The Elements on Earth

ÆDES CHRISTI
in Academia Oxoniensi
C.W.S. R.E. 1904.

The Elements on Earth

Inorganic Chemistry in the Environment

P. A. COX

Fellow of New College and Lecturer in Inorganic Chemistry,
University of Oxford

Oxford New York Tokyo
OXFORD UNIVERSITY PRESS
1995

Oxford University Press, Walton Street, Oxford OX2 6DP

Oxford New York
Athens Auckland Bangkok Bombay Karachi
Calcutta Cape Town Dar es Salaam Delhi
Florence Hong Kong Istanbul Karachi
Kuala Lumpur Madras Madrid Melbourne
Mexico City Nairobi Paris Singapore
Taipei Tokyo Toronto
and associated companies in
Berlin Ibadan

Oxford is a trade mark of Oxford University Press

Published in the United States
by Oxford University Press Inc., New York

A catalogue record for this book is available from the British Library

Library of Congress Cataloging in Publication Data
Cox, P. A.
The elements on earth: inorganic chemistry in the environment / P. A. Cox.
Includes bibliographical references and index. Acid-free.
1. Environmental chemistry. I. Title.
QD31.2.C682 1995 546—dc20 94-42633
ISBN 0 19 856241 1 (Hbk)
ISBN 0 19 855903 8 (Pbk)

Typeset by EXPO Holdings, Malaysia
Printed in Great Britain on acid-free paper
by the Bath Press

Preface

Most people probably regard chemistry as a largely artificial activity. Chemists themselves learn how to prepare and study compounds in the carefully regulated conditions of the laboratory or the industrial plant. Non-chemists may think of useful products or—more frequently nowadays—of unwanted pollution. Yet most chemistry on Earth happens under natural circumstances beyond our control. Human activity certainly has some influence, and the possibility of harmful effects needs to be understood and counteracted. This cannot be done however, without understanding the chemistry of the natural world. Such an aim involves fields of study which are often regarded as separate disciplines: geochemistry, biochemistry, atmospheric chemistry, environmental chemistry, and so on.

The aim of the present book is to present some of the findings of these studies in the context of the general chemistry of the elements. I have avoided the term 'environmental chemistry' in the title, as it often has a negative implication, being concerned largely with environmental problems caused by pollution. Some of these concerns are discussed here, but I have taken a larger view of the 'environment', to include natural processes as well as our own influence on them. The unusual format of the text is an attempt to reconcile two somewhat conflicting desires: in the first place to give a comparative account of the roles of different elements, and secondly to give each individual element its due. The result is a book in two parts, the first in a conventional chapter-by-chapter format, and the second with headings for each element in alphabetical order. Part I can be read as a self-contained account, or as an introduction to Part II. It is unlikely that many people will want to read completely the latter part in the order in which it is printed; but I have given numerous cross-references, both within and between the two parts, and I hope that readers who follow these may end up covering areas that they did not originally think of.

The book is intended to appeal both to chemists—who ought to know more about what happens in the world outside the laboratory—and to others interested in the scientific study of the environment. Some background in chemistry is necessary, but I have tried to keep this at a level equivalent to that covered by an 'A'-level course (in the UK) or a 'Freshman Chemistry' course (in the USA). No specialist knowledge of the inorganic chemistry of individual elements is assumed. Although I do not have a specific course in mind, the book could be used as a supplement to an undergraduate course on inorganic or environmental chemistry.

I would like to express my thanks to many people who have contributed in various ways. Some of the encouragement to write this book came from the favourable comments—from students, colleagues, and even complete strangers—on my previous one, *The elements: their origin, abundance, and distribution*. This feedback reassured me that there are many readers interested in aspects of the elements not covered in conventional chemistry courses. A number of people read all or part of the first draft, and made comments which were extremely helpful: they include Ann Chippindale, John Emsley, Richard Wayne, and Bob Williams. Finally, I must thank my family for their own support and encouragement.

P.A.C.

Oxford
May 1994

Contents

Part I

*The first Part of this book provides an account of the Earth, its physics and chemistry, its structure and dynamics, in a way which emphasises the varied roles played by the chemical elements in combination with one another. It can be read as an introduction to Part II (where elements are discussed individually) or as a self-contained treatment. Cross-references are provided: thus 'Chapter 3' or '§3.1' refer to sections in Part I, whereas a mention such as '*carbon' indicates that more details can be found under this element heading in Part II.*

Logically, the elements should come first. They were present in space as the raw materials from which the Earth formed, and their diverse chemical compounds determine every feature of our environment. Nevertheless, it is easier to appreciate the importance of the chemical elements with some idea of the physical structure and dynamics of the Earth in mind. Chapter 1 describes some relevant aspects. The remaining chapters are more directly concerned with the chemical elements. Chapter 2 provides a brief introduction to the periodic table, to the chemical and radioactive properties of elements, and to their abundance. Chapter 3 is concerned primarily with the types of chemical compound found on the Earth, and Chapter 4 with the chemical reaction and dynamics of the Earth's surface. Finally, Chapter 5 discusses the ways in which elements are used by us, and some of the consequences of those uses.

1

The physical Earth

This chapter describes briefly how the Earth originated, its place in the solar system, and its physical structure. The consequences of energy flows, coming from the sun and from radioactive elements within the Earth, are also discussed.

1.1 The origin of the Earth and the solar system

Around 4600 million years ago an enormous, slowly rotating cloud of gas began to contract under its own gravitational attraction. Composed largely of hydrogen and helium formed at the origin of the Universe, this gas also contained small proportions of other elements 'cooked' by nuclear reactions in stars, and spewed out into space by supernova explosions. The contraction of the cloud may possibly have been initiated by such an explosion of a nearby star. The greatest concentration of gas built up in the centre of the cloud, and the gravitational energy released by its contraction caused it to heat up. As the temperature in the middle climbed to some 10 million degrees, nuclear reactions started, converting hydrogen into more helium, and releasing tremendous amounts of energy. This source of energy stabilised the central region against further contraction, and thus was born a new star, our sun, the centre of the solar system.

Meanwhile, chemical reactions were taking place in the outer, cooler parts of the gas cloud. Atoms of different elements combined together to form molecules (such as water) and solid dust particles (such as metallic iron and oxides of silicon and magnesium). Frictional forces on the dust particles travelling through the gas caused them to settle into the plane of rotation, forming a disc. Gravitational forces then aggregated the dust, first into bodies called planetesimals a few kilometres in diameter, and finally into larger bodies, the planets of the solar system. This last stage was slow, taking perhaps 100 million years, and during this stage the young planets were subjected to a tremendous bombardment as they swallowed up most of the planetesimals. Even now, 4.6 billion years later, the process is not complete, and some of the original planetesimals survive as comets and meteorites, which roam the solar system and occasionally collide with planets. Other smaller bodies in the solar system, the asteroids and many types of meteorite, are the remains of one or more planets that started to form, but later broke up as a result of collisions. The Moon may have been split from the Earth in some such collision, although the details of its formation are still controversial.

Sometime during the aggregation of the solid dusts into planets, the remaining gas was blown away by radiation emitted from the newly glowing sun. Some gas was picked up by the gravitational pull of the giant planets Jupiter and Saturn, and gave them massive atmospheres of hydrogen and helium. The Earth and its neighbouring planets Venus and Mars were too small for this to happen. Their atmospheres came later, partly from volatile material contained in the impacting planetesimals, but mostly from volcanic outgassing from their hot interiors, causing the chemical decomposition of rocks and the release of gases trapped in the dust particles from which they formed.

This account of the origin of the solar system is rather speculative, as obviously no-one was there to see it happen.[1] But although many details are still argued over, the general outline represents the best guess about how it happened. The age of the Earth, the Moon, and meteorites which fall on the Earth, can be estimated quite precisely by measuring the products of the radioactive decay of some elements such as *uranium[2] (see also *lead). These measurements, and other evidence, are consistent with the idea that all the bodies of the solar system, including probably the sun, were formed at around the same time.

Table I.1 shows some data on the major bodies in the solar system: the sun itself, the nine principal planets, and the Moon. The sun is a thousand times more massive than the largest planet, Jupiter, and indeed contains over 99 per cent of the total mass of the system. One important consequence of this fact is that the orbits of planets, including the Earth, are largely controlled by the gravitational pull of a single central body, the sun. Most planets keep a nearly constant distance from the sun. As a result, the solar radiation they receive, which falls off with distance according to the inverse square law, remains nearly constant, and so do their average surface temperatures. If there were two or more major centres of attraction for the planets, their orbits would be much less regular. Surface temperatures would fluctuate wildly, and our environment would be very different and not conducive to the existence of life as we know it. The figures in the table show that surface temperatures of the planets generally fall with increasing distance from the sun, although not in a very regular

1 A nicely illustrated account of these astronomical details can be found in Andouze and Israel (1985). Dorman and Woolfson (1989) present an alternative view of the origin of the solar system. Cox (1989) discusses the origin of the elements, and the chemical reactions involved in the formation of the planets.

2 An asterisk indicates that more detailed discussion can be found under the entry in Part II for the corresponding element.

Table I.1 Major bodies of the solar system

Name	Distance from Sun (10^8 km)	Mass (10^{24} kg)	Radius (10^3 km)	Average density (10^3 kg m^{-3})	Surface temperature (K)
Sun	–	1.98×10^6	695	1.41	5600
Mercury	0.58	0.33	2.44	5.44	620
Venus	1.08	4.87	6.05	5.27	740
Earth	1.50	5.98	6.37	5.52	290
Moon	1.50	0.07	1.74	3.34	–
Mars	2.28	0.65	3.39	3.96	220
Jupiter	7.78	1900	70	1.3	170
Saturn	14.3	570	60	0.7	140
Uranus	28.7	87	26	1.2	80
Neptune	45.0	100	25	1.7	80
Pluto	59	0.7	3 (?)	3 (?)	80

way. As discussed later, there are internal sources of energy in addition to solar radiation (§1.3); the atmospheres of planets also play an important role in regulating surface temperatures (§1.4).

The masses of the planets can be determined from the gravitational forces they exert on their satellites and on other planets, and hence the average density of each planet can be estimated, as shown in Table I.1. Planets fall into two major groups: the inner planets, comprising Mercury, Venus, Earth, and Mars, which have higher densities, and the less dense outer planets, Jupiter, Saturn, Uranus, and Neptune. (Pluto remains something of a mystery, as it is so small and far away that its size is difficult to judge.) These densities are not easy to compare directly with those of familiar substances, as there are enormous pressures at the centres of planets, which lead to a considerable compression. Nevertheless, the densities in the table suggest that the inner planets are similar in structure and composition to the Earth (see §1.2), whereas the outer planets are very different: 'ices', including solid methane and ammonia, may be major constituents, and Jupiter and Saturn also have massive atmospheres of hydrogen and helium.

Among the inner planets, the Earth seems to be typical in many ways. The unique features of our environment, including the existence of life, are partly a result of the distance between the sun and the Earth, which is just right to give a surface temperature in the range where water is liquid. The consequences of this are immense, not just for life but for nearly all the chemical processes taking place at the Earth's surface. Water is by far the most mobile component of the environment, and as it is also an excellent solvent it entrains many other substances with it. The chemistry of other planets, where no liquid water exists, is dull in comparison with that on Earth.

1.2 The structure of the Earth

Our terrestrial environment has a complex physical structure. The tenuous but essential atmosphere; the rivers, lakes, and seas of liquid water; rocks and soils of diverse form; and living organisms are all familiar to us. The chemical composition of these surface regions, and the chemical reactions that take place in them, form a major part of the subject matter of this book. Many pieces of evidence show that the Earth also has an internal structure. Its overall density is too large, even allowing for compression at the centre, for it to be made entirely of the kinds of rocks which are found near the surface. More detailed information comes from the science of seismology, the study of how shock waves generated by earthquakes and underground explosions travel through the Earth. Beneath the crust of surface rocks is a somewhat denser mantle, which itself overlies the dense metallic core. It is believed that the core has an outer part which is liquid, and a central solid region.[3]

The main features of the Earth's structure are summarised in Fig. I.1, and some data on the different parts are shown in Table I.2. The most interesting regions for us—the 'biosphere' composed of living matter, the atmosphere, and the oceans—form only a very small fraction of the total. But the other parts are important. The crust is the direct source of nearly all the chemical components present in the outer regions, and in particular of most of the elements essential for life and of use to human technology. Material in the crust is derived from the mantle, by physical and chemical processes driven ultimately by heat coming from radioactive decay (§1.3). The chemical composition of the crust and mantle are well known, and will be discussed in later sections. The core is not so directly relevant to us, and its chemical composition can only be guessed at, largely on the basis of observations on other parts on the solar system (§2.2). Metallic iron is almost certainly the major constituent.[4]

As with the formation of the Earth itself, our knowledge of how its different parts originated is somewhat speculative. It is thought that the solids making up the inner planets, including the Earth, contained a mixture of oxides together with denser metallic solids. As the planets formed, their interiors were heated up by the energy liberated from the radioactive decay of some elements. The metals melted, and being denser than the 'rocky' oxides, sank to form the central core. The heating process also decomposed some of the rocks, liberating volatile molecules such as water, nitrogen, and carbon dioxide. When the surface was cool enough, water vapour condensed to form the oceans. Some of the controversy concerning the composition of the early atmosphere is discussed under *carbon and *oxygen.

---

3 See, for example, Smith (1981).

4 Brown and Musset (1981) discuss the problems involved in understanding the internal composition of the Earth.

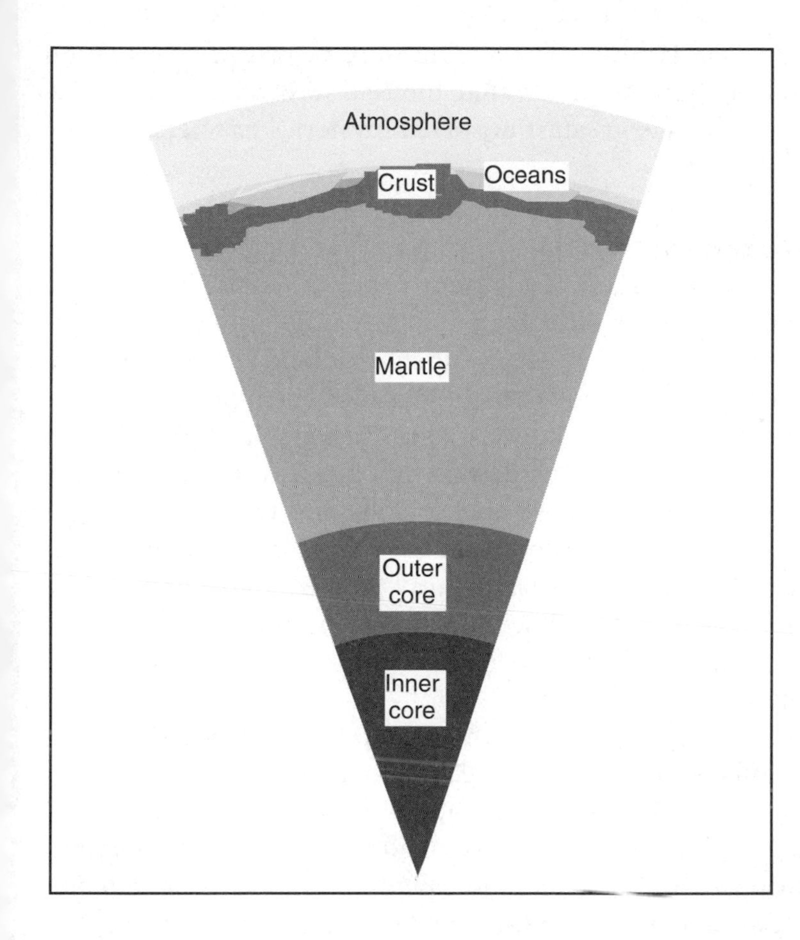

Fig. I.1 The structure of the Earth (not to scale). The different zones are described in more detail in Table I.2.

Table I.2 The structure of the Earth

Component	Average thickness (km)	Mass (10^{21} kg)	Average density (10^3 kg m^{-3})
Atmosphere	–	0.005	–
Biosphere	–	0.00001	1
Oceans	4	1.41	1.03
Crust	17	24	2.8
Mantle	2883	4016	4.5
Core	6371	1936	11.0

The early history of the crust is particularly uncertain. Radioactive dating shows that the earliest rocks still in existence were formed 3800 million years ago, some 800 million years later than the Earth itself. It is now known (§1.3) that much of the crust is being continually formed and recycled into the mantle. But precisely when and how this process began is controversial. One very surprising observation is that even the earliest rocks show evidence that life had already started.[5] The origin of life must count as one of the greatest unsolved mysteries of science.

1.3 Energy flows and dynamics

The world around us is in a state of constant change. Winds, rain, rivers, waves, tides, earthquakes, and volcanoes are some physical manifestations of the perpetual dynamics of our environment. With these, we should include our own physical and mental activity! Less immediately obvious are the complex chemical transformations involved in the cycling of the elements, discussed in Chapter 4. All this motion would quickly come to a halt if it were not driven by ceaseless flows of energy through the environment. A summary of these energy flows and transformations is shown in Fig. I.2. Inputs and outputs—mostly in the form of electromagnetic radiation—are shown on the left. These must balance fairly precisely, as otherwise the Earth would warm up or cool down rapidly. Successive columns in Fig. I.2 show different types of thermal (heat) energy, kinetic energy of motion, and finally chemical energy, including that involved in life processes.

By far the largest source of energy comes from the sun, in the form of radiation in the visible, ultraviolet, and infrared regions of the electromagnetic spectrum. About 30 per cent of this is directly reflected back into space, and the remainder is absorbed by the atmosphere and at the Earth's surface. A very small fraction initiates chemical reactions, both in the atmosphere and through photosynthesis by plants. The major part of the absorbed radiation is converted immediately to heat, and a significant fraction of this is in the form of latent heat of evaporation of water. A small additional source of heat comes from the radioactive decay of elements such as *potassium, *thorium, and *uranium within the Earth.

The conversion of heat—thermal energy—into kinetic energy of motion comes about because of the uneven temperature distribution. More energy is absorbed at the Earth's surface than in the atmosphere; the intensity of radiation is also greater near the equator than in the polar regions. Temperature gradients are therefore produced, both vertically in the atmosphere, and horizontally according to the latitude. These gradients cause convection currents. Warm air rises, cold air sinks, and winds are set up in the atmosphere. Water also circulates in the oceans, with currents caused partly by temperature differences between polar and tropical oceans, and partly by the effect of winds dragging surface waters. These motions are also strongly influenced by the rotation of the Earth, and the combined effects of this and the

5 Mason (1991).

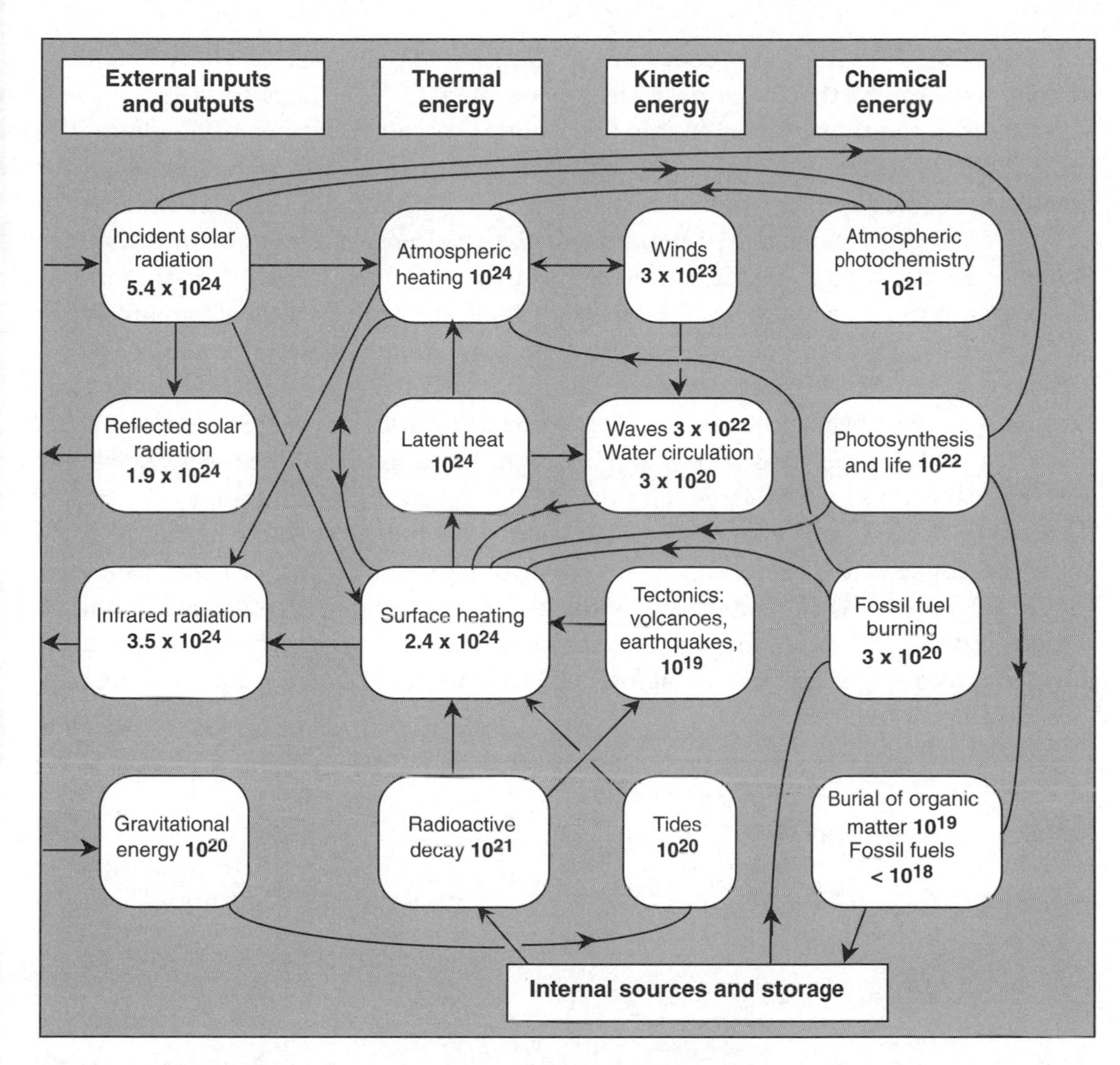

Fig. I.2 Energy flows in the Earth. The figure in each box shows the approximate flux, in joules per year, associated with the specified type of energy. The arrows show the transformation between different forms. Note the balance between the inputs and outputs of radiation (on the left), and the much smaller magnitudes associated with chemical energy (on the right). Most figures are order-of-magnitude estimates only, based on data from Smith (1981).

temperature gradients lead to the complex and ever-changing circulation that makes up the Earth's weather patterns.[6]

These physical circulations are crucial to the chemical transformations discussed in later sections. Chemical elements and compounds entering the atmosphere and

6 Henderson-Sellers and Robinson (1986).

oceans, either from natural processes or in the form of pollution, are transported and distributed around the globe on a time-scale ranging from days to years. Equally significant is the transport of energy from hotter to colder regions. One effect is to reduce the temperature differences between the tropical and polar regions. The greatest polewards energy flux occurs around 40° latitude, and corresponds to 3×10^{23} joules per year, nearly 10 per cent of the total energy received from the sun. Roughly 25 per cent of this is carried by warm water, 60 per cent by warm air, and the remaining 15 per cent by the latent heat of atmospheric water vapour which evaporates in warm regions and condenses and precipitates where it is cooler.

Water is one of the most significant constituents of the Earth's surface. Its large thermal capacity helps to even out the temperature differences between night and day, between the different seasons, and between the equator and the poles. The latent heat of evaporation is also large, and makes a substantial contribution to the energy fluxes as just discussed. Liquid water is essential for life as we know it, and is also a good solvent which is important for the distribution and cycling of many elements. Figure I.3 shows a simplified picture of the hydrological cycle, representing the distribution and flow of water through the environment. The annual fluxes involved are immense, totalling nearly half a million cubic kilometres of liquid water every year.

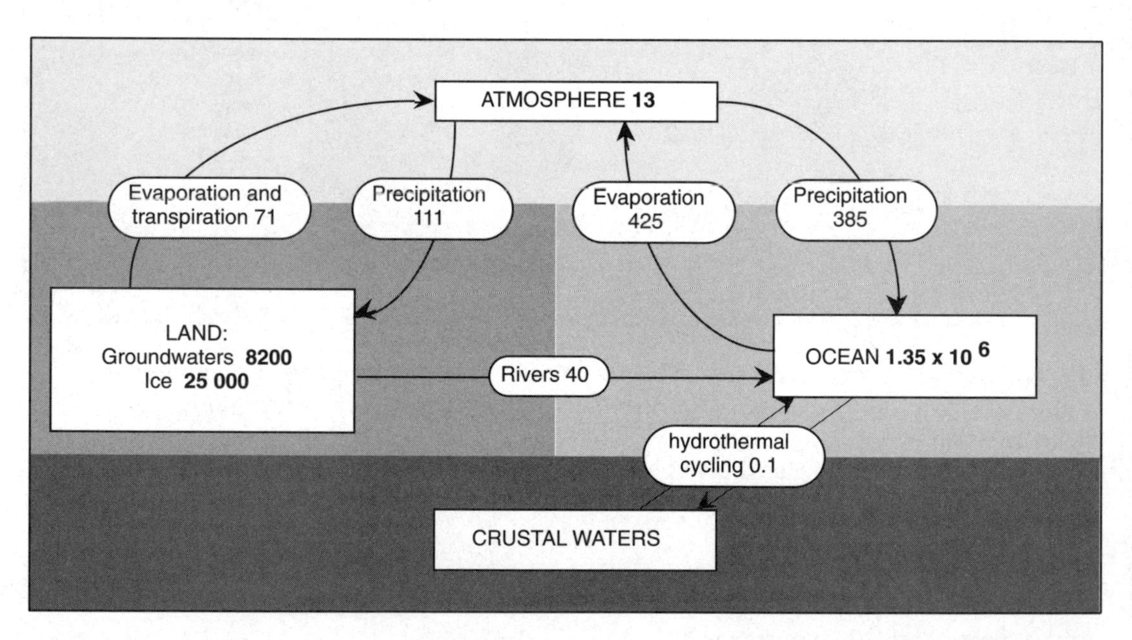

Fig. I.3 The hydrological cycle. Square-cornered boxes show the major reservoirs of water, and round-cornered boxes the annual fluxes, all in units of 10^{15} kg H_2O. 'Groundwaters' are fresh waters contained in lakes, rivers, and soils; 'crustal waters' are either fluids deep underground, or combined in rocks. Note the relatively tiny amount of water held as vapour in the atmosphere, but its huge importance in the cycle. (Based on Schlesinger, 1991.)

Not surprisingly, most evaporation occurs from the oceans, although there is a significant amount from land. Evaporation from lakes and rivers is supplemented by transpiration from plants, which makes an important contribution to the total. The net flow through the atmosphere transfers some water from the oceans to land, whence it ultimately returns to the oceans through rivers. A supply of water on land is crucial to the existence of life there, and the geographical distribution of rainfall, determined largely by the circulation patterns in the atmosphere, has had a major influence on the spread of life, including human settlement. Water falling on land is also largely responsible for the weathering of rocks, causing chemical transformations, and dissolving some constituents which are carried into the sea in rivers, and concentrated there by evaporation. Some water gets underground, and in the crust under the sea, and also participates in important chemical changes. These consequences of the hydrological cycle are discussed in Chapter 4.

Although radioactive decay liberates very little energy compared with that coming from the sun, it is important as it forms a heat source deep within the Earth. This keeps the centre of the Earth hot—a fact that was found mysterious to nineteenth-century scientists who did not understand the source of the energy required. It is now known that the heat flow from the Earth's interior gives rise to other convection currents, much slower than those of the atmosphere and oceans, but nevertheless important over geological time-scales of hundreds of millions of years. The hot mantle behaves as a very viscous fluid, and the crust which 'floats' on it moves slowly, of the order of one centimetre per year. This is the theory of plate tectonics, now well established, which explains the phenomenon of continental drift, and is also important for understanding the origin and chemical composition of the crust.[7]

As mantle material rises towards the surface, the pressure on it decreases. Some of the minerals melt to form magma which rises faster. Solidifying again as it cools near the surface, this forms the rocks of the oceanic crust. New crust is being formed continuously in this way, along the mid-ocean ridges which pass beneath all the major oceans of the Earth. The crust moves outwards pushing the continents with it, so that America and Europe, for example, are slowly moving further apart as a result of new crust forming in the middle of the Atlantic Ocean. At a few places along the ridge, such as Iceland, the molten material even breaks through the surface to give volcanoes.

In some other places on the Earth, crust is drawn back into the mantle by the sinking part of the mantle convection current. Known as subduction, this process is happening along the west coast of both North and South America. Some of the rocks melt as they heat up, forming magma which rises and mixes with material of the continental crust. Both earthquakes (from bits of crust slipping past each other) and

---

7 Smith (1981).

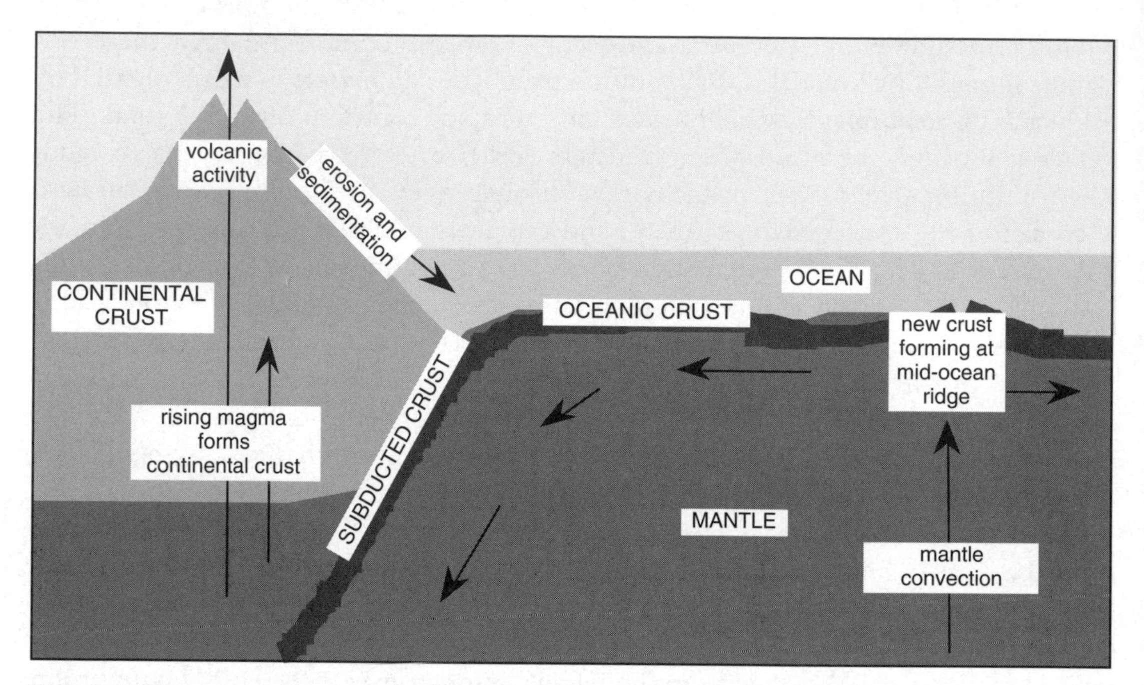

Fig. I.4 Plate tectonics, showing the convection currents in the mantle, and their role in the formation of the oceanic and continental crust. The surface erosion and sedimentation of rocks are driven by liquid water, and atmospheric constituents such as CO_2.

volcanoes are typical features of these regions. The continental crust which is formed there has a more varied and complex nature than the oceanic crust. Some of the chemical features of the formation of the crust from the mantle are described in §4.5; just as water is the most important chemical agent at the surface, so processes occurring deep within the Earth are dominated by silicate minerals (see *silicon).

A very small fraction of the solar energy received is converted by plants through photosynthesis into chemical energy, in the form of organic *carbon compounds and free molecular *oxygen. This is also of great significance to the Earth. Many chemical elements are involved in life, and the complex reactions involved make a major contribution to their cycling through the global environment. Even many supposedly 'inorganic' processes may be influenced. For example, the presence of plant life may break up rocks mechanically, and produces acids that may speed up their chemical weathering by a factor of a hundred. Biological processes in the sea are responsible for the formation of many sediments such as limestone. Most important of all, perhaps, is the presence of oxygen in the atmosphere, which is due almost entirely to photosynthesis. The chemical consequences are profound.

Most of the chemical energy stored by plants is passed along the 'food chain'. Oxygen and organic matter are recombined by air-breathing creatures, and by plants themselves in the dark; the energy appears ultimately as heat. A small amount of organic material is buried without reoxidation, however, and concentrated deposits of this constitute the fossil fuel reserves—coal, oil, and natural gas—on which contemporary human civilisation so much depends. Energy usage by humans is still much smaller than the total rate of biological production, but it is nevertheless much larger than the rate of burial. Fossil fuels are therefore being depleted. The exploitation of other energy sources—solar, wind, wave, tidal, geothermal, nuclear, etc.—has been partly stimulated by this problem. But in recent years another concern has arisen. The oxidation of fossil fuels is currently returning carbon dioxide to the atmosphere at a rate greater than it can be used in photosynthesis. This is only one form of pollution caused by human activities, but it does risk having global consequences by altering the physical properties of the atmosphere: the so-called 'greenhouse effect' is discussed briefly in the following section.

All the energy received by the Earth, whether converted temporarily into kinetic, chemical, or other types of energy, ends up eventually as heat. Heat energy in turn is radiated back into space, in the form of infrared radiation. All these transformations are governed by two important laws of nature, the First and Second Laws of Thermodynamics. The First Law asserts the conservation of energy. Any imbalance between the total energy received (including that from internal sources such as radioactivity) and the total amount emitted must change the internal energy of the Earth. This would lead to a change in the Earth's average temperature. To a good approximation, this is not happening, as the infrared emission closely balances the energy input. The way in which the surface temperature of the Earth is controlled by this balance depends on the properties of the atmosphere, as explained in the following section. The Second Law of Thermodynamics is more subtle, and deals with the types of energy conversion that can occur. All natural processes lead to an increase in entropy, a measure of the microscopic disorder present in all matter and in radiation.[8] Heat—the random motion of atoms and molecules—is the most disordered form of energy. One consequence of the energy flows through the Earth is a continual increase in entropy. A very small fraction of the energy is diverted into more ordered forms—the kinetic and chemical energy shown in Fig. I.3—but the local entropy decrease entailed by these conversions is greatly outweighed by the general increase associated with the degradation of solar radiation into thermal energy, and ultimately into infrared emission.[9]

---

8 Atkins (1984) gives a readable account of the Second Law. For a more technical account intended for chemistry students, see Atkins (1990).

9 The modern theory of irreversible thermodynamics shows how a global increase in entropy can sometimes drive such local decreases. See for example Prigogine and Stengers (1984).

1.4 Radiation and the 'greenhouse' atmosphere

Figure I.5 shows how the energy of solar radiation, and that emitted from the Earth, are distributed according to their wavelength in the electromagnetic spectrum. Solar radiation is concentrated in the visible region of the spectrum, although it extends somewhat into the infrared and ultraviolet. Radiation emitted from the Earth is all infrared. These differences are a consequence of the very different temperatures of the sun and the Earth. The curves approximate to so-called *black-body* distributions, corresponding to a temperature of 5780 K for the sun, and 255 K for the Earth. The theory of thermal radiation, which predicts these distributions, was developed by Planck in 1900, and formed the first application of the Quantum Theory.[10] Not only the spectral distribution, but also the total amount of energy radiated, depends strongly on the temperature; 255 K is the temperature necessary for the Earth to radiate away just as much energy as it receives from the sun, after allowing for the amount reflected. (As shown in Fig. I.2, other sources of energy are negligible in comparison.) This temperature is a chilly −18°C, much colder than the average surface temperature of the Earth, which is around 288 K (+15°C). The difference between these two values is the famous 'greenhouse effect', and is caused by the absorption of radiation by the atmosphere.[11]

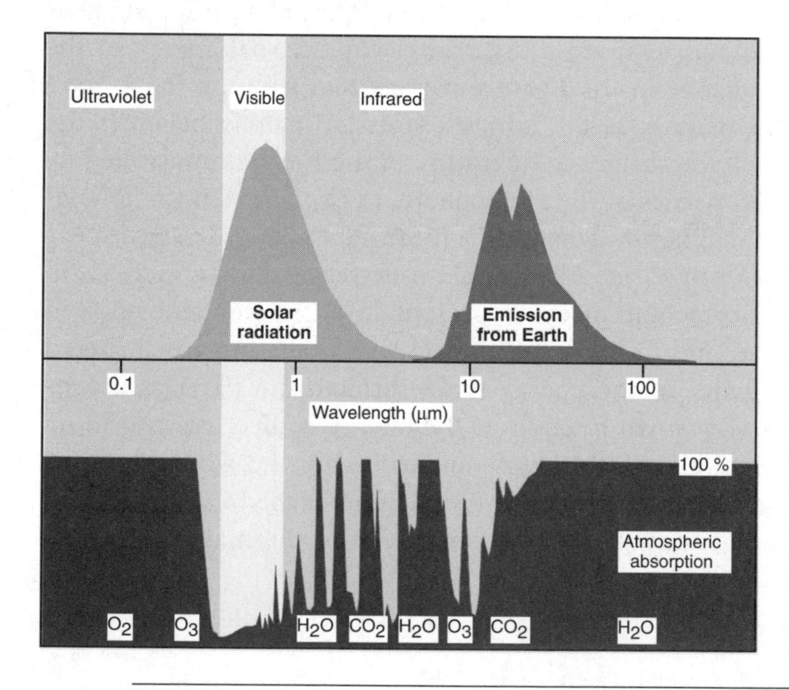

Fig. I.5 Radiation and the Earth's atmosphere. The upper curves show how the radiation received from the sun, and that emitted by the Earth, are distributed according to their wavelength in the electromagnetic spectrum. The lower curve, with the same wavelength scale, shows the degree of absorption of radiation in the Earth's atmosphere. The main absorbing molecules are listed at the bottom. (Based on Henderson-Sellers and Robinson, 1986.)

10 D'Abro (1951).

11 See Mitchell (1989) and Ramanathan *et al.* (1987) for reviews; the latter is more detailed and technical. As pointed out by many authors, the 'greenhouse' is an imperfect analogy, as real greenhouses work more by inhibiting convection than by absorbing infrared radiation.

The lower part of Fig. I.5 shows the extent to which radiation of different wavelengths is absorbed by the atmosphere, and the molecules responsible for most of the absorption.[12] Short-wavelength radiation is absorbed by increasing the energy of electrons in molecules. Longer-wavelength infrared radiation interacts with the atomic motions involved in the stretching and bending of chemical bonds (molecular vibrations). The principal atmospheric components, N_2 and O_2, do not absorb in this way, and most infrared absorption is due to water vapour (H_2O) and carbon dioxide (CO_2), although minor contributions come from other molecules, especially ozone (O_3), methane (CH_4), and nitrous oxide (N_2O).

It can be seen from Fig. I.5 that much more atmospheric absorption occurs in the spectrum of the radiation emitted by the Earth than in that of the incident solar radiation. The consequences are shown in Fig. I.6, which gives a more detailed view of the energy budget of the Earth's surface and the atmosphere. Some solar radiation is absorbed and scattered in the atmosphere, but a significant proportion reaches the ground, and about 50 per cent of the total incident radiation is absorbed at the surface. By contrast, very little of the infrared radiation emitted from the surface is directly transmitted into space. Radiation absorbed by the atmosphere, together with energy transferred from the surface in the form of sensible heat (that is, direct warming) and latent heat, must be balanced by radiation emitted by the atmosphere. This emission occurs in all directions, downwards towards the surface as well as upwards into space. The simplest way of thinking of the greenhouse effect is as the surface warming resulting from the infrared radiation from the atmosphere.

A better interpretation of the greenhouse effect makes use of the theory of black-body radiation. The Earth must have an effective temperature of 255 K in order for the radiation emitted to balance the total absorbed from the sun. This is the average temperature of the atmosphere at a height of about 5.5 km. At lower levels, the atmosphere is nearly opaque in the infrared spectrum. Most radiation is effectively trapped, as it is reabsorbed very soon after as it has been emitted. But both the density of the atmosphere and its temperature decrease with height. A decrease in the concentration of absorbing molecules leads to less absorption. The height of 5.5 km is the 'mean radiating level', above which radiation is able to escape into space. If we were to observe the Earth from space with infrared radiation of 20 μm wavelength, this level in the atmosphere would appear to be the surface. Below this level the atmosphere is warmed by the trapping of infrared radiation, so that the actual surface of the Earth is on average 33°C warmer than at an altitude of 5.5 km.

---

12 The physical principles involved in the absorption of radiation by molecules are described in Atkins (1990); Wayne (1991) discusses these in connection with the chemistry of the atmosphere.

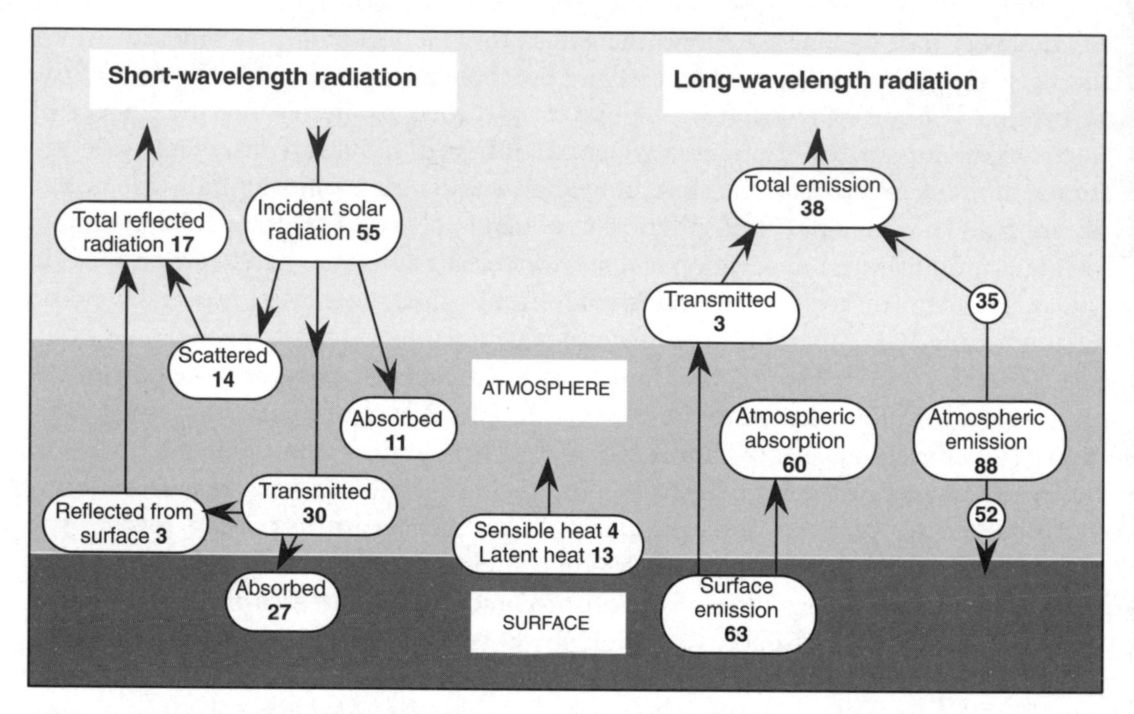

Fig. I.6 The Earth's radiation budget, showing the transmission, absorption, and reflection of short-wavelength solar radiation, and long-wavelength infrared. Energy fluxes are given in units of 10^{23} joules per year. Note the trapping of infrared radiation in the amtosphere. Radiation from the surface accounts for two thirds of the energy absorbed in the atmosphere, and a smilar fraction of the energy absorbed by the surface comes from infrared remitted in the atmosphere. Based on Henderson-Sellers and Robinson (1986).

This discussion shows how important the chemical composition of the atmosphere is in controlling the surface temperature. The greenhouse influence of each constituent gas depends on its chemical structure and on its concentration, but the relationship is not a straightforward one. At its present concentration of 345 parts per million (ppm), atmospheric CO_2 contributes about the third of the total 33°C greenhouse warming, most of the remainder coming from water vapour. If the CO_2 concentration were to double (which could happen in 200 years at the present rate of fossil fuel burning) the additional heating is estimated to be around 3°C: significant to the Earth's climate, but much less than double the current effect. One reason for this is that the present level of CO_2 is already sufficient to give total absorption of infrared radiation of appropriate wavelengths emitted from the surface. The main effects of an increased concentration would be at a height in the atmosphere where the absorption in less, and would lead to a slightly higher mean radiating level. Molecules such as methane, nitrous oxide, and ozone, which are present at concentrations around 1 ppm, naturally absorb much less radiation than CO_2. The effects of an increased concentration occur throughout the atmosphere, and are relatively more significant than with CO_2. This is especially true as they absorb infrared radiation at wavelengths different from CO_2 and water vapour, and so partly stop up the 'windows' in the spectrum of atmospheric transmission. The artificially produced chlorofluorocarbons are of concern for a similar reason, as their absorption peaks around 10 μm, where CO_2 and water vapour are nearly transparent.

Our neighbouring planets, Venus and Mars, have very different atmospheres from that of the Earth, and demonstrate some different aspects of greenhouse warming. Table I.3 shows some data. The first columns give the *solar irradiance*, the total radiation received per unit surface area, and the *albedo*, the fraction of this directly reflected by the planet. The combination of these two values determines the amount of solar radiation absorbed, and hence the effective radiation temperature required to maintain an energy balance. Solar irradiance falls with distance from the sun according to the inverse square law, and is almost double on Venus compared with the Earth. But Venus is covered with white clouds and has such a high albedo that less solar radiation is absorbed, so that its effective radiating temperature is less than that of the Earth. Venus also has an atmosphere composed almost entirely of CO_2, at a surface pressure 90 times larger than the Earth's. Infrared radiation is trapped up to a mean radiating level of 70 km. Because of the temperature gradient in the Venusian atmosphere, the surface is extremely hot, around 400°C. There can be no liquid water on Venus at present, but if there were originally, the planet may have suffered a 'runaway' greenhouse effect sometime in its early history: rising surface temperatures caused by the atmospheric CO_2 caused water to evaporate. Water vapour is an extremely effective absorber of infrared radiation, and so raises the temperature further, evaporating yet more water, and so on.

Mars has a very thin atmosphere, a rather small greenhouse effect, and a low surface temperature. Some of the surface features suggest that liquid water may have been present in the past. It is possible that the positive feedback loop just discussed may have operated in the opposite direction on Mars, giving a 'runaway freeze-out':

Table I.3 The greenhouse effect on Venus, Earth, and Mars. The solar irradiance is the energy per square metre received from the sun, and the albedo the fraction of this which is directly reflected. The 'greenhouse' effect is the difference between the temperatures in the last two columns. (From Mitchell, 1989)

Planet	Solar irradiance ($W\ m^{-2}$)	Albedo	Mean radiating level (km)	Effective radiation temperature (K)	Average surface temperature (K)
Venus	2613	0.75	70	232	700
Earth	1367	0.30	5.5	255	288
Mars	589	0.15	1	217	220

dropping surface temperatures would reduce the amount of atmospheric water vapour, thus lessening the greenhouse effect and causing further cooling.

This is a sobering context in which to think about the present situation on our own planet.[13] The atmospheric concentrations of greenhouse gases are currently increasing (see *nitrogen and especially *carbon). Water vapour is largely out of our control, because its concentration adjusts naturally according to the temperature. The different 'runaway' scenarios for Venus and Mars show however that its presence can a have a positive feedback effect, amplifying the changes caused by other gases. But there are very many other feedbacks, owing to possible changes in cloud cover (which both reflects more sunlight and absorbs more infrared), in the extent of the icecaps (highly reflective to sunlight), in the general pattern of atmospheric circulation, and so on. In fact the interaction between physical and chemical effects in the atmosphere is very complex and not yet fully understood. It is well established that mean temperatures on Earth have varied in the past; some of the evidence for this comes from the study of *oxygen isotopes in rocks, as well as from geological evidence of ice-ages. Such changes may have been influenced by alterations in the amount of solar radiation received, resulting from small periodic changes in the Earth's orbit. The chemical composition of the atmosphere, especially the CO_2 content, has also changed in response, however. One of the mysteries of the Earth's history is that the output of radiation from the sun is thought to have increased by around 25 per cent since the origin of the solar system, and yet the temperature on Earth has not changed nearly as much as would be expected. It is most likely that the atmospheric CO_2 has also decreased during this time, largely through the influence of living organisms. The idea that life may be able to adjust the environment to suit its need is one foundation of the *Gaia hypothesis*. This notion has been much criticised, but it is undoubtedly true that the existence of life has a crucial influence on our atmosphere: the chemistry of *carbon, *nitrogen, *oxygen, and *sulfur are all involved.[14]

Atmospheric absorption of short-wavelength radiation is important for other reasons. Very little absorption occurs in the visible region (0.4–0.8 μm) or in the neighbouring part of the ultraviolet spectrum (0.32–0.4 μm, known as UVA). Absorption of UVB (0.28–0.32 μm) is almost entirely due to ozone (O_3); at shorter wavelengths (UVC, less than 0.28 μm) normal dioxygen (O_2) molecules also absorb. Ozone is itself produced by the interaction of light with *oxygen. It is found principally in the *stratosphere*, at heights between 20 and 50 km above ground level, and is

13 A critical assessment of 'global warming' is given in Houghton *et al.* (1990).

14 For an account of the Gaia hypothesis by its chief exponent, see Lovelock (1982, 1988). Schneider and Boston (1991) have edited a collection of articles, not all favourable, presented at a scientific conference devoted to this topic.

important in many ways. The absorption of sunlight warms this region of the atmosphere, which prevents the convection currents responsible for the weather systems in the lower atmosphere, the *troposphere*, from rising further. The ozone layer therefore has an influence on the Earth's climate, and needs to be included in any detailed prediction of climatic change. The absorption of UVB radiation is also important for life, including ourselves, as this radiation is very damaging to cells. This is one reason for being concerned about the depletion of the ozone layer, caused partly by pollutant molecules containing *nitrogen and *chlorine.[15]

15 See Wayne (1991) for a detailed discussion of chemistry in the ozone layer.

2

The elemental constituents of the Earth

To know what the Earth is made of is an essential preliminary to understanding its chemistry. This chapter introduces the elements themselves, discusses their chemical classification according to the periodic table, and gives some examples of the chemical forms in which they appear in the Earth. The later sections discuss isotopes and radioactive properties, and some aspects of the relative abundance of elements. More details of the chemical compounds formed by the elements on Earth are given in Chapter 3.

2.1 The periodic table of elements

The discovery of the chemical elements was an enormous step in our knowledge of the world.[16] Ancient ideas on the nature of matter, such as the *earth, air, fire,* and *water* of the Greeks, were based on an obvious classification of familiar parts of the environment, but not on detailed experimenits designed to find their true constituents. The modern notion of a chemical element was first formulated in the seventeenth century, but it was only slowly that chemists learnt how to distinguish true elements from their compounds. One major stumbling block was the *phlogiston theory*, based on what we now regard as a confusion between two different aspects of nature, matter and energy. When this theory was abandoned, largely owing to the influence of Lavoisier, elements were discovered in quick succession.

At about the same time as Lavoisier's work, Dalton revived and popularised another idea essential to our modern understanding, the *atomic hypothesis*. All matter was supposed to be made of atoms, elements being made of one kind only, and compounds of more than one type. The proportions of different elements in a compound could be understood by assuming that all atoms of the same element have the same

16 Weeks and Leicester describe the discovery of the elements (1968). More general accounts of the history of chemistry are given by Ihde (1964) and Brock (1992).

mass, and that they combine together in simple ratios, expressed by their formulae such as NaCl and H_2O. Both of these assumptions are in fact wrong: many elements are composed of *isotopes* with different masses, and the composition of most of the minerals found in rocks is far from fixed. These problems will be discussed later (§2.3 and §3.2, respectively). In spite of these difficulties, Dalton's ideas allowed a great deal of sense to be made out the increasing chemical observations.

By the middle of the nineteenth century it was obvious that there were rather a large number of chemical elements. In spite of their differences in mass, considerable similarities were found between some of them, for example sodium and potassium, or chlorine and bromine. Various chemists therefore tried to arrange the known elements in such a way as to display their similarities. Early attempts at this were unsuccessful, but they culminated in the *periodic table* published by Mendeleev in 1869. One great advance made by Mendeleev was to allow for the possibility of elements that were not yet discovered. Indeed, he predicted not only the existence but also the properties of some new elements, such as gallium and germanium. Their subsequent discovery (in 1875 and 1886, respectively) did much to support his table. But Mendeleev's version was still incomplete and inaccurate in various ways. Some elements were wrongly placed, and a whole group of elements, the noble gases, was unknown to him. One problem was that he ordered elements according to their relative atomic mass (RAM). However, the RAMs of elements do not form a regular sequence, and it was impossible to know whether there were gaps between the known elements. In a few cases, the order of RAM is not the same as what we now know as the *atomic number* of the elements.

The basis of the atomic number is the *nuclear model* of the atom, first proposed by Rutherford and his collaborators in 1911. Most of the mass of each atom is concentrated at the centre, the positively charged nucleus, which is orbited by much lighter, negatively charged electrons. The atomic number is the number of electrons in a neutral atom, where their total charge just balances that of the protons in the nucleus. The first firm experimental basis on which to determine the atomic number came in 1913 from the spectroscopic work of Moseley. Only then was it possible to put the elements in an unambiguous sequence, and to be sure whether there were any gaps left. The last naturally occurring element is uranium, number 92, but not all the previous ones are found naturally, because they are radioactive and decay quickly. They have been made artificially, as have several *transuranium elements* with atomic numbers greater than 92 (see §2.3).

Table I.4 gives a list of the elements up to number 103. Some are known beyond this, but they have extremely short lives, and have been made in quantities of only few atoms, so that they have no relevance to the present account.

The modern form of the periodic table is shown in Fig. I.7. Elements are arranged according to their atomic number in horizontal *periods*, so as to fall in the vertical

Table I.4 The elements in order of atomic number, from Shriver *et al.* (1990). The RAM values for many elements are known to a greater precision than those quoted; see Greenwood and Earnshaw (1984) and especially IUPAC (1983) for a critical discussion. A table of the elements in alphabetical order of their symbols is shown inside the back cover of this book

Atomic number	Symbol	Name	Relative atomic mass
1	H	hydrogen	1.008
2	He	helium	4.003
3	Li	lithium	6.94
4	Be	beryllium	9.01
5	B	boron	10.81
6	C	carbon	12.01
7	N	nitrogen	14.01
8	O	oxygen	16.00
9	F	fluorine	19.00
10	Ne	neon	20.18
11	Na	sodium	22.99
12	Mg	magnesium	24.31
13	Al	aluminium	26.98
14	Si	silicon	28.09
15	P	phosphorus	30.97
16	S	sulfur	32.06
17	Cl	chlorine	35.45
18	Ar	argon	39.95
19	K	potassium	39.01
20	Ca	calcium	40.08
21	Sc	scandium	44.96
22	Ti	titanium	47.90
23	V	vanadium	50.94
24	Cr	chromium	52.01
25	Mn	manganese	54.94
26	Fe	iron	55.85
27	Co	cobalt	58.93
28	Ni	nickel	58.71
29	Cu	copper	63.54
30	Zn	zinc	65.37
31	Ga	gallium	69.72
32	Ge	germanium	72.59
33	As	arsenic	74.92
34	Se	selenium	78.96
35	Br	bromine	79.91
36	Kr	krypton	83.80
37	Rb	rubidium	85.47
38	Sr	strontium	87.62
39	Y	yttrium	88.91
40	Zr	zirconium	91.22
41	Nb	niobium	92.91

Table I.4 Continued

Atomic number	Symbol	Name	Relative atomic mass
42	Mo	molybdenum	95.94
43	Tc	technetium	99
44	Ru	ruthenium	101.07
45	Rh	rhodium	102.91
46	Pd	palladium	106.4
47	Ag	silver	107.87
48	Cd	cadmium	112.40
49	In	indium	114.82
50	Sn	tin	118.69
51	Sb	antimony	121.75
52	Te	tellurium	127.60
53	I	iodine	126.90
54	Xe	xenon	131.30
55	Cs	caesium	132.91
56	Ba	barium	137.34
57	La	lanthanum	138.91
58	Ce	cerium	140.12
59	Pr	praseodymium	140.91
60	Nd	neodymium	144.24
61	Pm	promethium	147
62	Sm	samarium	150.35
63	Eu	europium	151.96
64	Gd	gadolinium	157.25
65	Tb	terbium	158.92
66	Dy	dysprosium	162.50
67	Ho	holmium	164.93
68	Er	erbium	167.26
69	Tm	thulium	168.93
70	Yb	ytterbium	173.04
71	Lu	lutetium	174.97
72	Hf	hafnium	178.94
73	Ta	tantalum	180.95
74	W	tungsten	183.85
75	Re	rhenium	186.2
76	Os	osmium	190.2
77	Ir	iridium	192.2
78	Pt	platinum	195.09
79	Au	gold	196.97
80	Hg	mercury	200.59
81	Tl	thallium	204.37
82	Pb	lead	207.2
83	Bi	bismuth	208.98
84	Po	polonium	210
85	At	astatine	210
86	Rn	radon	222

Table I.4 Continued

Atomic number	Symbol	Name	Relative atomic mass
87	Fr	francium	223
88	Ra	radium	226.03
89	Ac	actinium	227
90	Th	thorium	232.04
91	Pa	protoactinium	231
92	U	uranium	238.03
93	Np	neptunium	237
94	Pu	plutonium	239
95	Am	americium	241
96	Cm	curium	247
97	Bk	berkelium	249
98	Cf	californium	251
99	Es	einsteinium	254
100	Fm	fermium	257
101	Md	mendelevium	258
102	No	nobelium	255
103	Lr	lawrencium	257

groups. These are numbered from 1 to 18, following the recent recommendations of the International Union of Pure and Applied Chemistry (IUPAC), designed to remove the confusion of previous numbering schemes.[17] One of the problems with this numbering, and indeed with the construction of the entire table, is that the numbers of elements in successive periods is not the same, but follows the sequence:

2, 8, 8, 18, 18, 32, (32)

the last one being incomplete.

These numbers can be explained by the quantum theory of atoms, and are a consequence of the electronic shell structure of atoms.[18] The modern 'long form' of the periodic table is constucted so as to take account of the different lengths of the periods. In some versions, the first two elements, hydrogen and helium, are not assigned to groups, although the ones shown here (1 and 18, respectively) are the most logical. The elements in the ten groups numbered 3–12 are known as the

17 See IUPAC (1989). In the most widely used previous scheme, Groups 1–10 were labelled IA–VIIIA, the latter including Groups 8–10; Groups 11–18 were labelled IB–VIIIB. In the alternative scheme, often used in the USA and especially in the geochemistry literature, the 'A' and 'B' suffixes were the other way round for all groups except I and II.

18 See Shriver *et al.* (1990) and Atkins (1990).

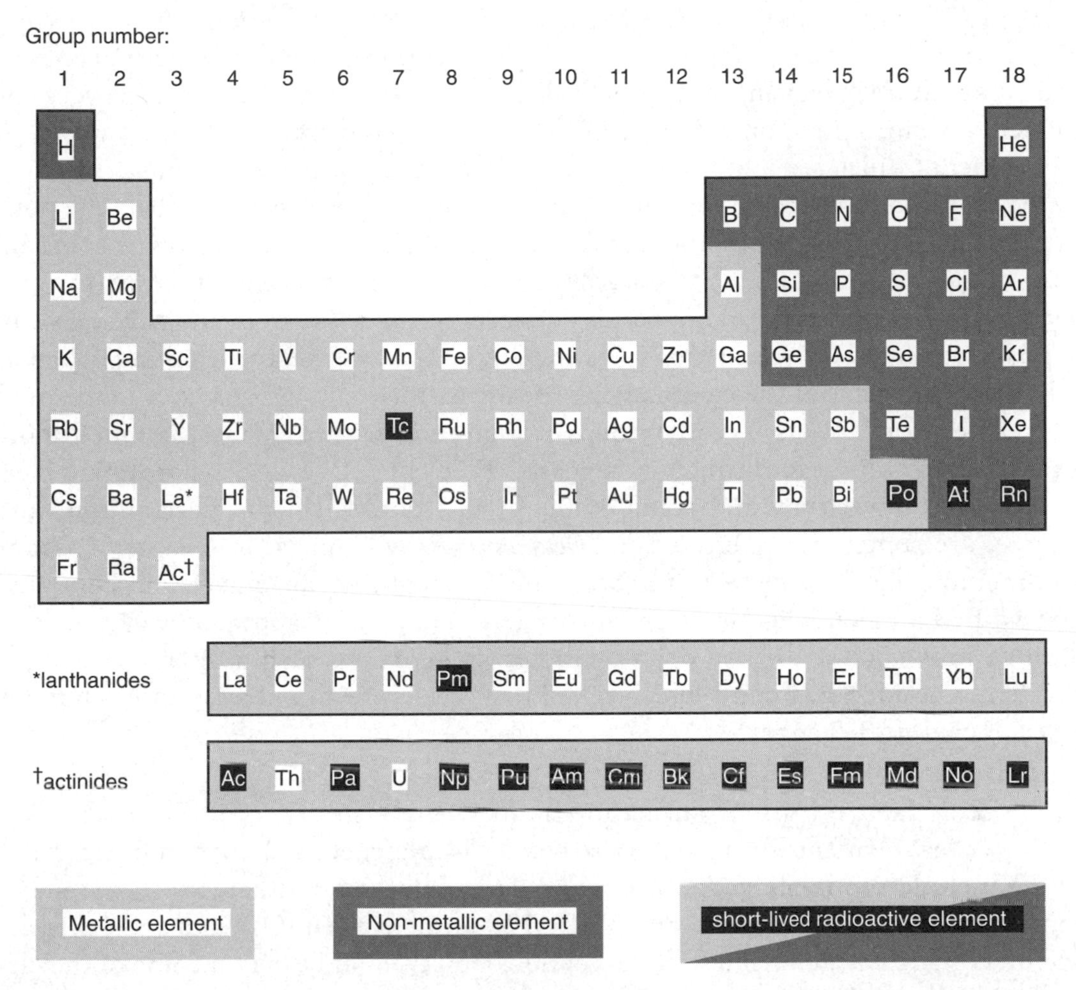

Fig. I.7 The periodic table of elements. The table is shown in the modern long form, with the new group numbers recommended by IUPAC. The 14 elements each in the lanthanide and actinide series immediately follow lanthanum (La) and actinium (Ac), respectively. The shading distinguishes metallic from non-metallic elements. Short-lived radioactive elements either do not occur naturally, or only occur in extremely small quantities.

transition elements, in distinction from the *main-group* elements of Groups 1, 2, and 13–18. Two extra series of elements, the *lanthanides* and the *actinides*, following lanthanum and actinium, respectively, are assigned to the same slot in Group 3 of the table. The lanthanide elements together with yttrium from Group 3 are often called the *rare earths*, and are very similar in their chemistry (see *lanthanum).

Apart from their numbers, some groups of elements are often referred to by name. The names sometimes used in this book are the *alkali metals* (Group 1, excluding hydrogen), the *alkaline earths* (Group 2), the *halogens* (Group 17), and the *noble gases* (Group 18).

Figure I.7 shows also how the elements are divided into *metals* and *non-metals*. This distinction is based formally on the physical properties of the pure elements, and is not entirely relevant to their behaviour in the natural environment, as very few elements occur in this form. It does however reflect some trends in the chemistry of the elements which are important. Metallic elements are generally *electropositive*, that is, they tend to form compounds in which their atoms lose electrons to become positive ions. For example sodium occurs as Na^+ in solid NaCl and in aqueous solution. Non-metallic elements are more *electronegative*, tending to gain electrons to form negative ions such as Cl^-. Compounds between non-metallic elements have covalent bonds, and often form small molecules such as CO_2 and H_2O; these are important as volatile constituents of the environment. However, the trend in behaviour in the periodic table, from metals in early groups to non-metals in later ones, is much more gradual than this clear distinction suggests. Elements close to the borderline have intermediate character, and some of them, like boron, silicon, germanium, and arsenic, are sometimes referred to as *metalloids*. As will appear in Chapter 3, many non-metallic elements close to the borderline normally occur in nature in combination with the electronegative element oxygen. They never appear however as free positive ions such as B^{3+} or P^{5+}, but rather as *oxyanions* such as BO_3^{3-} or PO_4^{3-}, covalently bonded fragments which retain their structure and identity through many types of chemical change.

2.2 Some typical compounds of the elements

A few elements occur naturally on Earth in a chemically uncombined form. These include the noble gases of Group 18 (helium, neon, argon, etc.), the major constituents of the atmosphere dinitrogen (N_2) and dioxygen (O_2), and some unreactive metals such as gold. These are exceptions however, and even for some of the elements just listed the uncombined forms are relatively uncommon. By far the majority of elements in the outer layers of the Earth are chemically combined, sometimes as molecules such as H_2O and CO_2, more often in solid minerals such as NaCl, SiO_2, and FeS_2. Table I.5 shows some typical chemical forms displayed on Earth by the first 30 elements, hydrogen to zinc, in order of atomic number. The second column shows the normal oxidation state (or oxidation number) of each element. Sometimes this is constant, but some elements such as carbon, nitrogen, sulfur, and iron can have different oxidation states in different conditions (see §3.1); in this case only the extreme range is generally shown. The following columns show only a selection of the chemical compounds found as solid minerals in the crust, as soluble species present in sea water, or as volatile molecules in the atmosphere.

One of the most striking features illustrated in Table I.5 is the preponderance of oxide minerals in the crust. This fact may be understood partly from the high abundance of oxygen (see §2.4), and partly from its strong affinity for many elements,

Table I.5 Typical chemical forms of the elements 1–30 found on Earth; only a few examples are shown for each element

Element	Oxidation state	Typical chemical forms in: Rocks	Sea water	Atmosphere
H	+1	$Al(OH)_3$	H_2O	H_2O vapour
He	0	–	He	He
Li	+1	$LiAlSi_2O_6$	Li^+	–
Be	+2	$Be_3Al_2Si_6O_{18}$	$Be(OH)_2$	–
B	+3	$Na_2B_2O_5(OH)_4.8H_2O$	$[B(OH)_4]-$	–
C	–4 to +4	$CaCO_3$, C, CH_4	HCO_3^-	CO_2, CH_4
N	–3 to +5	KNO_3, also NH_4^+	N_2, NO_3^-	N_2, N_2O
O	–2, 0	SiO_2, Fe_2O_3, $CaCO_3$	H_2O, O_2	O_2, H_2O, O_3
F	–1	CaF_2	F^-	CF_2Cl_2[a]
Ne	0	–	Ne	Ne
Na	+1	NaCl, $NaAlSi_3O_8$	Na^+	–
Mg	+2	$MgCa(CO_3)_2$, $Mg_3Si_2O_5(OH)_4$	Mg^{2+}	–
Al	+3	$Al(OH)_3$, $Al_2Si_2O_5(OH)_4$	$Al(OH)_3$	–
Si	+4	SiO_2, $MgSiO_3$	$Si(OH)_4$	–
P	+5	$Ca_5(PO_4)_3F$	HPO_4^{2-}	–
S	–2 to +6	FeS_2, PbS, $CaSO_4$	SO_4^{2-}, HS^-	SO_2, H_2S, $(CH_3)_2S$
Cl	–1	NaCl	Cl^-	CH_3Cl, CF_2Cl_2[a]
Ar	0	–	Ar	Ar
K	+1	$KAlSi_3O_8$	K^+	–
Ca	+2	$CaCO_3$, $CaAl_2Si_2O_8$	Ca^{2+}	–
Sc	+3	$Sc_2Si_2O_7$	$Sc(OH)^3$	–
Ti	+4	TiO_2, $FeTiO_3$	$Ti(OH)^4$	–
V	+3 to +5	$Pb_5(VO_4)_3Cl$	HVO_4^{2-}	–
Cr	+3, +6	$FeCr_2O_4$	$Cr(OH)_3$	–
Mn	+2, +4	$MnCO_3$, MnO_2	Mn^{2+}	–
Fe	+2, +3	FeS_2, Fe_2O_3	$Fe(OH)_3$	–
Co	+2	$CoAs_2$, also in Ni sulfides	Co^{2+}, $CoCO_3$	–
Ni	+2	Ni_9S_8, also Ni^{2+} in silicates	Ni^{2+}	–
Cu	0 to +2	'native' Cu, $CuFeS_2$, Cu^{2+} carbonates	CuCl+, $CuCO_3$	–
Zn	+2	ZnS	Zn^{2+}, $ZnCO_3$	–

[a]CFC compounds not of natural origin.

discussed in more detail in §3.1. These oxides are often very complex, and contain ions such as carbonate, silicate, borate, and so on. The high abundance of silicon as well as of oxygen means that silicate minerals, often containing many other elements in varying proportions, make up the major fraction of the Earth's crust and mantle.

The crust also contains halides of electropositive metals, such as NaCl and CaF_2, but these are less common because the halogen elements of Group 17 are much less abundant than oxygen (see Fig. I.9 in §2.4).

The later elements listed in Table I.5, for example cobalt and nickel, show a different pattern of behaviour. They are more often found in combination with sulfur, and other non-metals which are less electronegative than oxygen, such as arsenic. Copper occurs most commonly in sulfides, but it can also be found in 'native' form, that is as the uncombined metal.

Figure I.8 shows how these patterns of chemistry are distributed in the periodic table. Elements towards the top, the left, and the extreme right-hand sides (excepting the inert elements of Group 18) are most often found in oxide or halide minerals. These are sometimes known as *lithophilic* ('rock-loving') elements. Many elements on the lower right-hand side commonly occur in sulfide minerals, and are known as

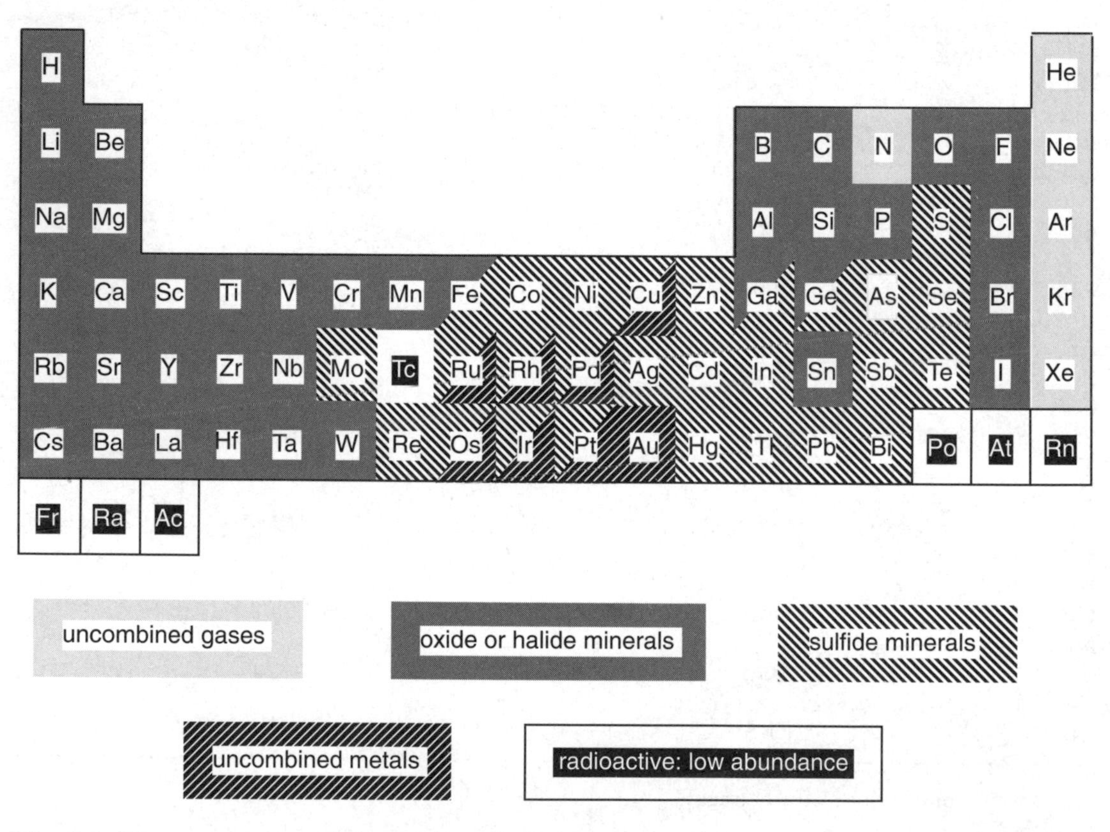

Fig. I.8 The occurrence of elements at the Earth's surface, showing the main types of minerals or other chemical forms for each element. Oxide minerals include a great variety of complex oxides, such as silicates and carbonates of metallic elements. The lanthanides, and the naturally occurring actinides thorium and uranium, follow the pattern of Group 3, and form oxide minerals.

chalcophiles. A few of the later and heavier transition metals can be found in uncombined, native, form; these are *siderophiles*.[19] As can be seen in Fig. I.8, many elements show a mixed or intermediate character. It is also important to note that chalcophilic and lithophilic elements can occur as oxides at the Earth's surface, when they are in contact with atmospheric oxygen.

Although this classification is based on solid minerals in the crust, it is significant for many other aspects of the chemistry of the elements concerned. For example, many chalcophilic elements are poisons, including the notorious 'heavy metals' such as *cadmium, *mercury, and *lead, and these facts are connected (see §3.5). The siderophilic elements are also the ones that on Earth are strongly concentrated in the core, so that their abundance in the crust is often very low (see §2.4).

Some of the patterns of chemistry shown in minerals also carry over to the aqueous environment—oceans, rivers, and lakes. Electropositive elements are present as positive ions such as Na^+, electronegative ones as negative ions such as Cl^-. Oxidised forms of non-metals form oxyanions such as bicarbonate, HCO_3^-, and phosphate, present as a mixture of HPO_4^{2-} and $H_2PO_4^-$. Many metal ions do not exist in simple form in solution, but as complexes, effectively in combination with other ions. Highly charged ions form hydroxides, such as $Al(OH)_3$, although they are often very insoluble in water (see §3.4). The chalcophilic elements also form strong complexes with ions such as chloride or carbonate.

Turning to the atmosphere in the final column of Table 1.5, we find only a limited number of elements. Others are present as solid dusts, blown from arid soils or derived from the evaporation of sea water, but only the non-metals of the top right-hand portion of the periodic table (plus hydrogen) form volatile molecules stable in the Earth's atmosphere (see §3.3).

2.3 The abundance of the elements

As the elements were discovered, it became apparent they are distributed very unequally in the Earth. Some elements, such as oxygen and silicon, are major constituents of nearly all minerals; others, such as helium or gold, are rare.[20] The first major survey of the relative abundance of elements in the crust was undertaken in the early twentieth century by the American geologists Clarke and Washington, who performed chemical analyses of thousands of rock specimens. These estimates have been refined in subsequent work, taking into account the relative amounts of

19 This geochemical classification of elements was first introduced by Goldschmidt on the basis of limited information in 1923. For a good modern account of geochemistry, see Mason and Moore (1982).

20 Cox (1989) gives an account of the abundances of elements and the factors that control them.

different types of rock, and the differences between the crust of the continents and that underlying the oceans. Figure I.9 shows a modern estimate of the abundance of elements in the Earth's crust. Elements are plotted against their group in the periodic table, the vertical scale showing the relative masses present in parts per billion (10^9). The solid line is drawn as a guide to the eye, and connects the successive elements of the first long period (potassium to krypton).

One noteworthy feature of this plot is the enormous range of abundance found, spanning a range of more than ten powers of ten. In other words, there is over ten billion times as much of the commonest elements, oxygen and silicon, as of the rarest ones such as osmium, tellurium, and xenon. Some very familiar elements such as gold are extremely rare overall: so much so that if they were uniformly distributed in the Earth they would be regarded as insignificant impurities, and very sophisticated

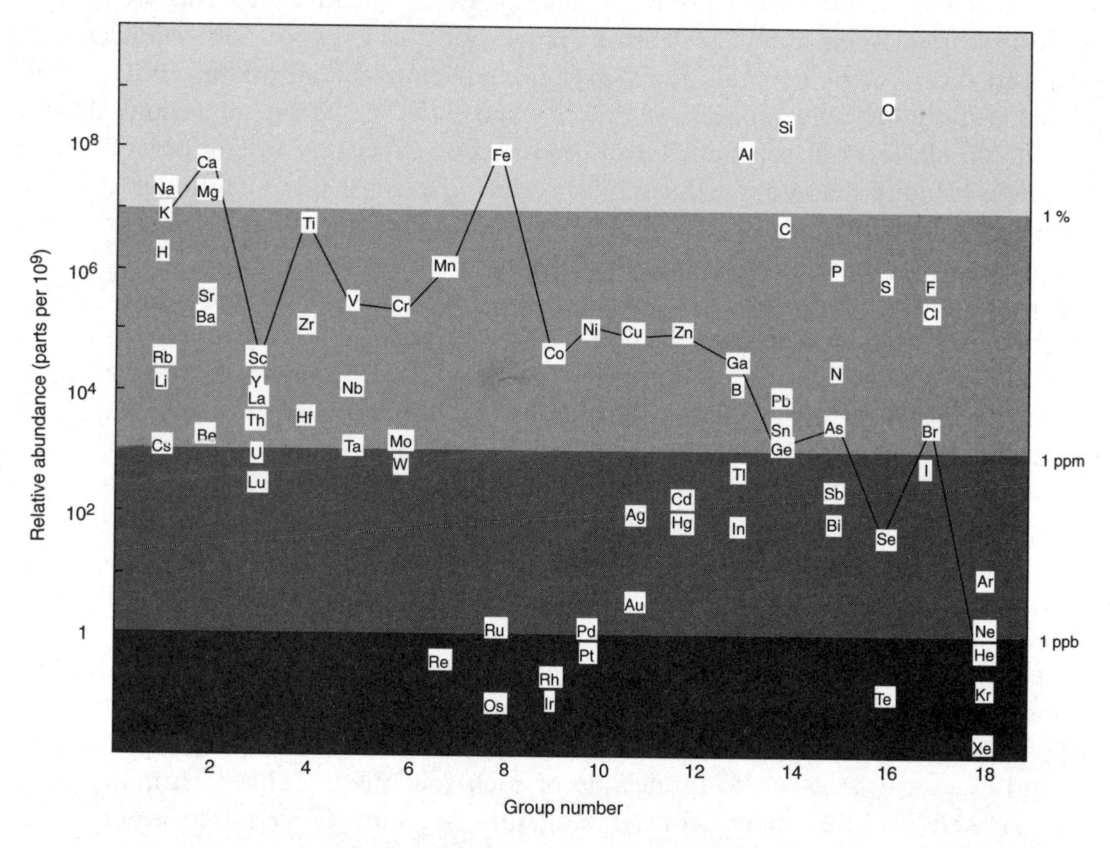

Fig. I.9 The abundance of elements in the Earth's crust, shown as the parts per billion (10^9) by mass on a logarithmic scale, with each division corresponding to a factor of 10 different from the neighbouring one. The horizontal scale is the group number in the periodic table. The line connects elements of the first long period (potassium to krypton) and is drawn as a guide to the eye. The different shading shows levels of abundance of one percent, one part per million (ppm) and one part per billion (ppb).

methods of chemical analysis would be needed even to detect them. Gold was known to the ancients only because it occurs in nearly pure form, albeit in small deposits. The exploitation of most elements depends on their relatively high concentration in specific minerals. The chemical composition of these, and the geochemical processes that produced such concentrated deposits, will be described briefly in some later sections (see §§3.2 and 4.4–4.6).

As mentioned in §1.2, the surface crust is not typical of the Earth as a whole. To ascertain the chemical composition of the underlying mantle and the central core is much harder, and many uncertainties still remain. Estimates of the mantle composition can be obtained from rocks which are thought to have been derived from deep within the Earth. Oxygen and silicon are dominant constituents as in the crust, but there are some differences, elements such as magnesium and chromium being relatively more abundant in the mantle, and others such as the alkali metals much less so. Information about the composition of the core is less direct.[21] Geophysical evidence shows that it is dense and metallic. Most estimates of its composition rely heavily on meteorites coming from space, especially the metallic types that are thought to be derived from the core of one or more small planets that formed and subsequently broke up. Iron is certainly the dominant element, together with a substantial fraction of nickel and other transition elements. But there must also be non-metallic elements present in the core to account for its density and other properties. It is most likely that sulfur is the predominant non-metal, although other possibilities, such as carbon or silicon, have been considered.

Putting together these admittedly uncertain estimates of the core and mantle composition enables us to calculate the overall chemical composition of the Earth. This is shown in Fig. I.10, plotted in the same way as the crustal composition of Fig. I.9. There are many detailed differences between the two plots, reflecting the way certain elements have been concentrated either into the crust or into the inner parts of the Earth. For example, the heavy transition metals, especially those of the *platinum group, are relatively much more abundant in the core, and so fall much lower in Fig. I.9 than in Fig. I.10. The chemical differences that lead to these patterns of abundance in different zones of the Earth are discussed later, especially in §3.1 and §4.5.

If crust is not typical of the Earth as a whole, it is also true that the Earth's composition is very different from that of the complete solar system. Most of the mass of the solar system is contained in the central body, the sun; the composition of this is known from the spectrum of sunlight, and the observation of spectral lines typical of different elements. One element, *helium, was indeed first discovered in this way, and named after the Greek word for the sun (*helios*). A few elements are hard to

21 See Brown and Musset (1981).

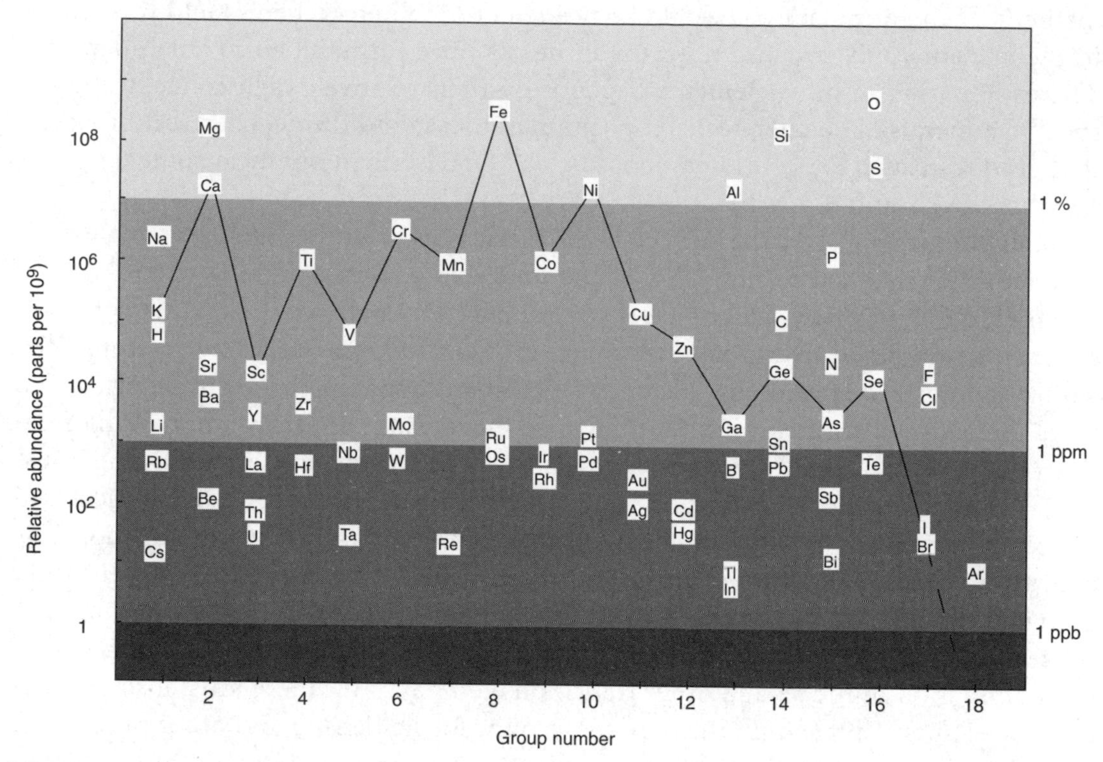

Fig. I.10 The abundance of elements in the whole Earth. This plot should be compared with Fig. I.9, which is drawn in the same way. Some elements (e.g. K, Br) are lower in the present plot because they are preferentially concentrated in the crust; others are higher, because of their concentration either in the mantle (e.g. Mg, Cr) or in the core (e.g. Ni, Se).

detect in the sun's spectrum, and much confirmation of the composition of the solar system comes from meteorites: in this case, a special type known as *carbonaceous chrondrites*, which as their name suggests contain a significant amount of carbon. Unlike most types of meteorites, which seem to have come from planets that subsequently broke up, the carbonaceous chondrites are thought to be derived directly from the planetesimals involved in the formation of the planets (§1.1). Their composition therefore gives direct information about the material from which planets such as the Earth were made.

Figure I.11 shows the abundance of elements in the solar system, plotted in the same way as in the previous two figures. Hydrogen and helium, the elements produced at the inception of the Universe, are by far the major constituents. Nearly all the remaining elements have been made by nuclear reactions in stars. Helium nuclei are fused together to make carbon and oxygen, the next most abundant elements. Subsequent nuclear processes make heavier elements, in proportions which are controlled, not by chemical properties but by the relative stability and reactivity of the

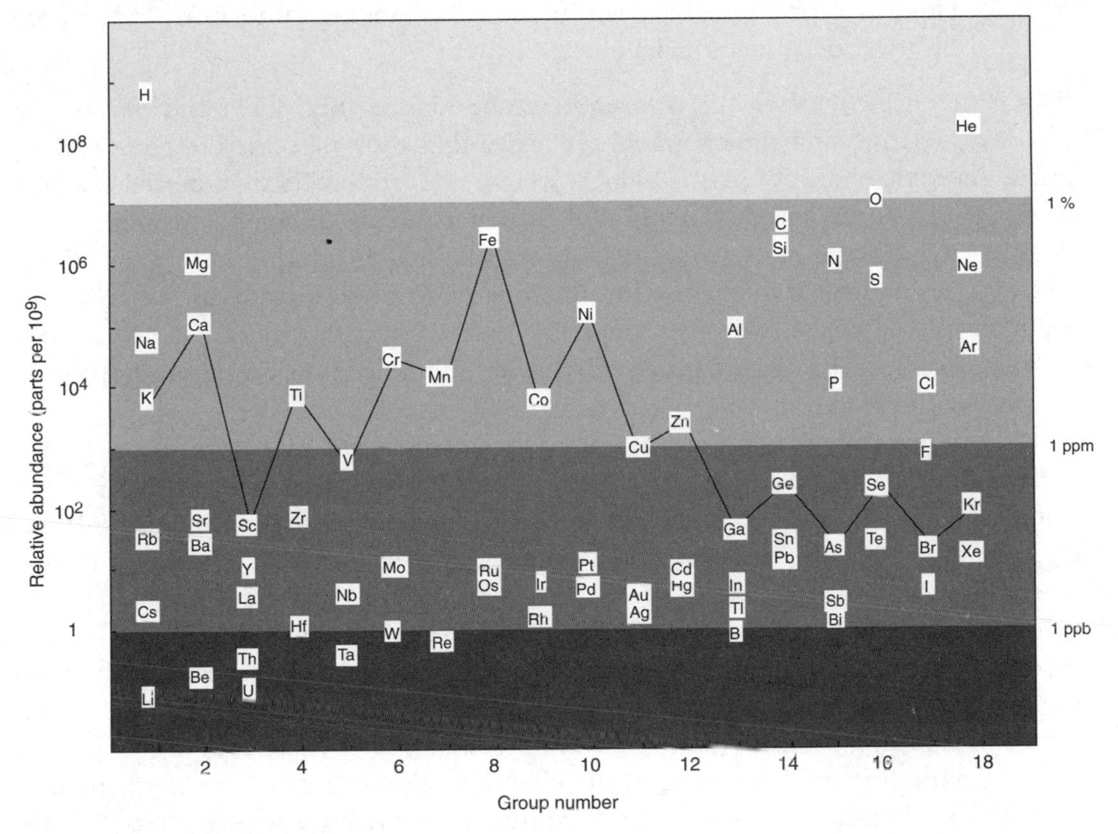

Fig. I.11 The abundance of elements in the solar system. Compare with Fig. I.9, and note the enormous preponderance of H and He in the present plot, as well as the much higher abundance of C, N, and the noble gas elements of Group 18. The higher abundance of elements with even, rather than odd, atomic number is a consequence of differences in nuclear stability, which control the proportions in which elements have been synthesised by nuclear reactions in stars.

atomic nuclei. Several features of the solar-system abundance are also apparent in the previous plots:

1. There is a general decline in abundance with increasing atomic number. The three light elements lithium (Li), beryllium (Be), and boron (B) are an exception, as they are among the rarest elements overall. This is because their rather unstable nuclei are not made in stars; their production in space is largely due to high-energy collisions involving cosmic rays, although small quantities were also made in the 'Big Bang'.
2. There is an alternation in abundance, even-numbered elements being commoner than their odd-numbered neighbours. This is a result of the relative stability of nuclei with odd and even numbers of protons.

3. There is a pronounced peak in the abundance around iron (Fe), which has the most stable nuclei of any element.

The differences between the solar-system abundance (Fig. I.11) and that of the Earth (Fig. I.10) reflect the chemical processes that took place as the planets were formed. As explained in §1.1, the planets were made from solid dust particles which condensed out of the gaseous material of the solar nebula. Elements forming stable solid phases, either metals (iron, nickel, etc.) or oxide compounds (aluminium, silicon, etc.), condensed efficiently into the dust, and predominate on Earth. Those present as volatile gases (hydrogen, the noble gases, and carbon and nitrogen as CH_4 and NH_3, respectively) have considerably lower abundance on Earth in comparison with the starting material.

Our discussion has shown that the abundance of elements at the Earth's surface is determined by a combination of very different processes. The nuclear synthesis of the elements formed the raw materials, reflected in the solar-system abundance of Fig. I.11. Elements were then fractionated according to their chemical behaviour, first to give the Earth its overall composition, and then within the different zones of the Earth to give the surface crust. Subsequent chemical fractionation has continued, providing the different compositions of the atmosphere and the oceans, and the concentrations of elements found in mineral deposits. Although the abundance of many elements has been altered by these chemical processes, certain features of the original pattern remain, and are important in the chemistry of the Earth. The elements that dominate the chemical and physical environment are mostly ones—such as carbon, oxygen, silicon, sulfur, and iron—which were made in high proportions in nuclear reactions, and so are generally abundant in the Universe.

2.4 Isotopes and radioactivity

Although most of this book is concerned with chemistry, the properties of atomic nuclei are also important. As mentioned in the previous section, nuclear properties controlled the proportions in which different elements were made, and are apparent in the relative abundance of the elements on Earth. Nuclear properties are of significance on the Earth today, through the existence of isotopes, and through the radioactive properties of certain elements.[22]

Isotopes are forms of an element with the same atomic number but with different atomic masses. For example oxygen has three stable isotopes, ^{16}O, ^{17}O, and ^{18}O. Since the discovery of the neutron by Chadwick in 1931, it has been understood that different isotopes of a given element have the same number of protons in the nucleus, but different numbers of neutrons. In the case of the oxygen isotopes, there are eight protons in

22 See Cox (1989) for an account written for chemists.

all cases, but eight neutrons in ^{16}O, nine in ^{17}O, and ten in ^{18}O. The chemical properties of an element are very largely determined by the nuclear charge, although the mass does have a small effect. For most purposes we can ignore the fact that oxygen is a mixture of three different isotopes, but the chemical equilibrium and rate constants for compounds containing these isotopes are slightly different. This means that the proportions of the isotopes, although determined largely by their nuclear synthesis in stars, are not exactly the same in different compounds. These proportions also depend on the temperature and pressure at which the compounds were formed, and this is one reason why earth scientists are very interested in studying the isotopic composition of different minerals. Not only do they give clues as to the conditions under which minerals were made, but they can also give important information about the Earth's changing environment, for example past climates, ice-ages, and so on.[23]

Variations in isotopic composition are usually given in δ values, expressing the parts-per-thousand (permil) differences between the abundance ratios in different samples. For example, if R is the ratio of $^{18}O/^{16}O$ abundance, then by definition

$$\delta^{18}O = 1000\,[R_{\text{sample}} - R_{\text{standard}}]/R_{\text{standard}},$$

where a convenient standard sample is taken for R_{standard}. In a similar way, one can define δ values for other isotopes, for example δ^2H for the variation in deuterium (^2H) abundance relative to normal hydrogen, and $\delta^{13}C$ relative to the more abundant carbon isotope ^{12}C. Variations of several hundred permil are found with δ^2H, because the mass difference between isotopes of *hydrogen is a factor of two, and has an unusually large effect on the properties of isotopes. Variations with *carbon and *oxygen isotopes are an order of magnitude smaller, although still significant.

Radioactivity is a consequence of the instability of some atomic nuclei, causing them decay into more stable ones through the emission of highly energetic particles. Three types of radiation are found:

1. *Alpha* (α) *radiation* consists of ^4He nuclei, containing two protons and two neutrons. The 'parent' nucleus gives a 'daughter' with an atomic number decreased by two units, and an atomic mass decreased by four units. For example, ^{238}U decays to ^{234}Th, which is itself radioactive, and decays by a succession of steps, ultimately to ^{206}Pb (see *uranium).

2. *Beta* (β) *radiation* is electrons. The result of β decay is a daughter with the same mass number as the parent, but an increase by one unit in the atomic number; effectively, one neutron changes into a proton to conserve charge. For example, ^{87}Rb decays to ^{87}Sr. Other types of β decay are possible. For

23 There is a well-illustrated chapter in Smith (1981); Faure (1986) is much more detailed.

example ^{40}K can decay to ^{40}Ar, in which the atomic number has decreased by one unit, by *electron capture* (EC) by the nucleus.

3. *Gamma* (γ) *radiation* often accompanies α or β decay, as the daughter nucleus loses excess energy. It consists of high-energy photons, like light but of much shorter wavelength (generally much less than 1 nm, compared with several hundred nm for visible light).

The radioactive decay of a nucleus is a random process governed by statistical laws. According to the quantum theory it is impossible, even in principle, to predict when a given ^{238}U atom will decay. The *probability* that it will decay in a given time is however known. (It is extremely small for ^{238}U: one chance in 2.05×10^{17} of decaying in the next second.) Very remarkably, this probability is the same today as it was billions of years ago when the atom was first made in a star. The statistical process leads to a definite law when a large number of radioactive atoms are present. If $N(0)$ is the number of atoms originally present, the number $N(t)$ after a time interval t is

$$N(t) = N(0)\ e^{-kt},$$

where k is a constant which gives the probability of decay per unit time for a given type of atom. It is more normal to quote the *half-life*, $t_{1/2}$, rather than the decay constant k; they are related by the equation

$$t_{1/2} = \ln 2/k.$$

The half-life gives a useful guide to how long the radioactivity of a given substance will last, but it is a feature of the law of exponential decay that it will never disappear completely. After 10 half-lives about a thousandth of the original material will remain, and after 20 half-lives about a millionth.

All elements have radioactive isotopes, but most of these have such short half-lives that they are never encountered in nature. For some elements on the other hand, all isotopes are radioactive. This is the case with all the elements beyond bismuth (atomic number 93), which decay by a succession of α and β processes to an isotope of lead. Two lighter elements, technetium (number 43) and promethium (number 61) are also unstable in this way. They have been made artificially in nuclear reactors and accelerators, as have many very heavy elements.

Table I.6 shows a selection of radioactive isotopes found on Earth. They are classified into three categories. L represents naturally occurring isotopes with very long half-lives, comparable to the age of the Earth (10^9 years or more). These have been present since the origin of the Earth, although originally in larger amounts. They include the radioactive elements *thorium and *uranium, and isotopes of a few lighter ones such as *potassium and *rubidium. The existence of such long-lived radioactive isotopes is important to us in several ways. They provide a heat source

Table I.6 Some radioactive elements in the environment, classified in the last column as follows: L, long-lived isotope present in significant natural abundance; S, short-lived isotope produced naturally in radioactive decay or by cosmic-ray bombardment; A, artificially produced isotope. Only a selection of the most important isotopes in each category is included

Element	Isotope	Decay mode	Half-life	Class
H	^{3}H (T)	β	12.3 years	S, A
C	^{14}C	β	5730 years	S, A
K	^{40}K	β, EC	1.25×10^{9} years	L
Co	^{60}Co	β	5.3 years	A
Kr	^{85}Kr	β	10.8 years	A
Rb	^{87}Rb	β	4.9×10^{10} years	L
Sr	^{89}Sr	β	51 d	A
	^{90}Sr	β	28 years	A
Tc	^{99}Tc	β	2.1×10^{5} years	A
I	^{129}I	β	1.7×10^{7} years	A
	^{131}I	β	8.0 d	A
Xe	^{133}Xe	β	5.2 d	A
Cs	^{137}Cs	β	30 years	A
Rn	^{222}Rn	α	3.8 d	S
Ra	^{226}Ra	α	1600 years	S
Th	^{232}Th	α	1.4×10^{10} years	L
U	^{235}U	α	7.0×10^{8} years	L
	^{238}U	α	4.5×10^{9} years	L
Pu	^{239}Pu	α	2.4×10^{4} years	A
	^{240}Pu	α	6.6×10^{3} years	A
Am	^{241}Am	α	433 years	A
Cm	^{244}Cm	α	18 years	A

within the Earth, which although small compared with the input of solar energy, is nevertheless responsible for fuelling many of the geological processes that have shaped the Earth's surface (see §1.3). They also contribute, both directly and through their shorter-lived radioactive daughters, to the background radiation to which all life is subjected. The decay of radioactive elements, at a rate which is constant and essentially unaltered by temperature, pressure or states of chemical combination, also serves as an important 'clock' which can be used by earth scientists to determine the ages of rocks, and even of the Earth and the solar system (see especially under *lead).

Isotopes marked *S* in Table I.6 are also natural, but have shorter half-lives so that they cannot be left over from the origin of the Earth. They are much rarer than the long-lived isotopes, and come from two sources: as intermediates in the decay of the long-lived elements *thorium and *uranium, and from nuclear reactions resulting from the bombardment of the Earth by high-energy cosmic rays. The most important element in the former category is ⋆radon, an element of the noble gas Group 18, which is a radioactive health hazard. The many products of cosmic-ray bombardment include the ⋆carbon isotope ^{14}C, which is widely used in archaeology for radioactive dating.

The final class in Table I.6 is *A*, denoting isotopes that do not occur naturally on Earth, but which are produced, either deliberately or inadvertently, by artificial means. They mostly come from nuclear reactors or explosions, either as fission products from the splitting of ⋆uranium nuclei, or as a result of neutron bombardment of non-radioactive elements. Although these artificial radioactive isotopes are relatively very rare, and are responsible for much less radioactivity than we receive generally from other sources, they are the ones giving rise to most environmental concern. Some of the issues involved here are discussed later, in §5.5 and under the heading of the relevant elements.

3

Chemical compounds in nature

The different types of compounds formed by each element obviously reflect its characteristic chemistry, as well as the conditions present when the Earth was formed, especially through the relative abundance of elements initially present. These issues were discussed briefly in Chapter 2; their consequences for the chemistry of the elements in Earth are described in more detail in the present chapter. §3.1 is concerned mostly with the combination of elements with oxygen. The following sections give a survey of the compounds found in different parts of the environment.

3.1 The oxidation states of the elements

The important role of oxygen has already been mentioned (§2.2). The dominance of this element arises from a combination of its abundance and its high electronegativity, so that it is found on Earth in combination with a majority of the other elements. In the solar nebula from which the Earth formed (see §1.1), most oxygen was present in combination with the most abundant element, hydrogen, in the form of water vapour. As the nebula cooled, many elements started to react with this and form solid oxides, through reactions such as

$$M + H_2O \rightarrow MO + H_2.$$

We would expect that the elements which ultimately combined with oxygen would be those with the greatest chemical affinity for it. This is largely found to be true. A quantitative measure of the chemical affinity of an element for oxygen is the *Gibbs free energy of formation*, ΔG_f, per mole of oxygen consumed in the formation of an oxide from the elements:[24]

$$(1/n)\ M + (1/2)\ O_2 \rightarrow (1/n)\ MO_n.$$

---

24 See Atkins (1990) and Johnson (1982) for discussions of chemical thermodynamics.

Such ΔG_f values depend on conditions such as temperature and pressure; furthermore most elements are not present as simple oxides, but in combination with other elements such as silicon in complex oxides. A complete set of data for all the relevant compounds would be difficult to assemble, but a reasonable measure of oxygen affinities can be obtained by taking water as the environment, and considering for each element the most stable oxidised form occurring at neutral pH (7). This is what has been done in Fig. I.12. The numbers plotted, against the periodic group number for elements, are ΔG_f for the most stable oxidised form of each element, at 25°C and

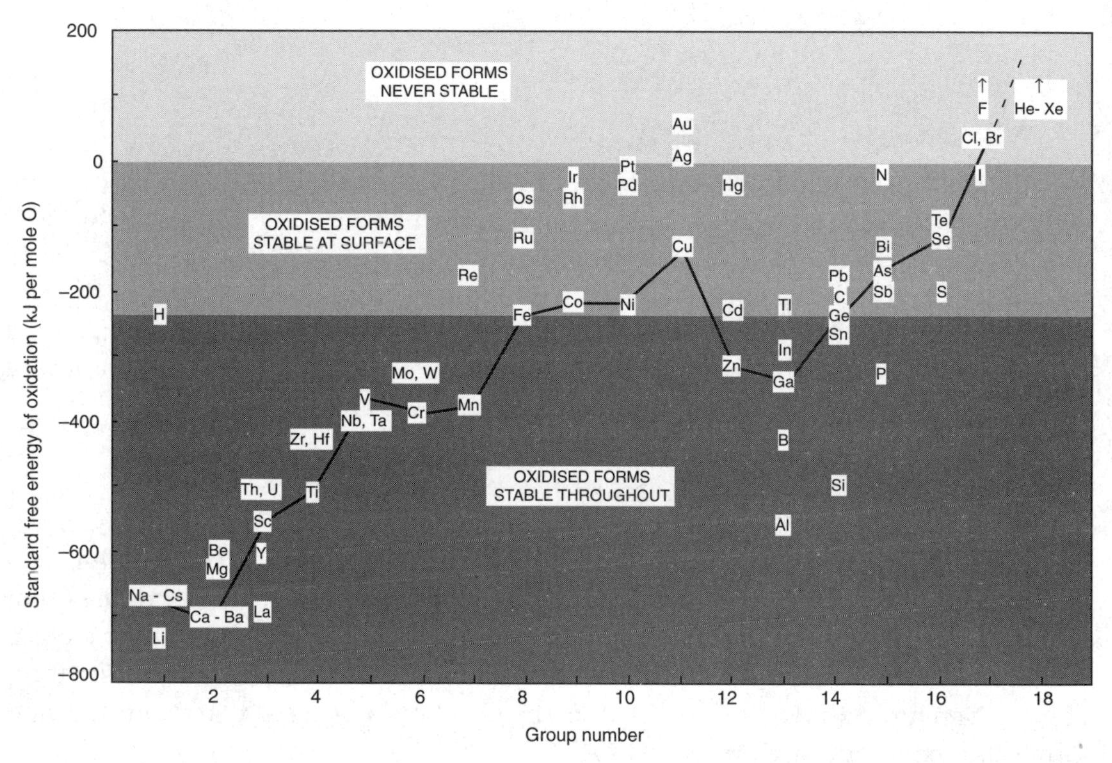

Fig. I.12 The affinity of elements for oxygen, showing the relative stability of oxygen compounds of the elements, the more stable ones being lower down in the plot. Stability is measured as ΔG_f per mole of oxygen consumed, for the formation of a compound from the element and O_2 at 25°C and one atmosphere pressure. In each case, the number refers to the *most stable* oxidised form of the element in water at pH 7, and is sometimes a simple oxide (e.g. TiO_2), but more often may be an ion in solution (e.g. Na^+, SO_4^{2-}) or a hydroxide (e.g. $Al(OH)_3$). The solid line is drawn as a guide to the eye, and connects elements of the first long period (K–Kr). The three differently shaded regions show: (top) elements where oxidised states are never predicted to be stable; (middle) ones where oxides are stable in the presence of atmospheric O_2; and (bottom) elements where oxides are predicted to be stable throughout the Earth. Note the controlling positions of H and Fe on the borderline between the lower regions.

1 atmosphere pressure.[25] Elements lower down in this plot have a stronger affinity for oxygen. The solid line is drawn as a guide to the eye, through elements of adjacent atomic number from potassium to krypton across the first long period. The main trend in the plot is change from electropositive elements on the left, which have very stable oxides, to electronegative ones on the right, where the oxides are much less stable. There is some irregularity in the later part of the transition series (e.g. chromium to copper).

Three regions are shown with different shading in Fig. I.12. The lowest one contains elements such as sodium, magnesium, and silicon, which have a higher affinity for oxygen than does hydrogen. They should react with water to liberate hydrogen, and so be found always as oxides. Nearly all the elements in this category are indeed found as oxides (including complex ones such as silicates) throughout the Earth, although a few such as zinc have a higher affinity for sulfur, and so occur as sulfides.

Elements falling in the top region in Fig. I.12 have oxides with a positive free energy of formation. This means that they are not stable even the presence of oxygen at one atmosphere pressure. Only a few elements are in this category: gold, the halogens fluorine, chlorine and bromine, and the noble gas elements of Group 18. These elements are almost never found in oxidised form. Gold occurs as the native (uncombined) element, the halogens in reduced form as F^-, Cl^-, and Br^- anions, and the noble gases as uncombined elements.

The middle region in Fig. I.12 shows elements with oxides stable in contact with atmospheric oxygen, but unstable in the presence of hydrogen. As the solar nebula condensed, these elements would not have been expected to form oxides, and indeed most of them do not occur predominantly in this form on Earth. They include the siderophilic metals, concentrated in the core, and some chalcophilic elements found in the crust as sulfides. The presence of free oxygen in the Earth's atmosphere, which is anomalous from the thermodynamic point of view and is due entirely to the presence of life on our planet, does however allow elements in this middle category to occur sometimes in oxidised states at the Earth's surface. For example, we occasionally find copper in the form of carbonates, and nitrogen and sulfur oxidised as nitrates (NO_3^-) and sulfates (SO_4^{2-}). The metallic elements in this group are often found as sulfides in the crust, although they tend to be concentrated as siderophilic elements in the Earth's metallic core.

Two elements should be singled out in this discussion, because of their marginal positions in the plot, and their importance on Earth: these are *iron and *carbon. Carbon is in the middle region in Fig. I.12, suggesting that it should not have originated on Earth in oxidised form. It appears in fact, that most (80 per cent) of carbon on Earth is oxidised, in the form of carbonate (CO_3^{2-}). Reduced carbon in the crust,

---

25 Thermodynamic data taken from various sources, including Johnson (1982).

such as fossil fuels, is thought to be secondary in origin, and the consequence of photosynthesis. Many reactions of carbon compounds are very slow in the absence of suitable catalysts. The conditions of thermodynamic equilibrium implied in our discussion are often not attained therefore. (For further discussion, see §4.1 and especially *carbon.)

Iron is interesting in other ways. It is the second most abundant element on Earth (see Fig. I.10), and like carbon, is present in a variety of chemical forms: metallic iron in the core, Fe^{2+} throughout the mantle and in the deeper crust, and Fe^{3+} at the Earth's surface. Iron is very slightly below hydrogen in Figure I.12, which suggests that it should all be present on Earth in an oxidised form. The conditions of the solar nebula were not those of liquid water at 25°C, however, and the partitioning of oxygen between hydrogen and iron depends on the temperature. More detailed calculations suggest that the first solids formed, when the nebula cooled below about 1200°C, contained iron in the unoxidised metallic form. It started to oxidise by reaction with water vapour at around 300°C, although some iron sulfides were also formed. The Earth was made of dust which had condensed over a range of temperatures, and as the nebula cooled, the dust was already starting to gather together under gravitational attraction into larger bodies. Metallic iron in the centre of these would not have been exposed to the cooler gases which could have oxidised it. Furthermore, at some stage during this process the uncondensed gases—hydrogen, together with its volatile oxide, water—were dispersed and blown off into deep space by a 'wind' of high-energy atoms emitted by the sun. Iron was therefore left on Earth in a variety of chemical forms—metal, oxide, and sulfide. Other elements with a similar ΔG_f to iron in Fig. I.12, such as cobalt, nickel, germanium, and tin, are in a similar situation. Iron has more influence than these because it is much more abundant, making up some 36 per cent of the total mass of the Earth. The presence of such an abundant element as metal, sulfide, and oxide, means that other elements must form oxides and sulfides in competition with iron. It appears therefore that the overall chemical state of the Earth comes from a coincidental combination of two very different circumstances: iron is a very abundant element (a result of its *nuclear stability*, which determines how much was made in stars) and has also a similar affinity to hydrogen for oxygen (a *chemical* property having to do with the behaviour of electrons outside the atomic nucleus).

The arguments so far have been based on equilibrium thermodynamics, predicting the most stable compounds under given conditions. If the Earth were truly in chemical equilibrium, chemistry here would be much less interesting than it is in fact. The surface of the Earth, and especially the atmosphere, is definitely not in equilibrium. This has partly come about because of the complex way in which the Earth was made. It was originally composed of solids which formed from the nebula under a variety of conditions, and then separated into layers which could not react further. In common with other inner planets such as Venus and Mars, the surface is more oxidised than the interior. A unique feature of Earth however is the presence of life, which has released free O_2 by capturing solar energy in photosynthesis, and so has led to a much more highly oxidising atmosphere.

One of the most important manifestations of chemical disequilibrium is the occurrence of some elements in variable oxidation states. The transfer of oxygen between such elements, in oxidation/reduction or *redox* reactions, is a crucial feature of chemistry at the Earth's surface. It is involved in the environmental cycling of many elements (see Chapter 4), and is central to the chemical processes of life. Examples of variable oxidation states include iron, found in oxides as Fe^{2+} or Fe^{3+}; sulfur as sulfide (S^{2-}), elemental sulfur, or sulfate (SO_4^{2-}) (oxidation state +6); and carbon as CO_2 or carbonate (CO_3^{2-}) (oxidation states both +4), and as organic matter, which may be approximately formulated as CH_2O with an average carbon oxidation state of zero. These elements are in the intermediate region of Fig. I.12, but some from the lower region, where oxidised forms are always stable, also show variability, for example, Mn in +2 and +4 states, and U as +4 and +6.

The thermodynamics of redox reactions are usually discussed in terms of *electrode* or *redox potentials*. These potentials express the voltage of a electrochemical cell or 'battery' which could in principle be generated by the reaction. More positive potentials show stronger oxidising agents, that is, species which tend to take up electrons from reducing agents, which have less positive potentials. For example, O_2 and reduced oxygen in water are in equilibrium at a more positive potential (+1.28 V in acid solution at pH 0) than Fe^{2+} and Fe^{3+} (+0.77 V under the same conditions), so that atmospheric oxygen will oxidise Fe^{2+} to Fe^{3+}, being itself reduced to water. On the other hand, metallic aluminium and Al^{3+} have a potential of –1.68 V, which is much more negative than that of hydrogen gas and water (0 V), so that aluminium will reduce water, giving Al^{3+} and H_2. This last example shows one important limitation of these potentials: they express the thermodynamic tendency of a reaction to go in a given direction, but say nothing about how fast this will happen. It is indeed true that all aluminium in the natural environment occurs in the Al^{3+} form, but manufactured aluminium metal can be used with water, for example in cooking vessels and soft-drinks cans. This is because the initial oxidation creates a very stable and impermeable layer of aluminium oxide on the surface, which protects the metal against further oxidation. As well as this kinetic limitation, redox potentials depend on conditions such as pH, the presence of other elements (especially when these lead to formation of insoluble compounds), and concentrations. For all these reasons, the *standard electrode potentials* found in chemistry books are not directly applicable to elements in the environment. Figure I.13 shows values which are more appropriate to many natural situations. They refer to neutral water (pH 7), and the stable forms of the elements in different oxidation states under these conditions, and at concentrations likely to be met.[26]

26 Data from various sources.

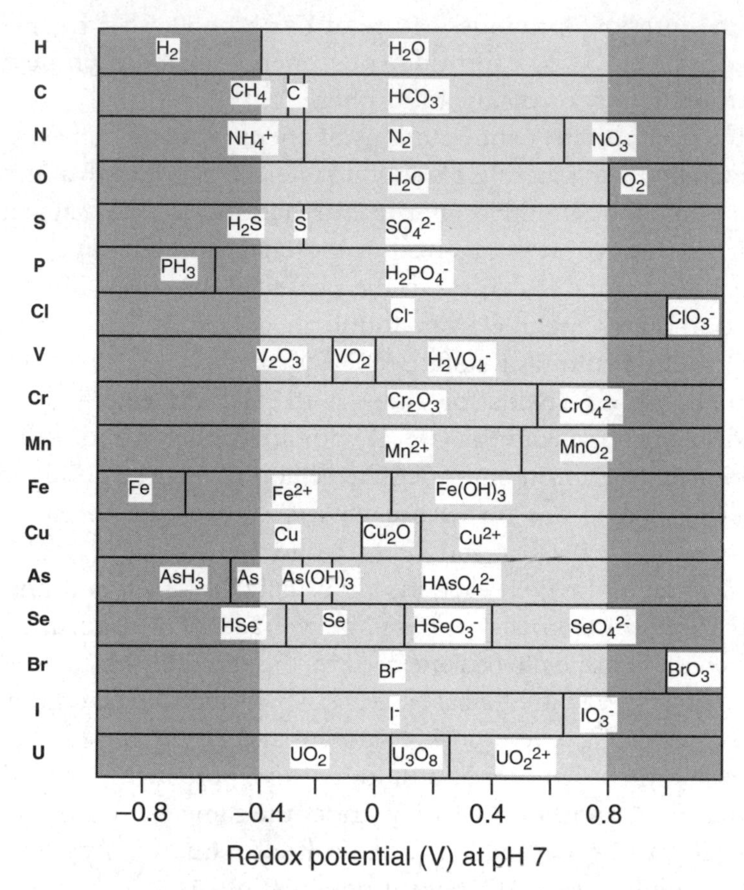

Fig. I.13 Redox potentials for some elements in the environment, showing the predominant species of each element, listed down the left-hand side, in water at pH 7. Each vertical line dividing two regions is the redox potential at which the species shown on either side are in equilibrium. Forms on the left are stable under reducing conditions, and those on the right where more oxygen is available. The darker shading shows potentials at which water should be reduced (on the left) or oxidised (on the right). Concentrations are appropriate to the natural environment, but nevertheless the potentials may vary considerably depending on other elements present in combination with those listed.

For each element, the species listed are the predominant ones in equilibrium with water at neutral pH; these may be soluble ions such as Fe^{2+} or Cl^-, gases such as O_2, or insoluble solids such as C or $Fe(OH)_3$. The vertical lines represent the electrode potential at which the species shown on either side are in equilibrium. A few elements are shown which do not normally occur in variable oxidation states, and are included to illustrate the limitations imposed by the stability of water. For phosphorus, the phosphate/phosphine $H_2PO_4^-/PH_3$ potential is lower than that for hydrogen/water, and so phosphine, if it existed naturally, would be able to react with water to give hydrogen. Phosphorus is found naturally always as phosphate. In the case of chlorine and bromine, the oxidised chlorate and bromate forms, ClO_3^- and BrO_3^-, are only stable at potentials above the oxygen/water one, and can therefore oxidise water to O_2. These elements are always found in nature in the reduced form as chloride and bromide, Cl^- and Br^-, respectively. Thus the darker shaded regions in Fig. I.13 represent 'forbidden areas', not normally found with naturally occurring forms of the element because they would be able to decompose water, either by liberating oxygen (to the positive side) or hydrogen (on the negative side).

One immediate conclusion from Fig. I.13 is that forms of the elements stable only towards the left-hand side should be oxidised by atmospheric O_2; e.g. methane and carbon to bicarbonate, sulfides and sulfur to sulfate, UO_2 and U_3O_8 to the soluble uranyl ion, UO_2^{2+}, etc. Many of these transformations do actually occur, and are important in the natural environment. When minerals containing reduced forms of the elements are exposed to the air and undergo weathering, iron oxidises to $Fe(OH)_3$, manganese to MnO_2, and UO_2 to UO_2^{2+}. However, some of the predicted reactions are extremely slow. The redox potentials suggest that atmospheric N_2 should be oxidised to nitrate, but the rate of this process is so slow as to be undetectable under normal conditions. A small amount of nitrogen is converted to nitrate in the atmosphere, through the action of lightning which generates reactive nitrogen oxides, but this natural conversion is now outweighed by the generation of nitrogen oxides during fossil fuel burning, and other parts of the *nitrogen cycle which involve oxidation and reduction are controlled by the action of life. This example illustrates the kinetic limitations, which are often especially important in the redox reactions of non-metals. Carbon is another example, as its most interesting compounds, those of life, do not appear in Fig. I.13 at all because they are not thermodynamically stable at any redox potential.

The redox potentials are important in spite of their limitations, as they show the thermodynamically favoured direction of change, and allow us to see some of the ways in which the environment is not in chemical equilibrium. The existence of reduced forms of the elements in the crust, together with oxygen in the atmosphere, is one of the most significant features of our environment on the Earth. Its presence is almost entirely due to photosynthesis by plants. If it were not for atmospheric O_2, the predominant redox states of each element would correspond to the equilibrium forms at a potential of about –0.2 V. This is very close to the potentials of three important redox couples, of *carbon (C/HCO_3^-), *nitrogen (NH_4^+/N_2), and *sulfur ($H_2S/S/SO_4^{2-}$); these are the elements which, along with *iron (Fe^{2+}/Fe^{3+}) largely control the redox chemistry of the Earth's surface. The environmental chemistry of some other elements appearing in Fig. I.13, especially *arsenic and *selenium, is also dominated by redox reactions.

3.2 The minerals of the crust

The solid compounds of the elements show an enormous diversity of composition and structure. Figure I.14 shows the arrangement of atoms in just three simple

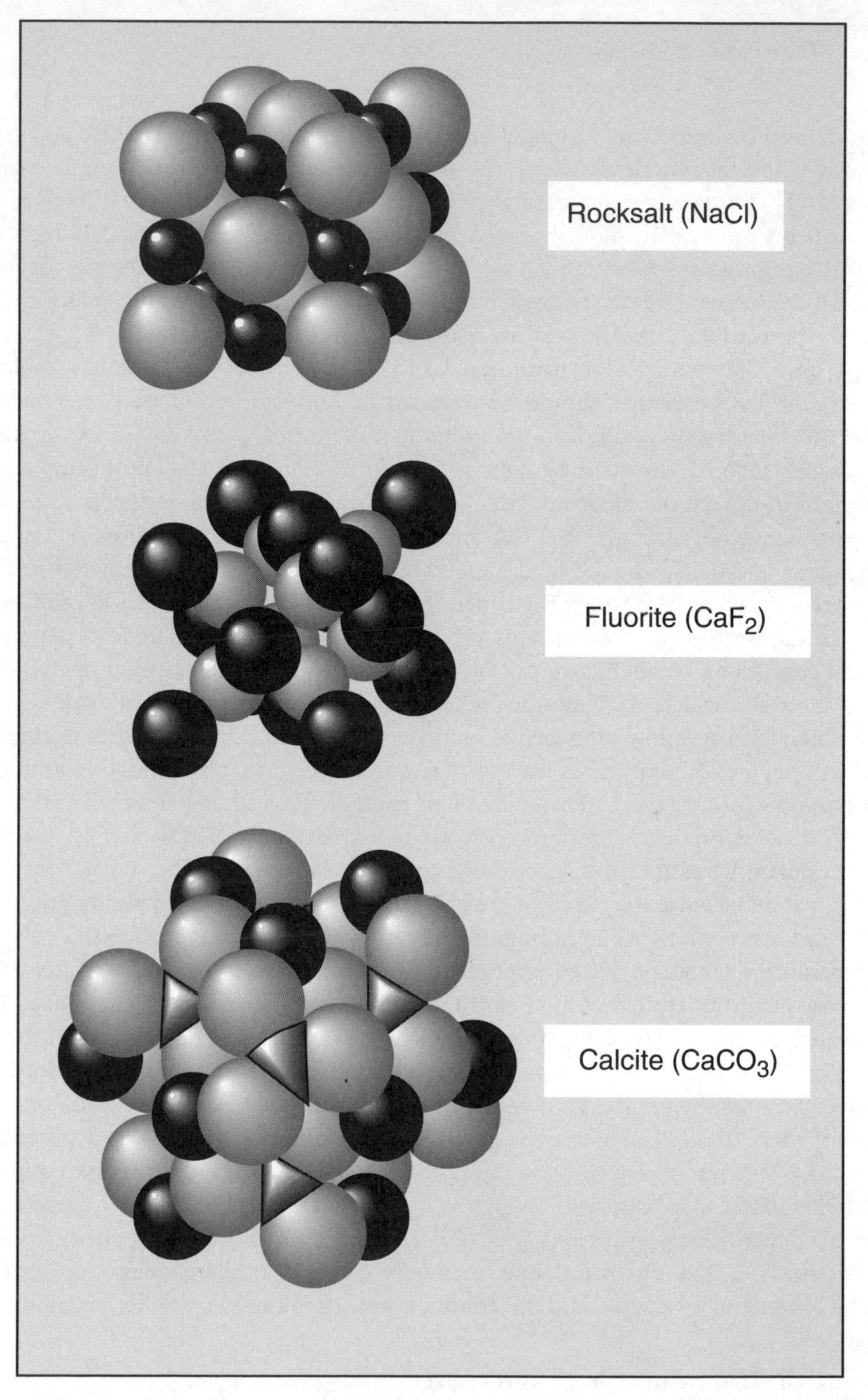

Fig. I.14 The crystal structures of three simple minerals. Anions are represented by light shaded spheres, cations dark. Note the triangular carbonate ions (CO_3^{2-}) in calcite.

structures, to illustrate the principles involved. *Rocksalt* (sodium chloride, NaCl) is one of the simplest, being found also in some other compounds, such as *galena* (PbS). *Fluorite* is CaF_2; a few oxides such as UO_2 have this structure. In these two structures simple cations (Na^+, Ca^{2+}) and anions (F^-, Cl^-) alternate so that ions of one kind are surrounded by ions of the other kind. The structure of *calcite*, the commonest crystal form of calcium carbonate, $CaCO_3$, illustrates a complex oxide where one can see the closely bound triangular carbonate ion, CO_3^{2-}. Complex ions, especially oxyanions such as carbonate, are a feature of many minerals; for example, the tetrahedral sulfate, SO_4^{2-}, and silicate, SiO_4^{4-}, ions are constituents of $BaSO_4$ and Mg_2SiO_4, respectively. Hydroxide, OH^-, is another common ion. Many minerals have structures much more complicated than those illustrated: silicates are especially complex, and discussed under *silicon.

The bonding between the atoms forming a complex ion, e.g. between oxygen and silicon in silicate, are largely covalent in nature; it is the directional and rather specific nature of these bonds that controls the structure of the ion, and maintains its integrity in a range of different compounds. But the electrostatic bonds between ions—whether simple or complex ones—do not require specific geometrical arrangements. For example, each Na^+ ion in rocksalt is surrounded by six Cl^- ions, and vice versa. This so-called *coordination number* of an ion is somewhat variable from compound to compound; thus Na^+ can have eight or even more near-neighbour oxygen atoms in silicates. There are however certain trends, which are commonly accounted for by attributing a definite size to each ion. Such *ionic radii* can be found by partitioning the observed distance between an anion or a cation, although there is no unique way of doing this; it must also be recognised that the apparent radius of an ion depends somewhat on its coordination number.[27]

Figure I.15 shows some typical ionic radii and average coordination numbers of elements in positive oxidation states found in oxide minerals. Coordination numbers are greater for larger cations, as it is possible to pack more anions around them than around small cations. It can also be seen that the largest ions, having the largest coordination numbers, are ones of low charge. When elements occur in different oxidation states, such as iron (Fe^{2+} or Fe^{3+}) or manganese (Mn^{2+} or Mn^{4+}), the more highly charged one is distinctly smaller. Some elements have been included even when they do not form simple cations. C^{4+}, P^{5+}, and S^{6+} should be regarded as constituents of carbonate, phosphate, and sulfate ions, respectively, but it is possible to assign a notional ionic radius to these species, by subtracting the appropriate radius for O^{2-} from the bond distances in the complex ions. The borderlines between the coordination numbers are quite flexible, and as mentioned already, radii depend on the coordination. For example Al^{3+}, on the 4/6 borderline in Fig. I.15, is found in

27 The values most widely used, and ones quoted in this book, are from Shannon and Prewitt (1969, 1970).

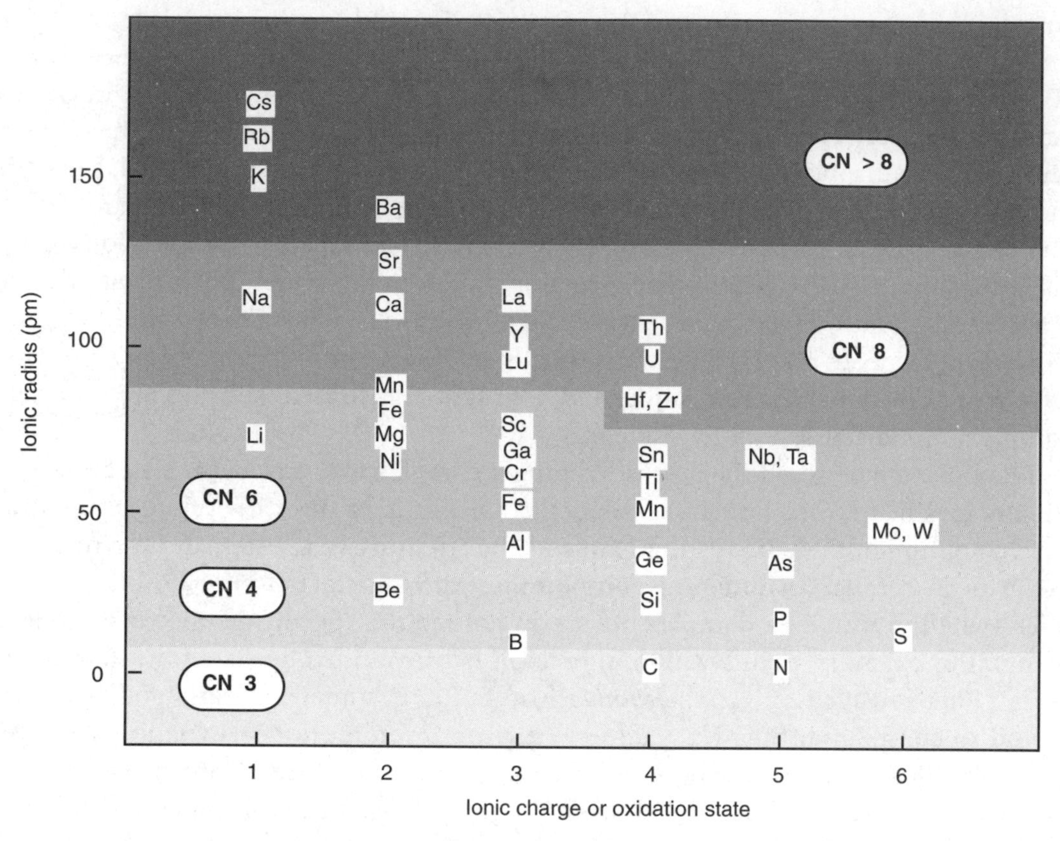

Fig. I.15 Ionic radii and typical coordination numbers (CN) of some cations found in oxide minerals. Note that both radii and coordination numbers are variable. (1 pm = 10^{-12} m = 0.01 Ångstrom Units.)

both octahedral (CN 6) and tetrahedral (CN 4) coordination, as well as less regular geometries. Its radius is distinctly less when the coordination number is smaller (e.g. 39 pm for CN 4, as against 53.5 pm for CN 6).

The ionic radii are useful not only in connection with mineral structures. Later sections will show how ionic charge and size are correlated with features such as solubility in water (§3.4) and the distribution of elements between the mantle and the crust (§4.5). For the moment, Fig. I.16 presents another interesting correlation. The types of oxide and halide minerals formed by elements are partly a function of the crustal abundance discussed earlier. But they must also reflect patterns of chemistry, and this is what appears in the figure. The predominant types of mineral in which an element is found are drawn as overlapping regions on a plot of radius versus charge. Very small or highly charged 'ions' (which do not have a real existence) are found as oxyanions, and occur in salts with common metallic elements, especially *calcium. Lower charged or larger ions occur frequently as silicates, a reflection on the high abundance of silicon. A wide range of other compounds can be found, however, and

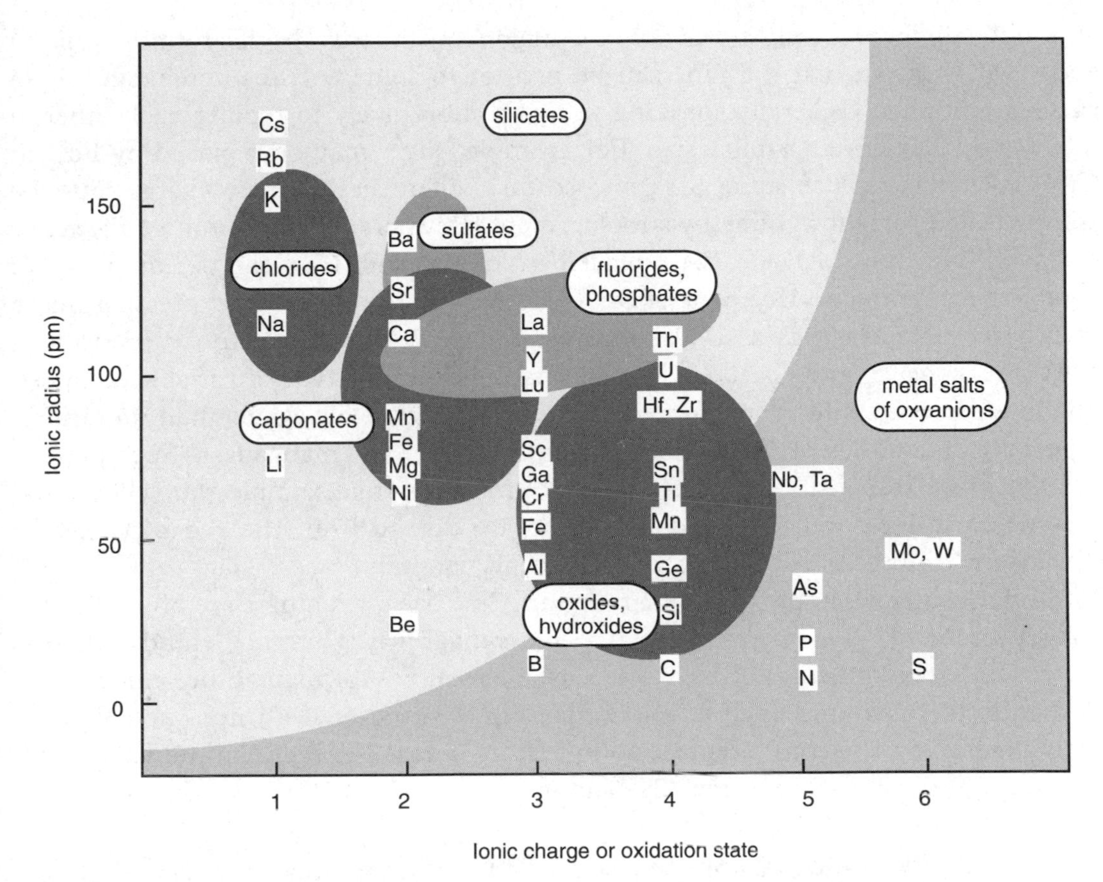

Fig. I.16 Oxide and halide minerals formed by lithophilic elements. The overlapping shaded regions show how the occurrence of different types of minerals is controlled by ionic charges and radii. Silicates are found for all ions in the upper left-hand region, and are generally more abundant than other types.

the diagram shows how the occurrence of these—oxides and hydroxides, fluorides, chlorides, carbonates, phosphates, and sulfates—is correlated with the ionic charge and size. Many of these are *sedimentary* minerals, either precipitated from water, or left as insoluble remains when other compounds have dissolved from rocks. The correlation expressed in Fig. I.16 therefore reflects the solubility of compounds in water, although the abundance of elements is also important.[28]

The processes involved in mineral formation are described briefly in §4.2. The separation of elements according to ionic charge and radius is far from perfect, and the simple formulae given to some minerals—$CaCO_3$, TiO_2, and so on—are often misleading. Silicates often contain a large number of elements, a formulation (of *hornblende*) such as $(Na,K)_{0-1}Ca_2(Mg,Fe,Al)_5[Si_{6-7}Al_{2-1}O_{22}](OH,F)_2$ being typical,

28 The relationships between solubility and ionic charge and size are rather complex. See Chapter 5 of Johnson (1982) for a discussion.

although still inexact as it shows only the major constituents. In this formula (Na,K) and (OH,F) mean that the ions can be present to some extent interchangeably in varying amounts. Generally speaking, ions are more likely to replace each other in this way if they have a similar size. For example, Mg^{2+} may be replaced by Fe^{2+} or Cr^{3+}, Al^{3+} may replace some Si^{4+}, and so on. Differences of ionic charge must be compensated somehow, often by a replacement elsewhere. In the common *plagioclase feldspars*, $(Na,Ca)(Si,Al)_4O_8$, the replacement of Na^+ with Ca^{2+} can be compensated by a corresponding replacement of Si^{4+} with Al^{3+}. Another way of expressing this is to regard plagioclase as a solid solution of the two ideal extreme compositions $NaAlSi_3O_8$ (*albite*) and $CaAl_2Si_2O_8$ (*anorthite*).[29] Even this type of variable composition is an oversimplification. Supposedly 'pure' minerals always contain impurities, especially of common elements such as iron, titanium, and manganese. Such impurities are often responsible for the colours of minerals; for example the yellows and browns of much sandstone and limestone are due to iron, the red of rubies to chromium, and the blue of sapphire to iron and titanium.[30]

Similar principles apply to sulfide minerals. The basic structures are often simpler, but replacements of elements take place in a similar way. Common sulfides such as *blende* (ZnS) and *galena* (PbS) are often used as sources for many other chalcophilic elements: metallic ones such as silver may replace zinc or lead; non-metallic ones such as arsenic or selenium replace sulfur. More details can be found under individual chalcophilic elements, especially *sulfur.

3.3 The volatile compounds of the atmosphere

In the composition of the rocks which make up its solid surface and interior, the Earth is fairly typical of its neighbours in the solar system: the Moon, Venus, and Mars. The origin of the Earth's atmosphere is thought also to be similar to those of Mars and Venus, by the *outgassing* of volatile materials combined or trapped in minerals during their condensation from the solar nebula (§1.1). In the subsequent development of atmosphere and oceans, however, the Earth is peculiar. The existence of liquid water in large quantities at the surface is one such unique feature, which is a consequence of its surface temperature being controlled within a suitable range by the factors discussed in §1.4. The other unique feature is the presence of a high concentration of free molecular oxygen, O_2, in the atmosphere; this comes almost entirely from life, which depends in turn on liquid water.

The volatile components of the atmosphere are small molecules, bound by covalent bonds, containing non-metallic elements. Chemists now know that volatile molecules may be formed by most elements; for example, so-called *organometallic*

29 For an account of mineral compositions and structures, see Deer, Howie, Zussman (1966).

30 See the article by Burns in Berry and Vaughan (1985).

compounds containing a metal atom and organic (carbon- and hydrogen-containing) groups. But although these compounds can be made in the laboratory, very few of them are stable for long outside carefully controlled environments. A few organometallic compounds, such as those of *mercury, are formed naturally and can be detected in the atmosphere. Generally speaking however, only the covalent compounds formed by non-metals have sufficient stability to exist in the natural environment. The most familiar such compounds are the hydrides and oxides, such as water, H_2O, carbon dioxide, CO_2, and methane, CH_4. There are also organic compounds of some elements, such as chloromethane (methyl chloride), CH_3Cl, and dimethyl sulfide, $(CH_3)_2S$. To these must be added the uncombined atoms of the inert gases such as argon, and the diatomic molecules formed by a few elements, especially H_2, N_2, and O_2.

Table I.7 shows some of these actual or potential constituents of the atmosphere, classified according to their solubility in water. This classification recognises the importance of water in the surface environment of the Earth. Compounds in the 'high' solubility column cannot be major components of the atmosphere. Some (such as HF and HCl) are formed within the Earth by the decomposition of rocks at high temperature and subsequently emitted from volcanoes; others (such as SO_3) are formed by the oxidation of other compounds in the atmosphere (COS, H_2S, and SO_2); all of them however combine rapidly with water and are precipitated in rain or snow.

Compounds in the 'low' solubility column are not taken up in water to an extent which depletes them from the atmosphere. Their solubility may not be totally negligible, and in the case of O_2 it is essential to the existence of life in rivers and the

Table I.7 Solubility of volatile compounds in water. The table shows the oxides, hydrides, and organic compounds of some non-metallic elements, classified qualitatively according to their solubility in water

Element	Low	Medium	High
H	H_2	–	–
C	CH_4, CO	CO_2	–
N	N_2, N_2O, NO	NH_3, NO_2	–
O	O_2, O_3	–	–
F	CF_2Cl_2, other CFCs	–	HF
P	–	–	P_2O_5
S	$(CH_3)_2S$, COS	H_2S	SO_2, SO_3
Cl	CH_3Cl, CFCs	–	HCl
As	$(CH_3)_3As$	–	–
Se	$(CH_3)_2Se$	H_2Se	–
Br	CH_3Br	–	HBr
I	CH_3I	–	HI

sea. Compounds in the intermediate 'medium' solubility column are more interesting in this respect: they dissolve in rain water, but can also evaporate from water. The most notable example is CO_2, which cycles extremely rapidly between the atmosphere and the oceans (see under *carbon).

Table I.8 lists the elements present as significant volatile constituents of the atmosphere. The major compounds are given, with some less important ones in parentheses. The composition is given (unlike those for other regions of the environment) in terms of the *mixing ratio*, which is the parts by volume, proportional to the number of molecules present.[31]

The origins of the molecules in Table I.8 are very diverse. They include: volcanic activity (N_2, CO_2, H_2S); biological activity (H_2, CH_4, CO_2, O_2, N_2O, and probably most of the organoelement, CH_3-containing, compounds); chemical reactions in the atmosphere (O_3); and pollution generated by human activities of some kind (chlorofluorocarbon compounds such as CF_2Cl_2, and substantial amounts of NO, NO_2, and SO_2).

The last column in Table I.8 introduces the concept of the *residence time*, explained more fully in §4.1. This represents the average life-time of the relevant

Table I.8 Constituents of the atmosphere. For each element, the predominant volatile compounds are shown, followed sometimes by less common ones in parentheses. The concentration is given as the mixing ratio (parts by volume), and refers to dry air apart from the (very variable) water vapour concentration given for H. The residence times are in some cases averages for several different compounds

Element	Species present	Total concentration	Residence time
H	H_2O (H_2)	0–4 %	10 d
He	He	5.2 ppm	10^6 years
C	CO_2 (CH_4, CO)	0.034 %	2 years
N	N_2 (N_2O, NO, NO_2, NH_3)	78.1 %	2×10^7 years
O	O_2 (O_3)	20.9 %	3000 years
F	CF_2Cl_2, other CFCs	0.6 ppb	100 years
Ne	Ne	18 ppm	*v. long*
S	COS, SO_2 ($(CH_3)_2S$, H_2S)	0.5 ppm	<1 year
Cl	CH_3Cl, CFCs	0.6 ppb	$\gg$ 1 year
Ar	Ar	0.93 %	*v. long*
As	$(CH_3)_3As$, $(CH_3)_2AsO(OH)$	*trace*	*short*
Se	$(CH_3)_2Se$	*trace*	*short*
Br	CH_3Br (CF_3Br, CH_2BrCH_2Br)	0.01 ppb	1 years
Kr	Kr	1.1 ppm	*v. long*
I	CH_3I	0.002 ppb	10 d
Xe	Xe	90 ppb	*v. long*
Rn	Rn	*v. low and variable*	3.8 d

31 Data from Ramanathan *et al.* (1987) and Wayne (1991).

constituent in the atmosphere, before it is taken up, into water or by a living organism, or converted by a chemical reaction into another compound. The residence times given are rough average values, as each compound may have its own characteristic residence time, depending on the rates at which it undergoes chemical reactions or dissolves in water. These figures do however give a useful insight into some factors that control the composition of the atmosphere. For example, compounds of *sulfur are emitted in substantial quantities, by volcanoes, by some living organisms, and as a result of vegetation and fossil fuel burning. Their concentration in the atmosphere is very small however, because the residence times are short, meaning that they do not stay long in the atmosphere. Not only solubility in water, but also chemical reactions such as oxidation are very important in the atmospheric chemistry of many elements. These are discussed in §4.2 and under the elements concerned, especially *carbon, *nitrogen, *oxygen, *sulfur, and *chlorine.

By contrast, the residence times for some elements are extremely long. Among the noble gases, helium is light enough to escape slowly from the upper atmosphere into space, and its residence time (a million years) is controlled by this loss. *Radon is radioactive, and its residence time given (3.8 days) is the half-life of the isotope ^{222}Rn. For the other elements in this group, the residence times are comparable to the age of the Earth itself. After the noble gases, the least reactive constituent of the environment is dinitrogen, N_2; in consequence it also has a long residence time, which partly accounts for its high concentration (78 per cent by volume) in the atmosphere. Oxygen has a much more rapid turnover (3000 years), and would disappear from the atmosphere quite quickly if the photosynthetic processes of life were to stop.

One important non-metallic element not found as a volatile compound in the atmosphere is *phosphorus. Its hydride (phosphine, PH_3) has rather low solubility in water, but is always unstable relative to the oxidised state in the presence of water or oxygen. The oxide P_2O_5 never occurs free, as it has an extremely high affinity for water: phosphate is formed, which is the sole chemical form known for phosphorus in the natural environment. The consequences for the cycling of phosphorus through the environment, and for its role as an essential element in life, are profound.

3.4 The aqueous environment

Liquid water is an excellent solvent, not only for some small molecules, but also for many of the ionic constituents of minerals. The high polarity of the water molecule enables it to solvate ions efficiently: we write simple formulae such as Na^+ or Cl^-, but in reality these are strongly associated with several molecules of water. The high dielectric constant of water also reduces the electrostatic forces between ions, and enables them to separate and move freely in solution. Thus although the major compound formed on evaporation of sea water is rocksalt ('common salt'), NaCl, this compound does not exist in solution; rather we have equal numbers of hydrated Na^+ and Cl^- ions.

This simple picture is only valid for ions of low charge. More highly charged ones such as Al^{3+}, and very small ones such as Be^{2+}, produce a strong polarisation of surrounding water molecules. In neutral solution such simple hydrated ions do not exist at all; rather they form hydrated complexes such as $Al(OH)_3$ and $Be(OH)_2$. These species are also quite insoluble, although their solubility increases in both strongly acid and basic solution.

Yet more highly charged 'ions' never exist as simple cations, but always as covalently bonded oxyacids or oxyanions, such as $Si(OH)_4$ (silicic acid), and $PO(OH)_3$ (phosphoric acid, more often written as H_3PO_4, and ionised in neutral solution to a mixture of $H_2PO_4^-$ and HPO_4^{2-}). These elements occur in minerals as salts of oxyanions, as discussed above in §3.1.

Figure I.17 shows how the solubility of an ion in an oxide mineral, and the type of species formed in solution, are a function of the ionic charge and radius. Large low-charged ions (including the –1 halide ions not shown in the figure) exist in simple hydrated form in solution. Their compounds tend to be very soluble. Small high-charged ions are also soluble as oxyanions. Ions of intermediate charge/size ratio are insoluble in neutral conditions. These considerations are important in understanding the relative concentration of dissolved ions in the major reservoir of water present on Earth—the oceans. The sea-water concentrations of some elements vary markedly, both between the Atlantic and Pacific Oceans, and especially between surface and deep-sea conditions. Figure I.18 shows some average values, plotted against the position in the periodic table.[32] As with the plots given earlier (Figs. I.8–I.10 in §2.4) a logarithmic scale has been used, each vertical division representing a factor of 10. The solid line has been drawn as a guide to the eye, through the elements of successive atomic number in the first long period (K–Kr). The high relative concentrations of elements forming low-charged ions (Na^+, Mg^{2+}, Cl^-) is notable.

There is an extraordinarily wide range of concentrations shown in Fig. I.18, from Na and Cl at concentration about 1 per cent, to some elements such as Be and Bi at levels below 1 part in 10^{12}. Some of these concentrations are so low as to be barely detectable and make severe demands on the most sophisticated methods available for sampling and analysis. The elements that predominate are the ones expected from our discussion above: cations from Groups 1 and 2, anions from Group 17, and non-metals such as boron, carbon, silicon, nitrogen, and sulfur that form oxyanions.

---

32 Data from Emsley (1991).

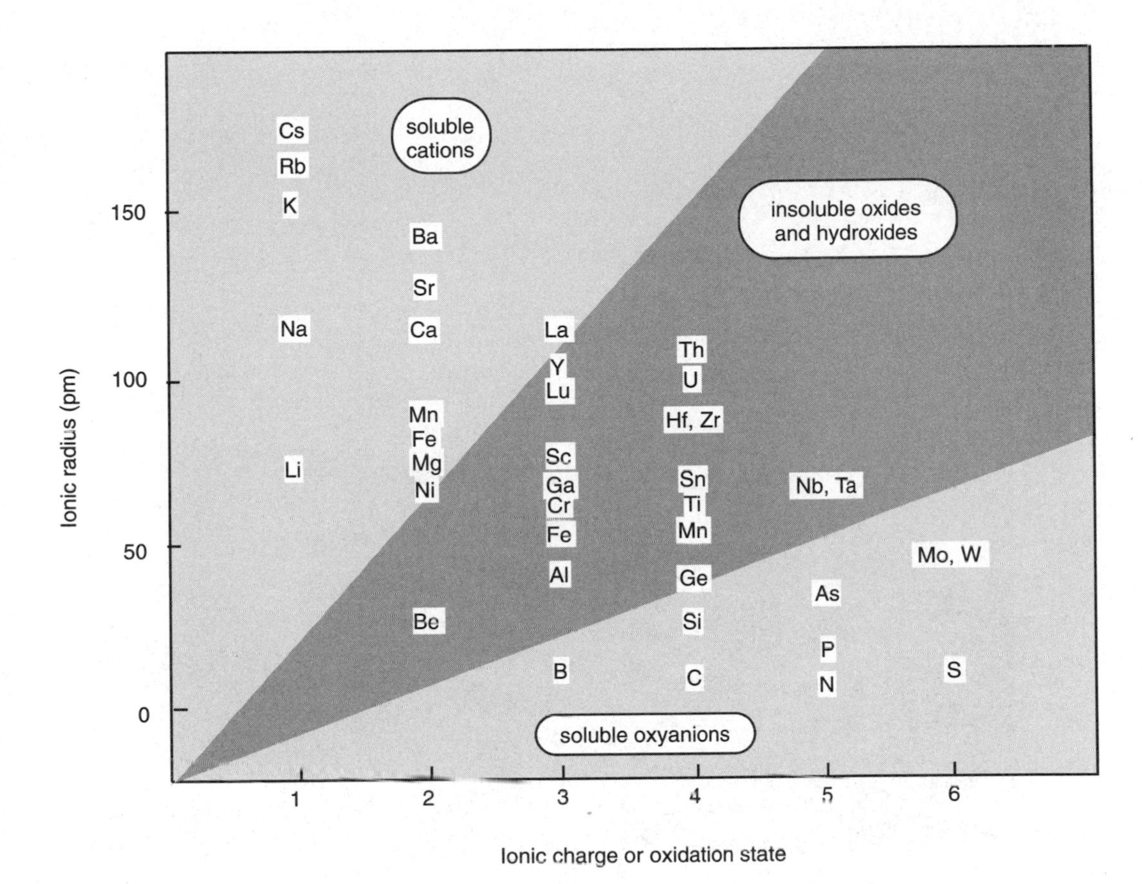

Fig. I.17 Solubility of lithophilic elements in water, showing the influence of ionic charge and radius. The division between 'soluble' and 'insoluble' species is only qualitative, and also depends on factors such as pH.

The detailed factors that control the concentrations of elements in sea water are much more complex than this broad picture implies. Most elements are derived from the *weathering of* rocks; that is, by the action of air and water, which leads to chemical breakdown of minerals and the solution of many elements, which are then carried to the sea in rivers. Other sources include biological fixation (nitrogen and carbon), dissolved volcanic gases (carbon, sulfur, chlorine and bromine), and sea-bed processes including *hydrothermal vents* from within the crust (manganese). It would be wrong however to think of the sea as merely a 'sink' in which these inputs gradually collect. Most elements are being actively removed from the oceans, by precipitation of insoluble minerals, by uptake into living organisms, and by reactions with sea-floor sediments. The concentration of any element therefore corresponds to a balance between the processes of addition and removal.

River-water inputs derived from weathering form the major immediate source for most elements in the ocean. Dissolved concentrations in rivers are extremely

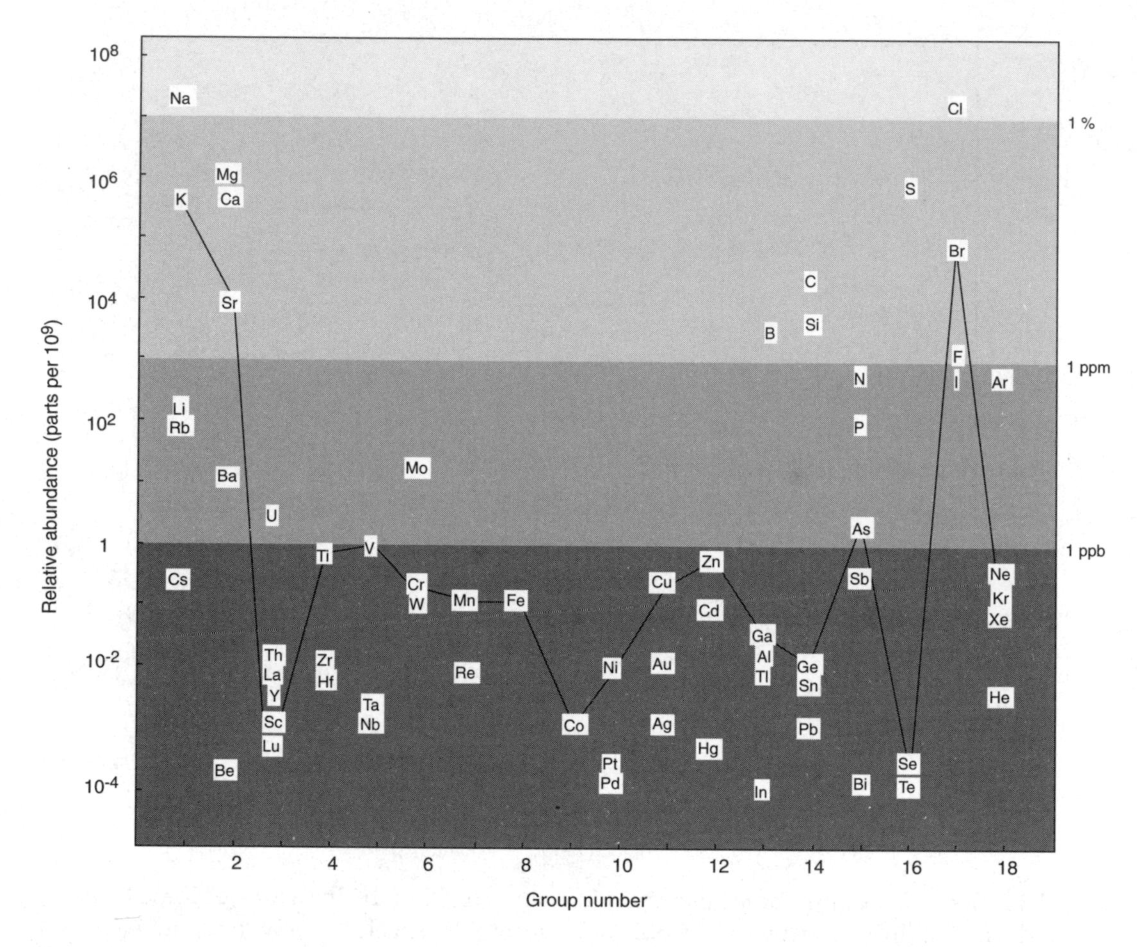

Fig. I.18 Abundance of elements dissolved in sea water. The concentrations in parts per 10^9 by mass (equivalent to μg per dm^3) are plotted on a logarithmic scale, against the group number in the periodic table. They are often very variable, and only average values are shown here. The solid line connects the elements K–Kr of the first long period, and is drawn as a guide to the eye. This plot should be compared with Fig. I.9, which shows the crustal abundance in the same way. Note the higher abundance of Cl and Br, and the much lower values for Al and Fe, in the present plot.

variable, depending on the types of rocks around, the rate of flow, and so on. Average values have been obtained for many elements.[33] Although soluble elements predominate, the commonest ones in rivers (C as HCO_3^-, Ca^{2+}) are not the same as in the sea. Figure I.19 shows some more detailed comparisons of crust/river/sea-water abundances.

The horizontal scale in Fig. I.19 gives the ratio of river to crustal abundance. Elements on the right-hand side in this plot (e.g. Cl, Br, C, S) are relatively common

33 Data from Mason and Moore (1982) and Henderson (1982).

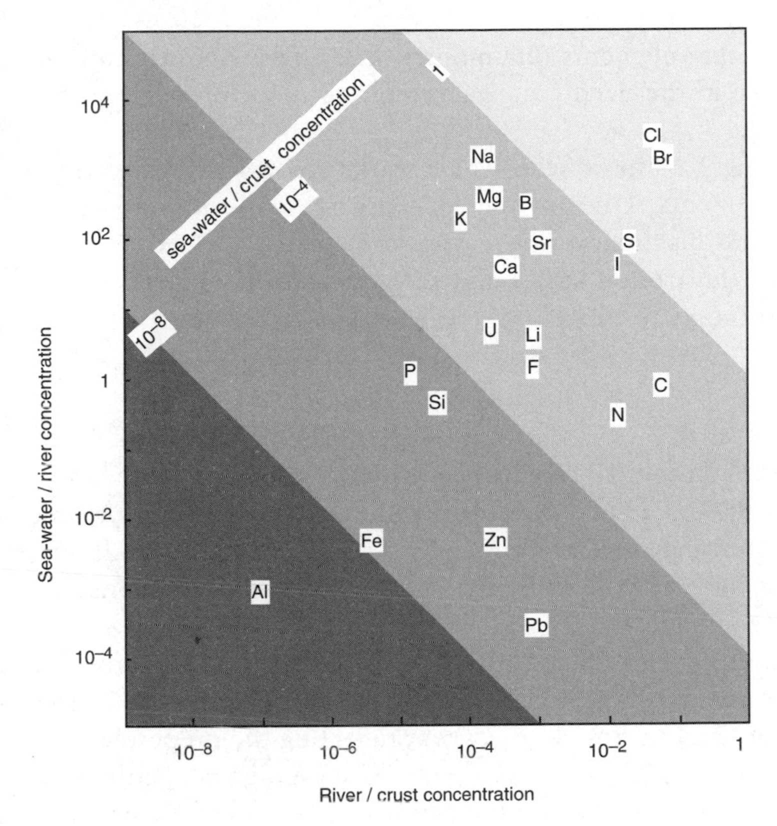

Fig. I.19 Crust–river–sea-water distribution of some elements. The horizontal scale gives the ratio of average river concentration to that in the crust; the vertical scale is the ratio of sea-water to river concentration. The diagonal scale gives the ratio of sea-water to crust concentration.

in rivers compared with rocks: they are ones very easily taken into solution, from rocks, from the atmosphere, or from living matter. Elements on the left-hand side (Fe, Al) are insoluble ones, relatively rare in rivers. The vertical scale in Fig. I.19 shows the ratio of sea-water concentration to that in average river water. Elements high up in the plot (Cl, Br, Na) are ones that become concentrated in the oceans, as sea water evaporates and leaves involatile materials in solution. On the other hand, elements low in Fig. I.19 (Pb, Al, Fe) have a very low sea-water concentration relative to that in rivers, and must therefore be rapidly removed from the ocean. The vertical scale in this figure is therefore correlated with the average residence times for elements in the ocean. Residence times for elements at the top, such as Cl and Br, are around 10^8 years, those at the bottom less than 100 years. We can see from Fig. I.19 that very soluble elements, on the right of the plot, tend to remain for a long time before removal from the ocean. But this correlation is by no means perfect (otherwise every element would lie on a straight line in the plot), and this reflects the diverse sources of addition and removal. Some of the elements (carbon, nitrogen, silicon, phosphorus, iron, zinc) are essential for life and often in short supply. Their concentrations are often less in surface waters, where life predominates. They are taken up by marine life, and to some extent recycled into the deep ocean when the dead organisms sink. Some other elements (aluminium, lead) have a higher concentration in surface waters than in the deep sea, suggesting that their presence comes partly from wind-blown dusts.

The diagonal shading in Fig. I.19 has a scale which shows the ratio of sea-water to crustal abundance of the elements. Ones at the top right are concentrated in the oceans relative to rocks. The discussion above shows that this depends on two factors, the rate of addition (horizontal scale) and that of removal (vertical scale). The cycling of elements through the hydrosphere is discussed in more detail in Chapter 4.

3.5 The elements of life

Life is often thought of (and presented in books) as essentially concerned with the chemistry of carbon. This idea is reflected in the use of the term 'organic chemistry' to describe the study of carbon compounds. In fact, as many as 25 elements may be essential for life, and the growing field of 'bioinorganic chemistry' attests to the importance of these.[34]

Figure I.20 shows the average abundance of elements in the human body; the right-hand scale shows how much is expected in a person of average weight (70 kg). The major elements are hydrogen, carbon, and oxygen, reflecting the preponderance of water and organic compounds. Eight other elements have average concentrations above one part in ten thousand, and are conventionally known as 'major elements' in contrast to 'trace elements' which have lower concentrations than one in 10^4.

Many of the elements present at low concentrations may be there 'accidentally', through their presence in food, drinking water, or as dust in the air. To decide whether a trace element is really essential for life may be difficult, and disputes remain about some, such as chromium, silicon, arsenic, and tin. Some elements, such as boron, are known to be essential for plants, but probably not for animals. Figure I.21 gives an idea of what is now known, with some uncertainties noted in essential elements.

Other elements have a well-established toxicity: these include the notorious 'heavy metals', *cadmium, *mercury, *thallium, and *lead, as well some lighter ones, such as *beryllium and *aluminium.[35] But again, the distinction between a toxic and a non-toxic element is hard to make, and depends on the amount. Some essential trace elements such as *selenium are definitely toxic in doses which may not greatly

34 Da Silva and Williams (1991) give a modern account of what is known about the role of elements in life. Useful discussions can also be found in special issues of *Journal of Chemical Education* **62** (Nov. 1985) and *Chemistry in Britain* (March 1982). Data on elemental abundance in the human body are given in Emsley (1991).

35 A fascinating historical and anecdotal account of trace elements is given by Lehnihan (1988).

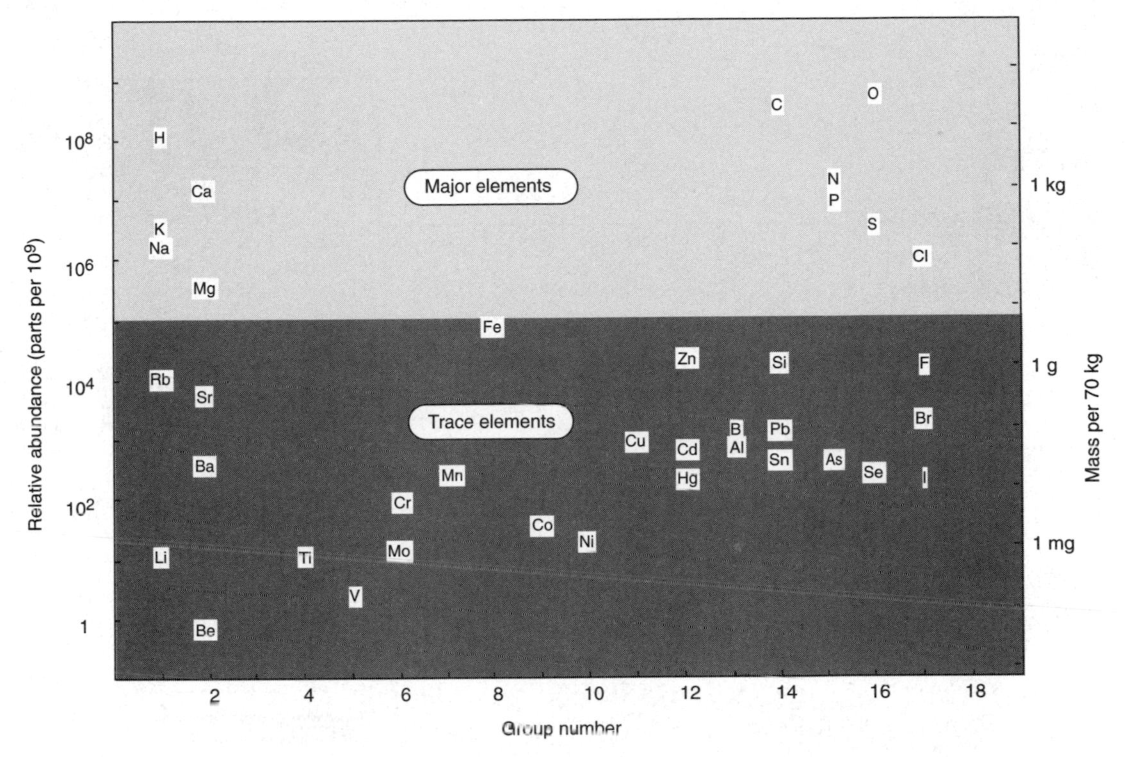

Fig. I.20 The concentration of some elements in the human body. The left-hand scale shows the abundance as in previous plots; the right-hand scale shows the total mass of each element in a person weighing 70 kg. The dividing line between major and trace elements is conventionally put at one part in 10^4. All the major elements are essential, but the trace elements shown in this plot include some inessential ones (see Fig. I.21).

exceed the necessary intake, and it is probably true that everything is harmful in excess; for example one can die from excess sodium chloride.

The types of chemical compounds formed by the different elements in biological situations are discussed in detail in Part II. They are obviously a function of the chemical behaviour of each element. Table I.9 gives a summary showing four different types of occurrence.

The first group of elements comprises non-metals bound by covalent bonds to organic molecules and polymers. The simplest organic molecules contain only hydrogen and carbon, but such hydrocarbons are quite rare in biological systems. Nearly all molecules important in life have other non-metallic elements, especially oxygen, often nitrogen, and sometimes sulfur. Figure I.22 shows the structures of a few important types of biological molecules, and is intended to illustrate some of the ways in which atoms bond together. Hydrogen, nitrogen, and oxygen are generally bonded directly to carbon. The same is often true of sulfur, although sometimes it is found in an oxidised form, as sulfate ($-OSO_3^-$), often attached to sugar molecules

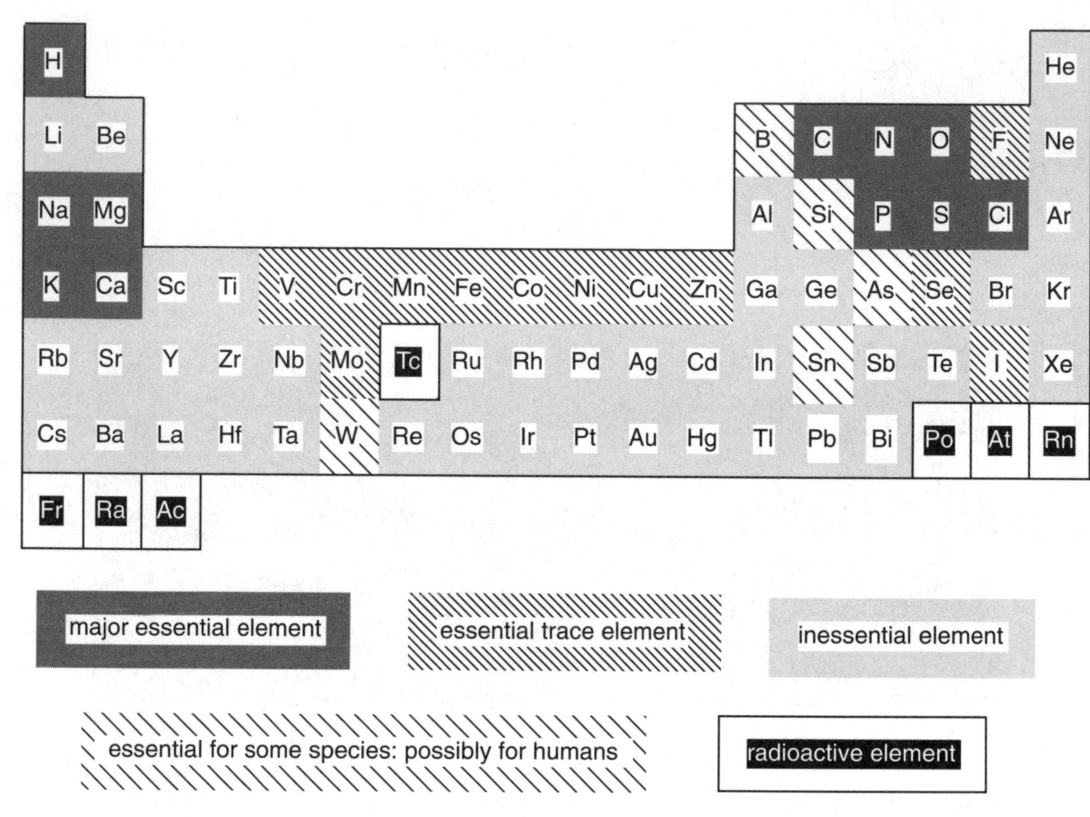

Fig. I.21 The biological significance of the elements. A few elements listed as inessential (e.g. Sr, Br) may be required by a few species, but not by humans.

Table I.9 The role of elements in biology

Chemical form and function	Elements
Covalently bound atoms in biochemical molecules and polymers	H, C, N, O, P, S, Se, I
Ions in aqueous solution and bound to charged organic groups	H^+, Na^+, Mg^{2+}, Cl^-, K^+, Ca^{2+}
Constituents of inorganic solids	C, O, F, Si, P, Ca, Fe (Sr, Ba)
Metals bound in enzymes and other specialised molecules	V, Cr, Mn, Fe, Co, Ni, Cu, Zn, Mo

such as glucose. Phosphorus is also an important constituent of many biological molecules, including DNA, but is not bound directly to carbon. It has a very high affinity for oxygen, and only the oxidised phosphate form is normally stable on Earth (see Fig I.13 in §3.2). This is true also in biology, where phosphate-containing molecules such as ATP play a crucial role in most energy-utilising reactions in cells.

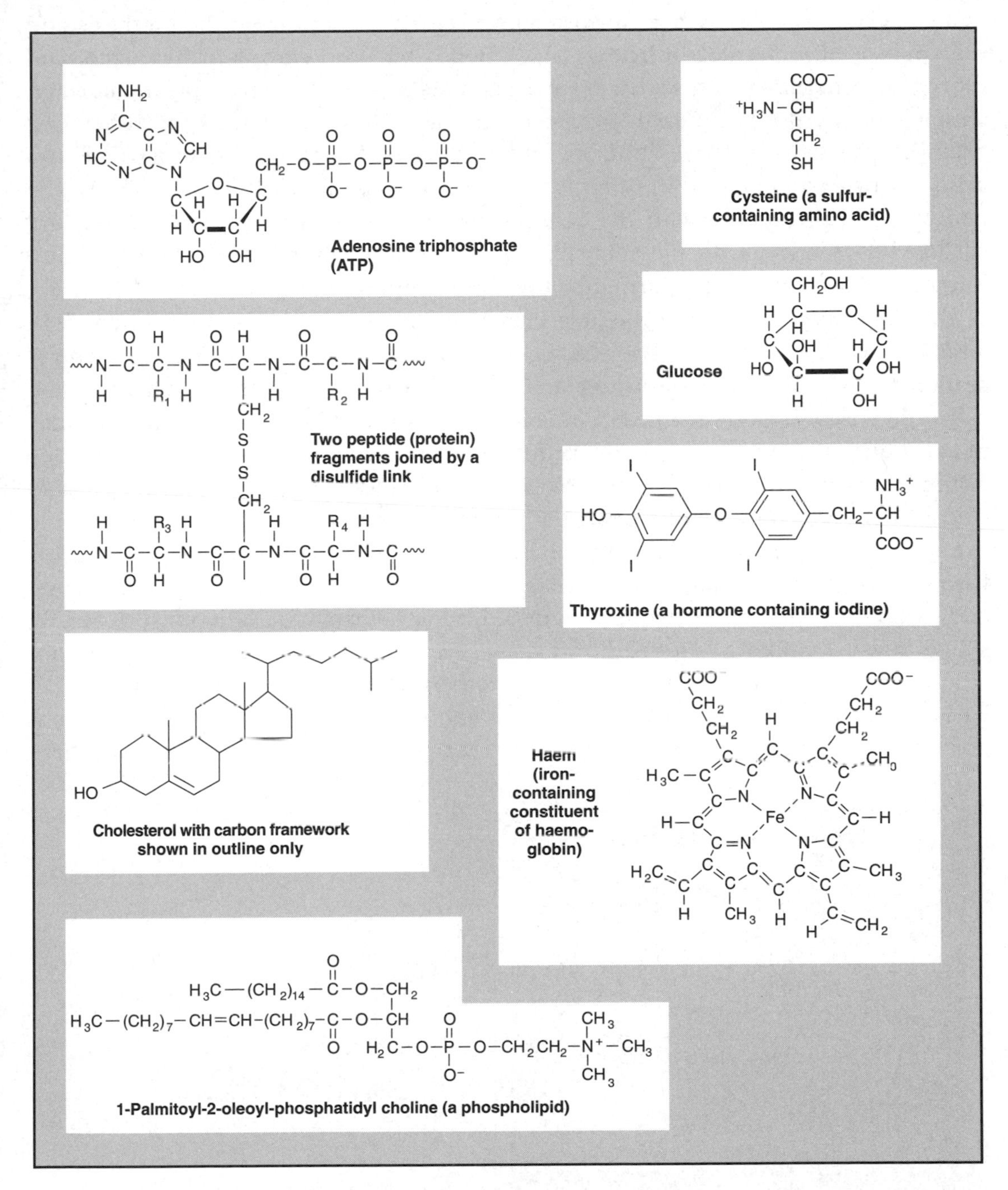

Fig. I.22 A selection of biological molecules, showing the typical modes of bonding of some elements important in life.

Elements other than carbon and hydrogen give functional groups, such as –OH, *CO, $-NH_2$, and –SH, which form reactive centres important in the synthesis and metabolism of molecules in living cells.[36] Bonds between carbon or hydrogen, and more electronegative atoms such as nitrogen and oxygen, also have a polar character, which enables the functional groups to interact physically with each other (for example, through *hydrogen bonds), and with water. Some of these groups are also acidic or basic and ionised in water, by the loss or gain of a hydrogen ion (H^+). The common acidic groups (with the charges appropriate to water at neutral pH) are carboxylate ($-CO_2^-$) and phosphate ($-OPO_3^{2-}$), although sulfate ($-OSO_3^-$) is also found. Positive charges are carried by the basic amino group ($-NH_3^+$ at pH 7).

One important function of the polar and ionic groups is to control the structure of large biological molecules, and indeed of whole cells and organisms. Non-polar hydrocarbon regions tend to avoid water and thus to aggregate together, a phenomenon called the *hydrophobic effect*. Specific interactions between polar groups, or more general interactions with water, then determine the arrangements of these groups. One example of this is the formation of biological membranes, which surround all cells and define their boundaries, thus controlling their shape and insulating them to some extent from their surroundings. Other membranes define specialised compartments within cells: the DNA-containing nucleus, chloroplasts (where photosynthesis occurs) and mitochondria (where energy is produced from the oxidation of food). Figure I.23 shows how a membrane is formed from a *lipid bilayer*. Molecules such as the phospholipid illustrated in Fig. I.22 have long hydrocarbon chains and polar groups at the end. The bilayer has the polar groups in contact with water on the outside, with the non-polar chains forming the interior. Real membranes are more complicated, and have

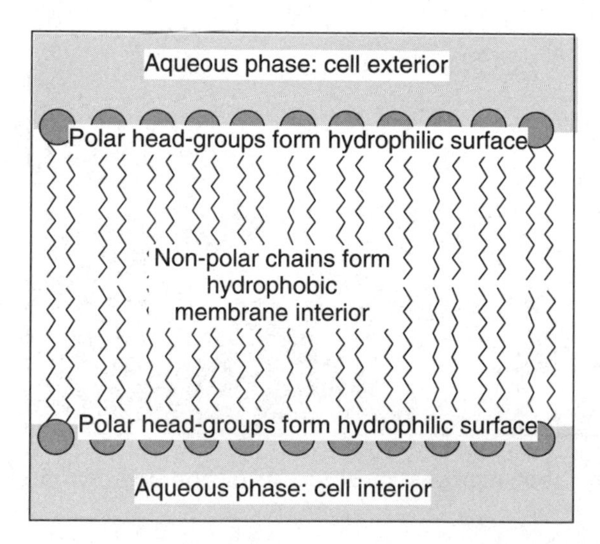

Fig. I.23 Schematic illustration of a lipid bilayer membrane.

36 Stryer (1988) gives an excellent account of biochemistry.

many other molecules—such as proteins and molecules specifically designed to pass particular constituents into or out of the cell—integrated into their structure.

The second group of elements in Table I.9 comprises ones which occur in living matter as hydrated ions dissolved in water. These include the commonest elements from Groups 1 and 2 in the periodic table, Na^+, K^+, Mg^{2+}, and Ca^{2+}, as well as the hydrogen ion H^+, and the anion Cl^-. The oxyanions bicarbonate (HCO_3^-) and phosphate (HPO_4^{2-}) are also part of the soluble components of the cell. One essential role of these ions is to provide an osmotic and electrolyte balance. Cations have a higher total concentration than anions, because biological molecules more often carry negative than positive charges. The presence of ions at an appropriate concentration is also important in modifying the interactions between charged groups in organic molecules, and thus in influencing their structure. The more highly charged ions Mg^{2+} and Ca^{2+} are often associated directly with particular anionic groups; for example many phosphate-containing compounds, including ATP, are present as magnesium salts. The Ca^{2+} ion also binds strongly to charged and polar groups such as carboxylate, and in doing so may change the shape of an enzyme or other molecule and modify its chemical properties. Thus *calcium acts as a control over many biochemical reactions.

Ions and highly polar molecules cannot pass directly through the non-polar constituents of biological membranes. Special channels exist to allow them to enter and leave cells. But this is not a purely passive process: most ions are in fact 'pumped' to maintain deliberate concentration differences between the interior of the cell and the outside world. The concentrations of Na^+ and Ca^{2+} are much less within cells than outside. One function of such ionic gradients is to maintain a correct osmotic balance. When a concentration gradient exists across a membrane, water will tend to pass through, from the side where dissolved substances are more dilute to that where the concentration is higher. It is important that the inorganic salt concentration is generally less inside the cell than outside, to compensate for the concentration of organic constituents inside cells. This is probably the primary reason why sodium—the commonest cation in sea water—was originally excluded from cells. There is an additional reason for excluding calcium: many of its compounds (such as phosphate) are very insoluble, and also as mentioned above it binds quite strongly to many charged groups.

Having arisen for various reasons, ion concentration gradients have come to have important functions in multicellular organisms: in nerve cells (*sodium), as chemical signals to cells such as muscles (*calcium), and in biochemical energy-producing reactions (*hydrogen).

Elements in the third group shown in Table I.9 are constituents of inorganic solids, forming shells, skeletons, teeth, etc. A surprising variety of living organisms manage without such solid constituents; for example the strength of trees is provided by biological polymers such as cellulose (a glucose polymer). But inorganic compounds are used to provide extra strength and protection against predation. Silica (SiO_2) is often used in plants, including single-cell marine diatoms. Calcium carbonate forms the shells of many marine animals; internal skeletons and teeth are more often made of apatite, a basic calcium phosphate of approximate composition $Ca_5(PO_4)_3(OH)$. Some other solids are also used in more specialised cases. Some marine organisms use strontium or barium sulfate. Iron oxide is also found in some cells, both for the storage of iron, and using the magnetic properties of Fe_3O_4 for internal 'compass needles' for navigation and orientation.

Many of the elements in the above categories are major ones. The final class in Table I.9 shows metallic trace elements that fulfil quite specialised functions, bound in enzymes and other types of molecules. The most abundant of these is *iron, and its binding in the haem molecule is shown in Fig. I.22. This is a constituent of haemoglobin, which carries oxygen in the blood, but is also found in some enzymes, especially ones which play a role in oxidation and reduction processes in metabolism. Transition metals are mostly bound to specially designed sites, with nitrogen, sulfur, and oxygen as the surrounding ligand atoms. Their role in biological systems is a function of the variety and flexibility of their typical chemistry: they have the ability to form complexes and to bind particular molecules or functional groups, and also have more than one stable oxidation state. It is indeed for these same reasons that transition metals and their compounds are often used as catalysts in the chemical industry, although the compounds involved are very different.

4

Chemical processes in the natural environment

Chemical compounds illustrate only one aspect of the behaviour of elements in the environment: chemical and physical change is also a feature of the world we live in. The dynamic aspects of the Earth's surface stem from the energy flows discussed in §1.3. Solar energy and internal sources (radioactive decay) fuel physical circulations which help cycle the elements through the environment. Some solar energy also acts more directly to cause chemical change: this happens in atmospheric photochemistry, and in photosynthesis by life. Although the energy fluxes involved in these processes are small overall, much less than 1 per cent of the total (see Fig. I.2), they have a disproportionately large effect. The oxygen in the atmosphere owes its origin almost entirely to photosynthesis, and its presence in turn gives rise to much chemical change.

4.1 An overview of element cycling

Figure I.24 shows some of the processes important in the environmental cycling of elements near the Earth's surface. This figure uses a convention that will be adopted in all the diagrams of element cycles shown in later sections (see also Fig. I.3): boxes with square corners represent 'reservoirs', the principal compartments of the environment where the elements reside in different compounds; round-cornered boxes represent 'fluxes', the processes that transfer the elements from one compartment to another. In Fig. I.24 only a selection of elements is shown, chosen to illustrate the most significant fluxes of each kind. The natural processes are discussed in more detail in the following sections of the present chapter, and those involving industrial extraction and pollution in Chapter 5.

Table I.10 gives some idea of the amounts of some elements transferred annually through environmental compartments. There are enormous differences in the magnitudes involved. Hydrogen and oxygen—as water—are the most 'mobile' elements; the processes of evaporation and condensation were shown in the hydrological cycle in Fig. I.3. Next in magnitude are fluxes of carbon and oxygen (as O_2) coming from

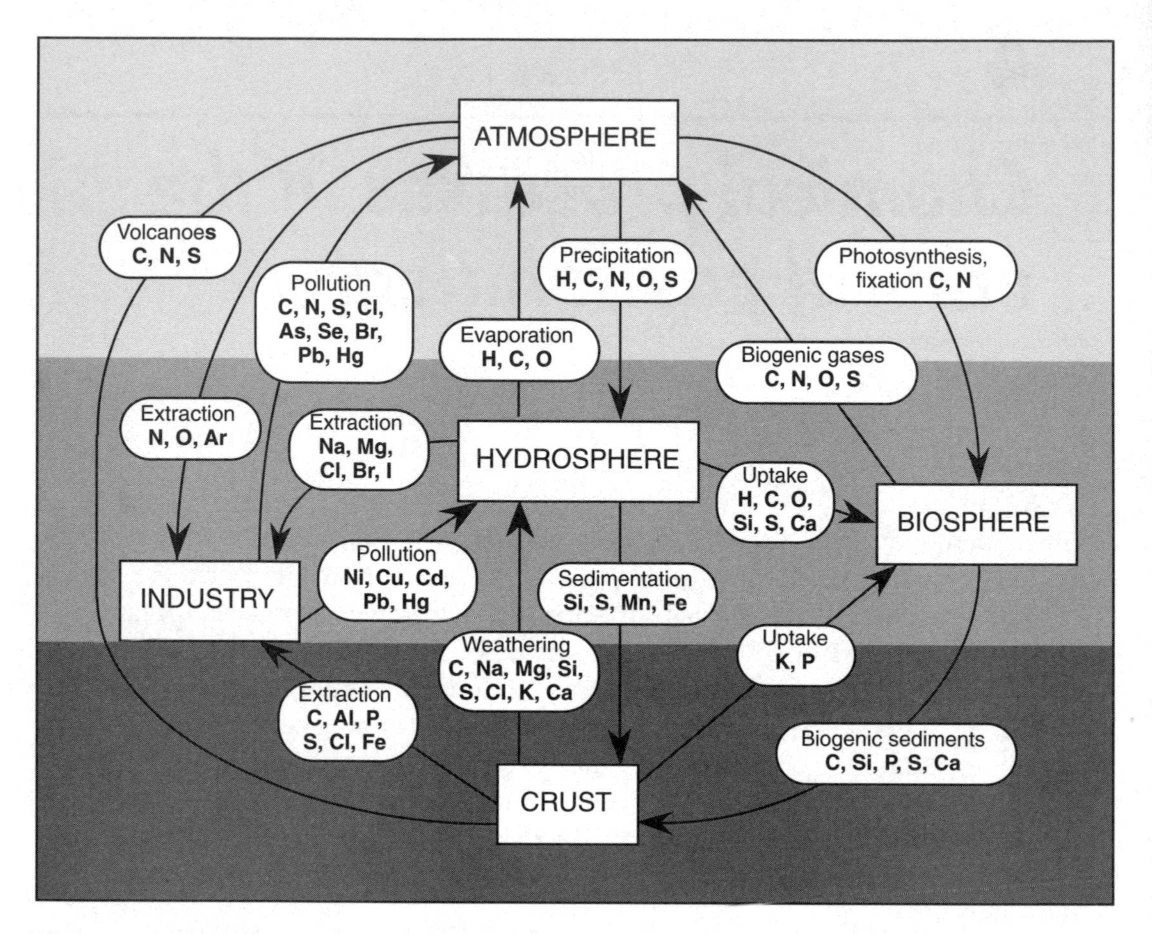

Fig. I.24 A summary of the processes important in the cycling of elements near the Earth's surface; only a selection of the elements involved in each type of transfer is shown.

photosynthesis, respiration, and decay of living organisms. The processes of life also mobilise other essential elements, especially nitrogen, phosphorus, sulfur, and calcium. Other fluxes through the hydrosphere are a consequence of the hydrological cycle, through the action of rainwater on rocks: *weathering* is a mixture of physical and chemical breakdown, producing both soluble constituents and insoluble remains; only the former are listed in Table I.10. Industrial processes contribute to the cycling of several elements. The largest flux of this kind is carbon converted to atmospheric CO_2 by fossil fuel burning. There are also appreciable anthropogenic contributions to the cycling of phosphorus, sulfur, and chlorine, and at smaller levels to toxic elements such as mercury and lead.

A different view of the mobility of various elements is shown in Fig. I.25. The vertical scale gives the annual transfer divided by the average crustal concentration of an element. The mobility of hydrogen (as water) appears even more remarkable in rela-

Table I.10 Annual cycling of elements through the atmosphere, hydrosphere, biosphere, and industry. The atmosphere and hydrosphere columns show the natural transport of selected elements in volatile and soluble forms only. There are also significant transports of dust through the atmosphere (at a level of 10^{-6} to 10^{-3} units) and of sediments through the hydrosphere (at a level up to 0.1 units). A dash indicates that no figure in available, not that the corresponding flux is zero

Element	Annual flux (10^{12} kg year^{-1}) through: atmosphere	hydrosphere	biosphere	industry
H (as H_2O)	6×10^4	6×10^4	*v. large*	*large*
C	200	100	150	8
N	0.25	0.1	6	0.1
O (as O_2)	300	1	300	0.1
(as H_2O)	5×10^5	5×10^5	*v. large*	*large*
Na	0	0.2	–	0.001
Mg	0	0.3	–	3×10^{-4}
Si	0	0.2	–	0.01
P	0	0.001	1	0.15
S	0.1	0.4	0.5	0.15
Cl	0.005	0.2	–	0.17
K	0	0.05	–	0.05
Ca	0	0.5	0.5	0.1
As	2×10^{-5}	4×10^{-4}	–	5×10^{-5}
Se	5×10^{-6}	8×10^{-6}	–	1×10^{-6}
Hg	5×10^{-5}	5×10^{-6}	–	8×10^{-6}
Pb	0	0.01 ?	2×10^{-4}	0.004

tion to its low crustal concentration, and the mobilisation of carbon, nitrogen, and oxygen by life can also be seen.

In addition to the soluble and volatile forms of elements appearing in Table I.10, large amounts of solid material are also transported by physical cycling of the atmosphere and hydrosphere. Rivers carry an estimated 2×10^{13} kg of insoluble sediments per year to the ocean, derived from the physical breakdown of rocks in weathering. Smaller quantities (maybe 10^{12} kg per year) of dust enter the atmosphere, from wind-blown soil and from the evaporation of sea spray. The metallic elements forming insoluble compounds (see §3.4) will thus be transported from land to the ocean in quantities proportional to their crustal concentration, and come at the 'baseline' level shown in Fig. I.25. A few highly soluble elements—especially chlorine and bromine as Cl^- and Br^-, respectively—have mobilities considerably above this level, but soluble metals such as the alkalis (e.g. sodium as Na^+) do not, because they are themselves derived from rock weathering; although they enter solution rather than remaining with the insoluble sediments, the physical breakdown of rocks is necessary to liberate them. A number of elements—copper, zinc, cadmium, mercury,

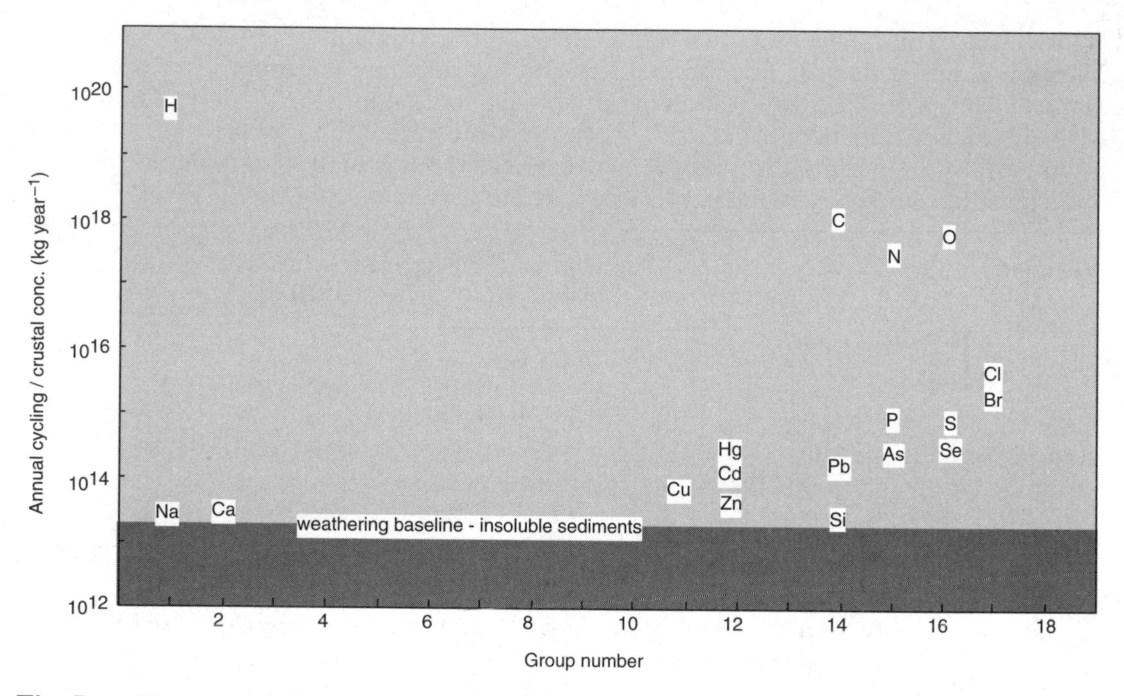

Fig. I.25 Element mobility in the environment. The vertical scale gives the annual transfer divided by the mean crustal concentration of each element. An estimated 2×10^{13} kg per year of sediments derived from rock weathering are carried each year by rivers to the ocean. These contain most elements in rough proportion to their crustal concentration, at the level of the 'weathering baseline' in the plot. Only elements with significantly larger fluxes are shown explicitly.

lead, selenium, and arsenic are shown in Fig. I.25—have mobilities enhanced above the baseline level by their industrial usage.

The presence of life has a profound effect on the chemical processes taking place at the Earth's surface. The natural cycling of several elements through the atmosphere, especially carbon, nitrogen, oxygen, sulfur, and the halogens, is strongly enhanced by *biogenic emissions*, that is, volatile compounds derived from living organisms in the sea and on land. The elements shown in Fig. I.24 under this heading are carbon, oxygen, nitrogen, and sulfur, and some of the processes concerned with this cycling are discussed below. Several other elements may also be involved in smaller quantities, including arsenic, selenium, and mercury.

Other essential elements of life may end up in *biogenic sediments*. Marine organisms produce solids such as calcium carbonate and silica, for shells or skeletons. When the plants or animals die, these sink to the ocean floor. Thus life makes a contribution to the general sedimentation processes discussed later.

A useful measure of the mobility of an element or compound in a particular part of the environment is the *residence time*. This concept was mentioned in connection with the atmosphere and oceans in §3.3 and §3.4, respectively (see also Table I.8). The definition of this quantity is

Mean residence time = (total content)/flux.

For example, the atmosphere contains 7.2×10^{14} kg carbon (mostly in the form of CO_2), and the estimated flux is 2×10^{14} kg per year, giving a residence time of 3.6 years. For a rigorous definition the rate of input should be equal to the rate of output, so that the flux can be unambiguously defined. Often this is only roughly true; for example the burning of fossil fuels leads to a current rate of input of carbon to the atmosphere which exceeds the output by around 2×10^{12} kg per year.

Residence times for elements vary enormously. In the atmosphere the values given in Table I.8 range from a few days to millions of years, reflecting large differences in the physical and chemical properties of elements and their compounds. For elements in the oceans wide ranges are also found, from a few years for elements such as aluminium and iron which have very insoluble compounds, to a hundred million years for the very soluble chloride and bromide ions.

Residence times need to be viewed in relation to another time-scale, determined by the physical mixing of different parts of the environment. Winds in the atmosphere are capable in principle of carrying constituents round the world in a few days. They are very unevenly distributed however, and may often be confined to a narrow range of latitudes. About a year is needed for complete mixing between the northern and southern hemispheres of the troposphere (the lower part of the atmosphere which extends up to 10–15 km above sea level). Mixing with the stratosphere (the upper atmosphere) is slower, requiring several years. These figures have some important consequences. Constituents with residence times measured in days will not be evenly mixed, but will be concentrated closer to their sources; an example is SO_2 from industrial pollution. Longer-lived constituents on the other hand will be evenly distributed throughout the lower atmosphere. The slow mixing between the stratosphere and troposphere is important for pollutants capable of damaging the ozone layer, such as *chlorine and *nitrogen compounds; as these are released at ground level, they must have long tropospheric residence times in order to reach the stratosphere in significant quantities.

The oceans are also stratified, with the surface layer being much more rapidly mixed than the deeper waters. Mixing with the deep ocean is only complete over a time of a few hundred years. This is particularly important in connection with the *carbon cycle and its perturbation through fossil fuel burning. Some of the increased atmospheric CO_2 is absorbed rapidly by surface sea waters, but the capacity of these is rather small. The time-scale for removal of most of the extra CO_2 is determined by its transfer to deeper waters.

The solid parts of the Earth move over much longer time-scales. The tectonic cycling discussed in §1.3 (see Fig. I.4) gives rise to a crust which is chemically differentiated from the mantle, but the mixing times concerned are of the order of the age of the Earth, i.e. a few billion years. Although these processes occurring deep within the Earth are not shown in Fig. I.24, they have important consequences at the surface, as they ultimately give rise to the minerals of the crust. They contribute to the overall cycling of elements in the atmosphere and oceans as well, through volcanic gases and hydrothermal processes.

4.2 Redox processes and life

Redox reactions (that is, oxidation and reduction) play a major role in the cycling of some elements. One reason for this is that a change in oxidation state of an element generally changes its volatility or solubility. For example, the most oxidised states of nitrogen and sulfur, nitrate and sulfate, respectively, are involatile, and so the presence of these elements in the atmosphere depends on less oxidised molecules (see Table I.7 in §3.3); when these are oxidised in the atmosphere they are washed out quite rapidly. In a rather similar way, iron and manganese are more soluble in water as Fe^{2+} and Mn^{2+} than in the higher oxidation states Fe^{3+} and Mn^{4+} (see Fig. I.17 in §3.4); if they are liberated from rocks in the reduced form, they will be precipitated on atmospheric oxidation. As discussed in §3.1, one of the most significant features of the Earth's chemistry is the presence of free O_2 in high concentration in the atmosphere. This comes almost entirely from photosynthesis by life. All living creatures require a continual supply of free energy to survive. Redox reactions of various kinds form the main source of this free energy, as well as being involved in other ways; for example, the supply of carbon and nitrogen to the biosphere requires conversion of the predominant inorganic forms (CO_2 and N_2, respectively) into the reduced compounds of life.

Figure I.26 shows some of the redox conversions performed by biology. Each reaction is represented by a pair of arrows, a right-pointing one representing oxidation, and a left-pointing arrow reduction. For example, in *aerobic respiration*, the ultimate energy source of air-breathing organisms such as ourselves, organic carbon compounds (C_{org} in Fig. I.26), are oxidised to CO_2, and oxygen is simultaneously reduced to water. If we formulate organic carbon as $[CH_2O]$, the empirical formula of carbohydrates such as starch or glucose, its oxidation may be written as

$$[CH_2O] + H_2O \rightarrow CO_2 + 4H^+ + 4e^-.$$

The reduction of oxygen is

$$O_2 + 4H+ + 4e^- \rightarrow 2H_2O,$$

and so combining these gives an overall reaction

$$[CH_2O] + O_2 \rightarrow H_2O + CO_2.$$

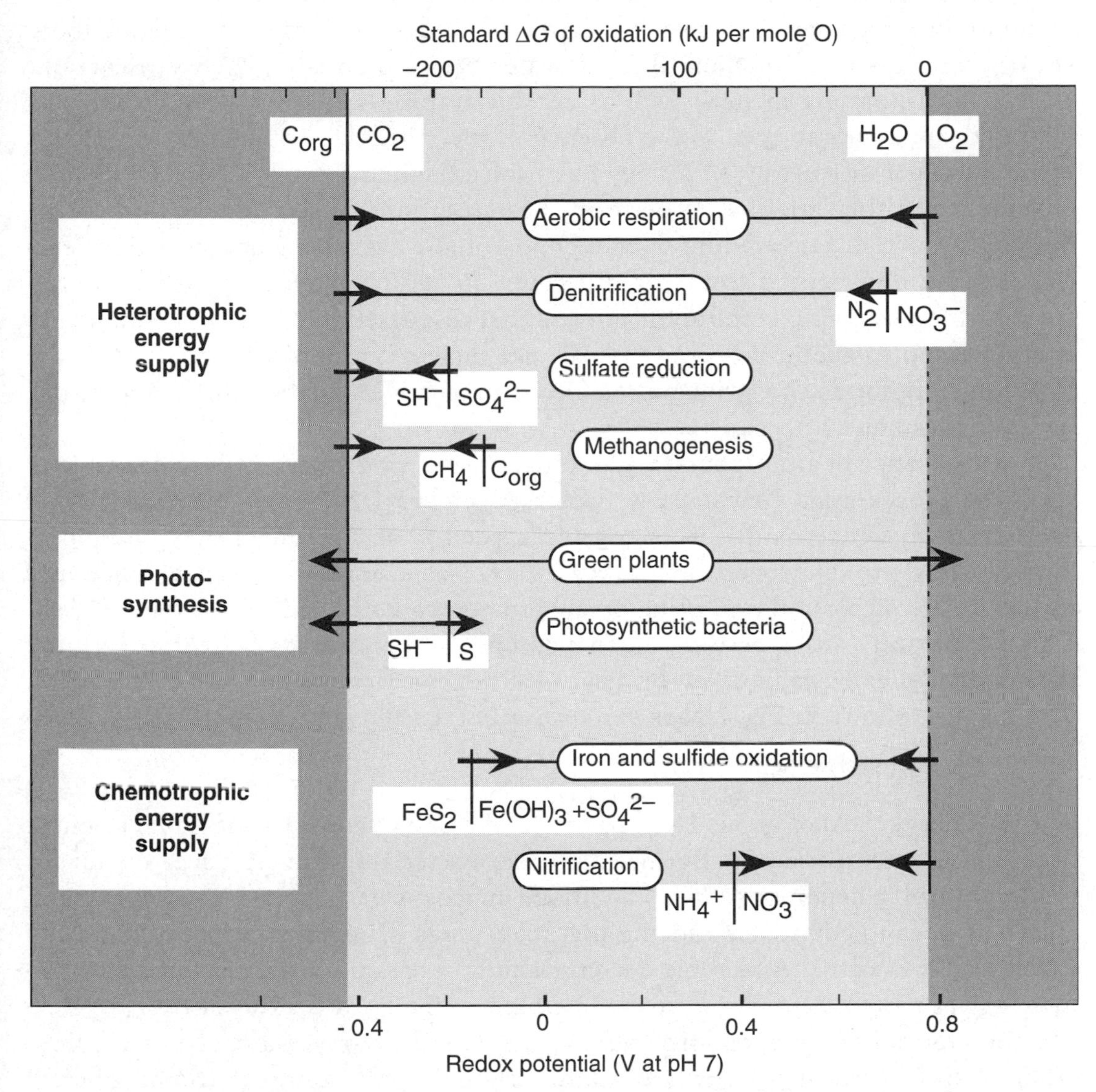

Fig. I.26 Some redox reactions important for the energy supply of biological systems. Each overall reaction consists of an oxidation (represented by a *right*-pointing arrow) and a reduction (*left*-pointing arrow). A combination of arrows pointing inwards corresponds to a possible source of free energy, whereas the photosynthetic reactions with outward-pointing arrows require an external source of free energy, for example from sunlight. The reactions are classified into those of *heterotrophic energy supply*, undertaken by organisms that require a source of organic matter as food, *photosynthesis*, in organisms which utilise sunlight to build up organic compounds, and *chemotrophic* reactions, supplying energy without needing organic matter for oxidation.

The horizontal scales in Fig. I.26 give the free energy change of the reaction (ΔG) and the redox potential, both at pH 7. For example, the oxidation of common organic compounds such as carbohydrates occurs at a potential of about –0.4 V, whereas the reduction of O_2 to water has a potential of +0.8 V. Reactions with inward-pointing arrows, such as aerobic respiration, are thermodynamically favoured. They occur even in the absence of life, although sometimes very slowly in the absence of a catalyst. Biology has evolved suitable catalysts, in the form of enzymes, and they are able to perform these reactions in such a way that energy may be extracted. The amount of energy potentially available from an overall reaction can be determined from the difference in redox potential of the two constituents. The aerobic respiration process just discussed yields a large amount of energy, equal to about 475 kJ (114 kcal) per mole of carbon oxidised. This is the reaction that forms the principal energy source for all air-breathing organisms, including humans.

In the absence of atmospheric O_2, alternative oxidising agents may be available to specialised organisms (particularly bacteria). Figure I.26 shows the examples of *denitrification*, where oxidation of organic carbon is accompanied by reduction of nitrate (NO_3^-) to dinitrogen, and *sulfate reduction* as a similar alternative. There is a further range of possibilities, using organic matter to both oxidise and reduce itself. This opportunity arises because organic compounds such as carbohydrates are not thermodynamically stable, even on their own in the absence of any oxidising agent. The reaction shown in Fig. I.26 is *methanogenesis*, written approximately as

$$2[CH_2O] \rightarrow CH_4 + CO_2,$$

with an energy yield of about 18 kJ per mole of carbon consumed. It is one example of a *fermentation reaction*, performed by some bacteria and other micro-organisms under anaerobic conditions. Such conditions include water-logged soils and marshes, lake and ocean sediments, and the digestive tracts of animals, especially grazing species such as cattle. Anaerobic decomposition of organic matter in these environments gives rise to the major source of methane in the Earth's atmosphere.

Other fermentation reactions include the familiar conversion of glucose into ethanol and CO_2, and there are many very complex examples. Some liberate dihydrogen, H_2, as in

$$3[CH_2O] + H_2O \rightarrow CH3COOH + CO_2 + 2H_2$$

which has an energy yield of about 30 kJ per mole of carbon. Other reactions use H_2 to produce methane.[37]

---

37 Levett (1990) gives an account of anaerobic bacteria, including a discussion of the variety of reactions used by them to obtain energy.

The examples just given show that fermentation is a much less efficient use of organic matter than aerobic respiration. Indeed, the pioneering chemist and microbiologist Pasteur first recognised that yeast needs about 10 times as much sugar to grow by fermentation than it does under aerobic conditions; this is directly related to the relative energy yield of the different processes. In the present-day environment, anaerobic reactions are only used when no other oxidising power is available, but presumably these were the first types of reactions from which early life obtained energy: there was no oxygen, and probably very little nitrate or sulfate. The possible sources of complex organic compounds suitable for the development of life are discussed briefly under *carbon. Whatever these were, the supply was probably limited, most carbon being in the oxidised form (CO_2 and carbonate). Thus the development of some kind of photosynthesis may have been stimulated by the shortage of organic 'food' for early life. Oxygen was an unwanted by-product. The photosynthetic reaction by green plants is essentially the reverse reaction of aerobic respiration, yielding organic matter and O_2 from CO_2 and water. In this case an energy *input* of 475 kJ per mole is needed, which comes from sunlight. (The reactions producing and using O_2 are described in more detail under *oxygen.) As shown in Fig. I.26, there is an alternative reaction, in which the oxidation of H_2S (or HS^- in water) to sulfur accompanies the reduction of CO_2. This reaction is still performed by some bacteria. It is energetically less demanding than oxygen-based photosynthesis, and it is thought that it may have developed first.

Most of the organic carbon produced by photosynthesis is reoxidised by respiration and aerobic decay processes mediated by micro-organisms. A high proportion of the remainder is consumed by anaerobic fermentation, giving products such as methane which are reoxidised in the atmosphere. Only a very small fraction escapes these decay processes, to be buried as reduced carbon, from which we derive our fossil fuels. Over the lifetime of the Earth, this has been significant, as around 20 per cent of the carbon in the crust is derived from organic remains. This reduced carbon is balanced by the production of atmospheric oxygen.

The advent of atmospheric O_2 not only gave rise to the possibility of its direct utilisation in respiration, but also opened up a variety of indirect possibilities for energy supply. Many of the reduced compounds of the crust and of the primitive Earth surface are oxidised by O_2. For example the common *iron mineral *pyrites* is oxidised to Fe(III) hydroxide and sulfate, SO_4^{2-}, and reduced nitrogen, formed from organic decomposition products as ammonium, NH_4^+, is oxidised to nitrate, NO_3^-. These reactions happen naturally under inorganic conditions, and have used up nearly all the oxygen produced by photosynthesis. As illustrated in Fig. I.26 however, the free energy produced can be utilised by specialised microbes as *chemotrophic* energy supplies, that is, ones giving chemical energy without needing a supply of organic carbon to be oxidised. The oxidised sulfates and nitrates, once formed, can be used in turn under anaerobic conditions as oxidising agents, as already mentioned.

The redox processes discussed here are not only essential for the maintenance of life; they also play an important role in the chemistry of several elements, including *carbon, *nitrogen, *oxygen, *sulfur, and *iron. The reactions of O_2 with other biogenic constituents of the atmosphere are discussed below in §4.3, and the importance of oxidation and reduction in mineral formation processes in §4.4.

4.3 Atmospheric chemistry

The Earth's atmosphere is unique among planets we know, in containing O_2 at a high concentration. This molecule is capable in principle of oxidising many of the compounds that enter the atmosphere as biogenic and volcanic emissions, but such reactions are often slow. Most atmospheric chemistry involves reactive species present at very low concentrations, generated in the first place from stable molecules by the photochemical action of sunlight. An example is

$$O_2 + h\nu \rightarrow 2O,$$

where $h\nu$ is the energy of one quantum of solar radiation (see below); in this reaction the light energy generates two highly reactive oxygen atoms. Other reactive species include *free radicals* which have an odd number of electrons. One of the most important is the hydroxyl radical (OH); the following sequence shows how, once formed, it can be regenerated in reactions involving other trace constituents of the atmosphere:

$$OH + CO \rightarrow H + CO_2;$$

$$H + O_2 \rightarrow HO_2;$$

$$HO_2 + NO \rightarrow OH + NO_2.$$

The atmospheric chemistry initiated by O and OH is discussed in more detail under *oxygen. The remainder of the present section deals with some general considerations important for all elements.[38]

Fig. I.27 shows the energies required to break some bonds that are important in atmospheric chemistry.[39] On the bottom scale the bond energy (or enthalpy) is given in kJ per mole. The bonds are shown in three rows—those to hydrogen at the top, to carbon in the middle, and to oxygen in the bottom row. In most cases the value given is an average for the type of bond concerned; for example the C–H bond energy is the average value in CH_4, which is not exactly the same as the energy required to remove the first hydrogen. In a few cases, however, Fig. I.27 shows values for specific bonds where these are important; O–O_2 and O–NO represent the energies required to remove an oxygen atom from ozone (O_3), and from nitrogen

38 See Wayne (1991).

39 From data quoted by Johnson (1982).

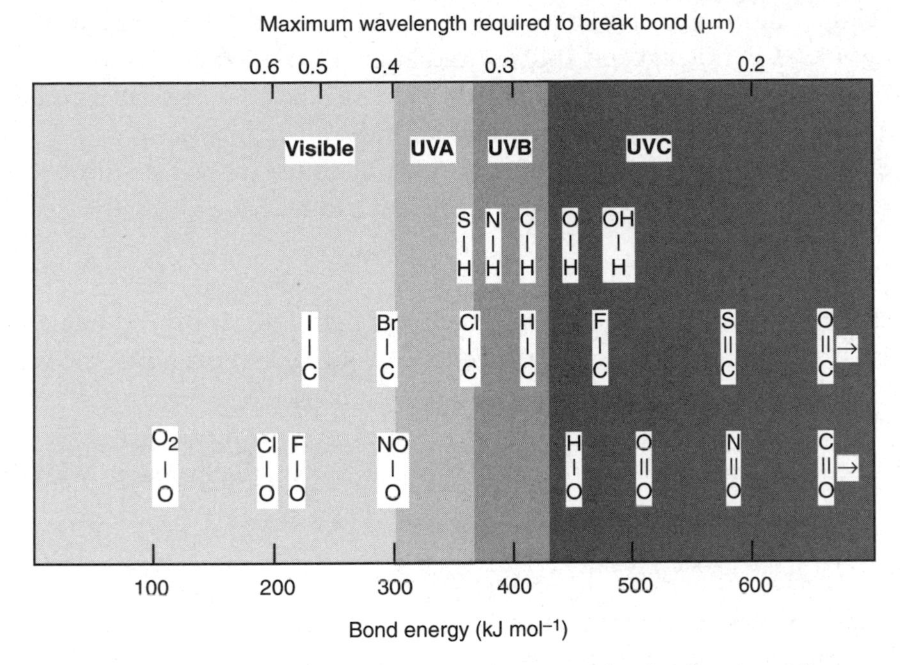

Fig. I.27 A selection of bond energies important in atmospheric chemistry (lower scale). The upper scale shows the maximum light wavelength which gives sufficient energy for one photon to dissociate a bond of the appropriate energy. The wavelength ranges for visible and UV radiation are shown by shading.

dioxide (NO_2), respectively; H–OH corresponds to the removal of the first hydrogen atom from water.

The upper scale in Fig. I.27 shows the wavelength of radiation appropriate to the corresponding bond energy. According to the quantum theory of radiation, energy 'packets' or *quanta* each have an energy

$$E = h\nu,$$

where ν is the frequency of the light and h is Planck's constant. Long-wavelength light has a low frequency, and the quanta may not have enough energy to break bonds. For each bond energy, therefore, there is a certain maximum wavelength, below which the light quanta have sufficient energy to break the bond concerned. For example, we can see from Fig. I.27 that light shorter than 0.4 μm in wavelength can break the first N–O bond in NO_2, but that a wavelength of 0.2 μm or less may be required to complete the dissociation of the NO molecule.

The spectral regions shown in Fig. I.27 give some idea of where in the Earth's atmosphere different molecules can be dissociated by radiation. Sunlight reaching the lower atmosphere (the troposphere) is concentrated in the visible region, although it extends into the UVA part of the spectrum. Most UVB is strongly absorbed by ozone at an altitude of 20–50 km, and UVC is absorbed by O_2, mostly at even higher altitudes. Thus we can see that O_3 and NO_2, as well as CH_3I containing a C–I bond, can be split by radiation in the lower atmosphere. On the other hand, a compound with C–Cl bonds must reach the ozone layer to be dissociated, and CO_2 is only affected at very high altitudes.

The relative bond energies also give an idea of the reactions possible for a given fragment formed by photodissociation. Consider the reaction

$$A + X\text{–}B \rightarrow B + X\text{–}A.$$

If the energy of the new bond (X–A) is much less than that of the old one (X–B), this reaction is strongly endothermic and unlikely to proceed. For example, the reaction

$$OH + CF_3Cl \rightarrow CF_3 + ClOH$$

does not occur, as the Cl–O bond energy is much too small compared with the C–Cl energy. On the other hand with

$$OH + CH_3Cl \rightarrow CH_3 + ClOH$$

the order of energies is different: this reaction is energetically favourable and does occur. Examples of other possible reactions that can be deduced from the data in Fig. I.27 are

$$OH + CO \rightarrow H + CO_2$$

and

$$O_3 + Cl \rightarrow OCl + O_2;$$

the latter reaction occurs in spite of the weak Cl–O bond because it requires even less energy to remove an oxygen atom from ozone.

Because of the different wavelengths of radiation involved, and also because many reactive species are first generated in the lower atmosphere, the chemical processes differ greatly at different heights. The *troposphere*—the lower region, strongly mixed by convection and weather systems—is dominated by the reactive properties of hydroxyl radicals. In the *stratosphere* reactions involving ozone and oxygen atoms are more important. Ozone is also important in the troposphere, but principally as a source of OH radicals. When its concentration increases under urban conditions it can become a serious pollutant, because it is toxic and contributes to the production of other harmful molecules. In the stratosphere it is essential to us, as it absorbs hard UV radiation and shields life at the Earth's surface from its harmful effects. More details about the ozone layer and its formation are given under *oxygen. Molecules which act as a threat to the ozone layer are those which survive in the lower atmosphere long enough to undergo the rather slow process of transfer into the stratosphere. This means that they must not be dissociated by visible or UVA radiation, and that they must not be attacked easily by OH radicals. The chlorofluorocarbons (CFCs) such as CF_2Cl_2 fulfil these conditions for the reasons discussed above; furthermore unlike pure fluorocarbons such as CF_4, which is only dissociated by UVA radiation above the ozone layer, they are readily split by UVB once in the stratosphere, giving free chlorine atoms which react with ozone. More details on CFCs are given under *chlorine.

Chemical processes in the lower atmosphere are described under *hydrogen, *carbon, *nitrogen, and *sulfur. The main feature is the oxidation of reduced compounds—to H_2O, CO_2, HNO_3, and H_2SO_4 for these four elements, respectively—mediated largely by OH radicals. Nitrogen compounds are important as a source of such radicals, as well as for contributing to some types of pollution. Carbon compounds are the main 'sink' for OH radicals, as well as being pollutants. Sulfur is important as the main source of acid rain, but sulfuric acid is significant for another reason, as in the stratosphere it can form a long-lasting aerosol 'haze', which has a net reflecting effect on sunlight and therefore a cooling influence on the Earth's climate.

4.4 Weathering and sedimentation

Liquid water is another unique feature of the Earth's surface. We owe its presence to the temperature range, controlled partly by the amount of solar radiation received at a distance of 150 million km, and partly by the influence of the atmosphere in modifying the output of infrared radiation (see §1.4). Water is by far the most mobile constituent of the environment, and has in addition some highly unusual properties, discussed under *hydrogen. Most importantly, it is an excellent solvent for many ionic compounds. The 'distillation' of water, by evaporation under solar heating and subsequent condensation and precipitation from the atmosphere, provides a powerful driving force for the chemical breakdown of rocks. *Weathering* is a combination of physical and chemical changes, and involves the action of other atmospheric constituents (O_2 and CO_2), as well as factors such as freezing and melting which promote physical breakdown. Some of the chemical consequences of weathering are summarised in Fig. I.28.

The most obvious action of water on rocks is to separate soluble and insoluble constituents (see §3.4). This is rarely a purely physical process, but involves some chemical breakdown. A typical sequence is the progressive decomposition of the common crustal mineral *potassium feldspar*, giving rise to the clay minerals *mica* and *kaolinite*, and eventually *gibbsite* (aluminium hydroxide):

$$\underset{\text{feldspar}}{3KAlSi_3O_8} + 2H^+ + 12H_2O \rightarrow \underset{\text{mica}}{KAl_3Si_3O_{10}(OH)_2} + 2K^+ + 6Si(OH)_4;$$

$$2KAl_3Si_3O_{10}(OH)_2 + 2H^+ + 3H_2O \rightarrow \underset{\text{kaolinite}}{3Al_2Si_2O_5(OH)_4} + 2K^+;$$

$$3Al_2Si_2O_5(OH)_4 + 15H_2O \rightarrow \underset{\text{gibbsite}}{6Al(OH)_3} + 6Si(OH)_4.$$

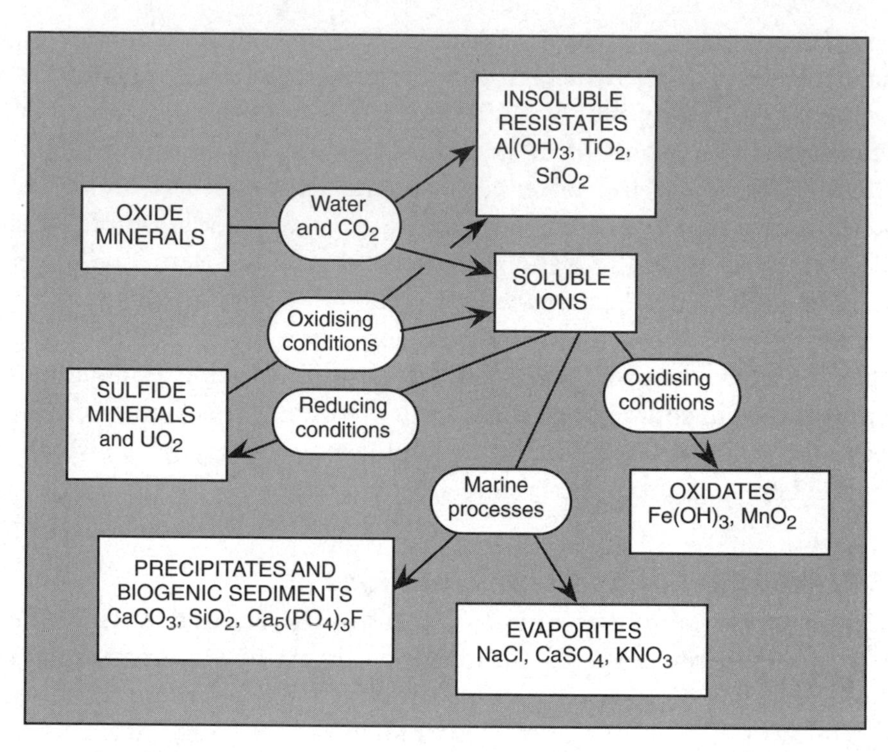

Fig. I.28 A summary of some of the processes involved in weathering and sedimentation. A few examples of sedimentary minerals formed by different processes are shown.

Soluble potassium ions (K^+) and silicic acid ($Si(OH)_4$) are produced, and the insoluble aluminium-containing minerals remain as important constituents of soils. How far this sequence of reactions goes depends on the conditions. Complete weathering is facilitated by high rainfall and temperature, and is a feature of tropical soils; under more temperate conditions mica and kaolinite may be the major products. Another prerequisite is some acidity (H^+ in the above reactions). Normally this is provided by atmospheric CO_2 dissolving in rainwater to form bicarbonate:

$$CO_2 + H_2O \rightarrow HCO_3^- + H^+.$$

Weathering may be accelerated by additional sources of acidity; these may come from the atmosphere, in the form of nitric and sulfuric acids formed by oxidation of nitrogen and sulfur compounds, or from the presence of vegetation. The decay of organic remains in soils can give high concentrations of CO_2 and also generates complex organic acids, known as *humic acids*. The growth of plant roots also facilitates the physical break-up of rocks. It has been estimated that these biological contributions may speed up weathering by a factor of a hundred, compared with rocks not covered with vegetation.

Insoluble minerals derived from weathering are known as *resistates*. They include aluminium hydroxide and the oxides of elements such as titanium and tin. Such minerals may be carried by fast-flowing rivers, and eventually deposited as sediments. Soluble constituents mostly pass into the sea, where they are concentrated by evaporation and also eventually precipitate. This happens over a wide range of time-scales, however, depending on the solubility and other properties of the element concerned. A high proportion of marine sediment is formed by the action of life in the ocean. Shells and skeletons are commonly made of calcium carbonate or silica, and these sink when the organisms die. These processes are important in the global cycling of *carbon, *silicon, and *calcium. More soluble elements such as alkalis (Na^+ and K^+) and halogens (Cl^- and Br^-) remain in sea water for much longer. They may eventually form *evaporite* minerals when sea-level changes or tectonic movements cause some sea water to become land-locked in a place where it can evaporate.

Figure I.28 shows that oxidation and reduction reactions are also important in sedimentary processes. Some elements may be liberated initially in a soluble form, which then oxidises on exposure to atmospheric oxygen to an insoluble compound. Examples are Mn^{2+} and Fe^{2+} (the common oxidation states of *manganese and *iron in igneous rocks), which oxidise to form deposits of insoluble MnO_2 and $Fe(OH)_3$. In other cases, oxidation may give a more soluble form. The insoluble *uranium mineral *uraninite* (UO_2 or U_3O_8) gives the soluble UO_2^{2+} ion on exposure to air. Sulfide minerals are often very insoluble, but can give soluble cations such as Cu^{2+} or Zn^{2+} when the sulfide oxidises to sulfate (see *sulfur for more details). Conversely, sulfide minerals may be deposited in some anaerobic conditions where bacterial reduction of sulfate occurs (see §4.2).

The brief sketch just given ignores a good deal of complexity in aqueous chemistry. The solubility and mobility of elements in water are influenced by many factors, including pH, complex formation, and adsorption. The pH of pure water is 7, and many natural waters have a pH not too far from this. As discussed under *hydrogen, however, there are several natural and anthropogenic compounds in the environment which give extra acidity. As mentioned above, acid conditions promote weathering. They may also change the solubility of some elements: in particular, the solubility of *aluminium increases with acidity, and this in turn influences the solubility of *phosphate, because of the formation of highly insoluble aluminium phosphate.

Complex formation is an association between soluble species, often (but not necessarily) an anion and a cation. As mentioned in §3.4, many elements are normally present in water as complexes. Metal ions are often complexed with natural inorganic anions such as hydroxide, carbonate, or chloride. For example, the predominant species containing aluminium, lanthanum, and gold in sea water are $Al(OH)_3$, $LaCO_3^+$, and $AuCl_2^-$, respectively. Other complexing agents are found in life, or come from pollution. The metalloproteins containing trace elements such as iron, copper, or zinc are essentially complexes, albeit of a very elaborate kind. Many elements are absorbed by living organisms by complex formation, either on the cell surface or through special molecules which are excreted in order to capture essential elements. An example is *iron, where proteins known as *transferrins* are able to complex the normally very insoluble Fe^{3+}, not only to get it into solution, but also to transfer it into cells. Dead organic matter is also important in complexing, and humic acids may bind trace metals and hold these in the soil.

Adsorption is analogous to complexing, except that a soluble ion binds to the surface of a solid mineral particle, not to another soluble species. Minerals often have electrically charged surfaces in water, resulting from an imbalance of ions of opposite charge. Cations such as H^+ and many metal ions may be attracted and held on the surface. The hydrated oxides of iron and manganese are good at this, the aluminosilicate clay minerals even more so. These solids are made up of layers with water and cations between them. Cations (such as K^+) released in the weathering reactions may be exchanged with others. This type of adsorption is especially important in soils, for retaining essential trace elements which would otherwise be washed away. Anions do not normally adsorb so well, but an exception to this rule is phosphate, which is quite strongly bound by aluminium and iron hydroxides.

Adsorption is undoubtedly important in the sedimentation and removal of elements from the oceans. Only the most abundant elements reach concentrations where they may form insoluble compounds such as $CaCO_3$. Some minor elements may be taken up by life in surface waters, and subsequently deposited when organisms die. But many trace elements are removed from the ocean principally by their adsorption onto sinking sediments. The strongest adsorption occurs with cations with charges +3 and +4, which also form very insoluble oxides and hydroxides (see Figs. I.17–I.19 in §3.4); this process contributes to very rapid removal of such ions from the ocean, and hence to their extremely low concentrations in sea water.

4.5 Igneous rocks

Igneous rocks, such as granite and basalt, are formed by melting and recrystallisation of magma—molten rocks—in the tectonic cycling described in §1.3. This crustal cycling is much slower than the circulation of the atmosphere and oceans, but has been equally significant over geological time-scales. As shown in Fig. I.4, slow convection currents in the underlying mantle rise towards the surface in some places—especially along the mid-ocean ridges—and sink again into the depths in others, the subduction zones. The decrease in pressure as mantle material rises allows some partial melting to occur, forming molten magma, which rises and solidifies again as it cools near the surface. Some melting also occurs in subduction regions, where the heating can also decompose rocks and give rise to volcanic emissions of volatile constituents. Elements can be separated by their different chemical behaviour in these tectonic cycles just as they can in the hydrological cycle although the active medium is not water, but silicates.

As mentioned in §3.2, most minerals have complex chemical compositions. Geochemists often express these, not directly in terms of individual elements, but as the proportions of binary oxides such as SiO_2, MgO, and Na_2O, which would be required to make up the compositions. This method is *not* intended to imply that these oxides are individually present. (Although small proportions of some elements may be in the form of a binary oxide such as silica, SiO_2, this is never the case for many elements, which only occur in complex oxide form.) For example, Mg_2SiO_4 may be expressed $2MgO.SiO_2$, although that is a very misleading description of the mineral structure of olivine. Viewed in this way, silica is the commonest constituent by far in crustal rocks, making up between 40 and 80 per cent by weight. Figure I.29 shows the distribution of major elements in igneous rocks, plotted according to their SiO_2 content. The upper plot shows the frequency of occurrence of rocks with different silica levels. It shows a 'bimodal' distribution with two peaks, one around 50 per cent and the other around 70 per cent SiO_2.

Rocks with a high proportion of SiO_2 are known as *acidic*, and those with low SiO_2 as *basic*. This terminology reflects the chemical fact that SiO_2, although only a very weak acid in aqueous solution, is different from the highly basic oxides such as Na_2O and MgO, formed by electropositive metals. The two peaks in Fig. I.29 represent rocks with a different origin. Basic rocks with 50–60 per cent silica are typical constituents of the oceanic crust, formed along mid-ocean ridges by melting and solidification of the underlying mantle (see Fig. I.4). More acidic rocks such as granite are common in the continental crust, which results from a more complex reworking of material producing further chemical fractionation. In line with this trend, the mantle itself is *ultrabasic* in chemical character, with relatively very low silica content, and especially high in magnesium; its composition approximates to a mixture of the magnesium silicates Mg_2SiO_4 (olivine) and $MgSiO_3$ (pyroxene).

The lower plot in Fig. I.29 illustrates how the average abundance of other common elements varies with the silica content: MgO can make up over 50 per cent of rocks with low silica, but is very much less common in those with high silica contents. Aluminium and the alkali elements, sodium and potassium, behave in the opposite way and are commoner in acid rocks. Calcium is commonest in rocks of intermediate character.

The progressive chemical fractionation of rocks from the mantle arises during the cycles of melting and solidification which accompany the formation of the crust. When a mixture of silicate minerals starts to melt, the liquid formed will initially be richer than the solid in silica, and in other elements such as aluminium and alkalis. Some other elements such as magnesium will tend to remain in the solid phase. The

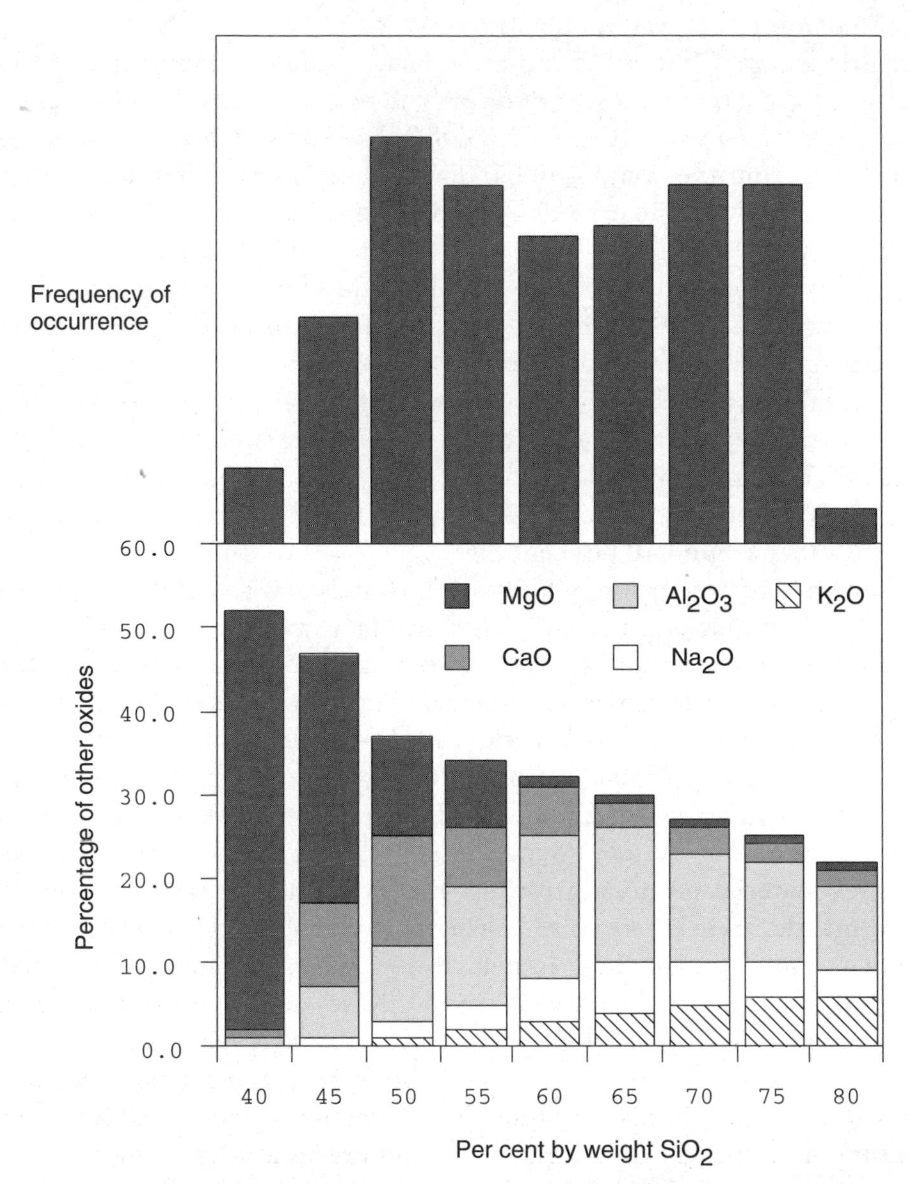

Fig. I.29 The composition of igneous rocks, expressed in terms of the binary oxides of major elements. (See text: this is *not* meant to imply that these oxides are present as individual compounds.) The top diagram shows the frequency of occurrence of rocks with different proportions of the major constituent SiO_2; note the 'bimodal' distribution, with peaks around 50% SiO_2 (basic rocks) and 70% SiO_2 (acidic rocks). The lower plot shows how the average proportions of other oxides in rocks change with the silica content. (Based on Mason and Moore, 1982.)

structural chemistry of silicates involved in this fractionation is discussed under *silicon in Part II.

The behaviour of minor elements is also interesting, especially as igneous rocks often form the most important mineral sources of these. Rarer elements may form crustal minerals in which they predominate, but they also occur in minerals of major elements by substitution (see §3.2). Mg^{2+} is very often replaced by the common Fe^{2+}, and by other ions of similar size such as Ni^{2+}. Substitution of ions with a different charge is also possible; for example Cr^{3+} can replace Mg^{2+} if another replacement is made, such as Al^{3+} for Si^{4+}, to compensate the charge. As Mg^{2+} is the commonest metal ion in the mantle, other ions that can easily replace it also tend to be concentrated in the mantle. Ions very dissimilar to Mg^{2+} pass easily into the molten phase when magma forms, and end up concentrated in the crust. Figure I.30 shows how the distribution of lithophilic elements (that is, ones normally occurring in oxide minerals) is correlated with their ionic charge and radius. Elements which form ions similar in size and charge to Mg^{2+} are mostly strongly concentrated in the mantle; these are known as *compatible* elements. *Incompatible* ones very dissimilar to Mg^{2+} are found in the crust at much higher concentrations than in the mantle. For some highly incompatible elements such as caesium, it is estimated that as much as a half of the total amount originally in the mantle has been concentrated into the crust during the geological history of the Earth.

Chemical differentiation continues as magma solidifies. Elements forming sulfide minerals (not shown in Fig. I.30) are quite insoluble in molten silicates. These minerals may solidify early, forming *layered deposits* which are important sources of some elements such as nickel and the *platinum metals. Most of the magma solidifies to form common minerals such as feldspars, but some highly incompatible elements are not incorporated easily into these major minerals; they include metals such as beryllium,with especially small ionic radii, and other such as niobium, tantalum, and thorium, with particularly large radii. Such elements become progressively concentrated in the liquid phase as solidification proceeds, and end up in rocks known as *pegmatites*, formed during the last stages of solidification of the magma.

4.6 Mineral formation

Our own exploitation of elements would be much harder if it were not for the presence of deposits where minerals rich in particular elements are concentrated. The formation of such deposits is one consequence of the chemical mobility and fractionation discussed in this chapter. Figure I.31 shows a summary of the processes which give rise to mineral and other sources of the elements. Many of these have been mentioned already; other processes are discussed in the present section.[40]

---

40 See Mason and Moore (1982) for discussion of chemical aspects, and Evans (1987) for more detailed geology.

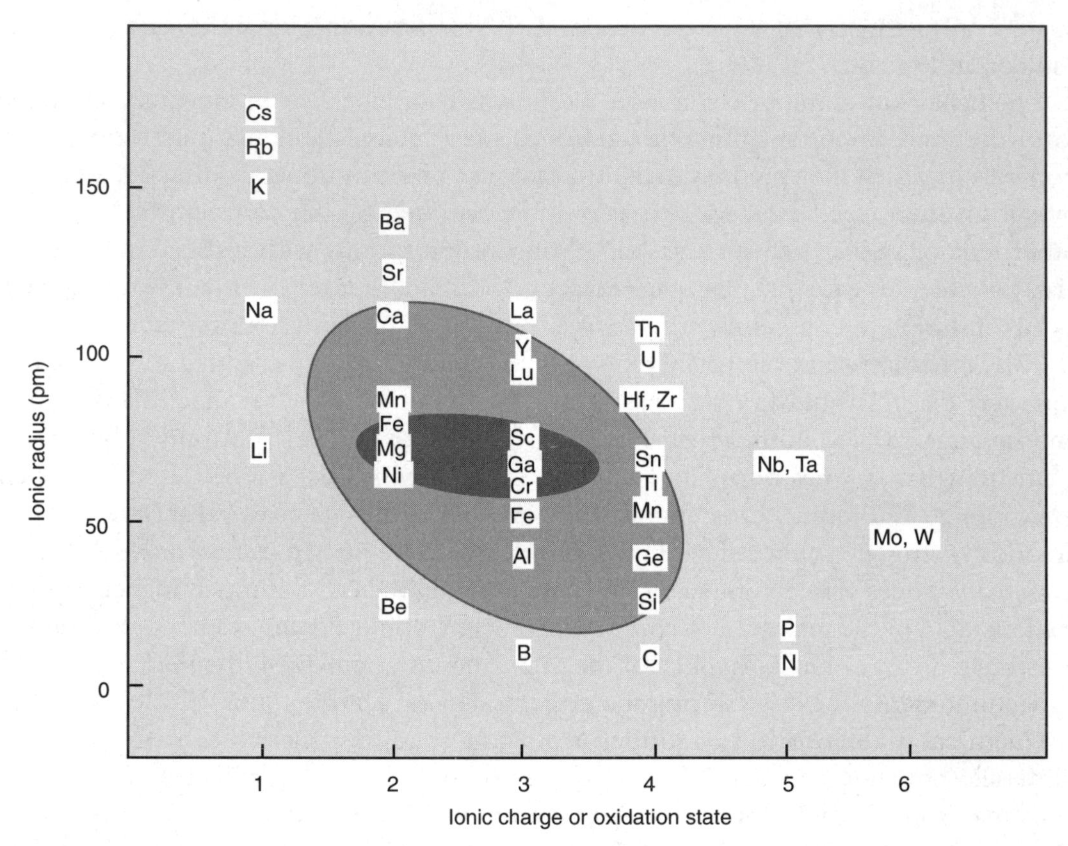

Fig. I.30 Crust–mantle distributions of lithophilic elements. Elements are plotted according to the charge and size of their important ions. The levels of shading show the degee of fractionation into the mantle or the crust. The central, dark-shaded region shows highly compatible ions that are strongly concentrated in magnesium silicates in the mantle. On the other hand, ions in the outermost region are highly incompatible with magnesium silicates, and concentrated in the crust. (Cox, 1989.)

The igneous processes discussed in §4.5 form the primary source of most minerals, but the subsequent action of water is often crucial. Many important minerals result from *hydrothermal* processes, on the sea floor or underground. Some water deep in the crust comes directly from the heating of rocks; other sources include rainfall which seeps in from the surface, and sea water which percolates through rocks in the ocean floor. Deep underground the pressure and temperature rise, and under these conditions many normally insoluble compounds can dissolve to an appreciable extent. Many elements such as copper, zinc, molybdenum, tin, and lead, are mobilised by the formation of complexes with anions such as fluoride and chloride, and are dissolved out of igneous rocks. Oxide or sulfide minerals may then precipitate through various reactions, for example on mixing with cooler water, or by chemical reaction with sedimentary carbonate rocks.

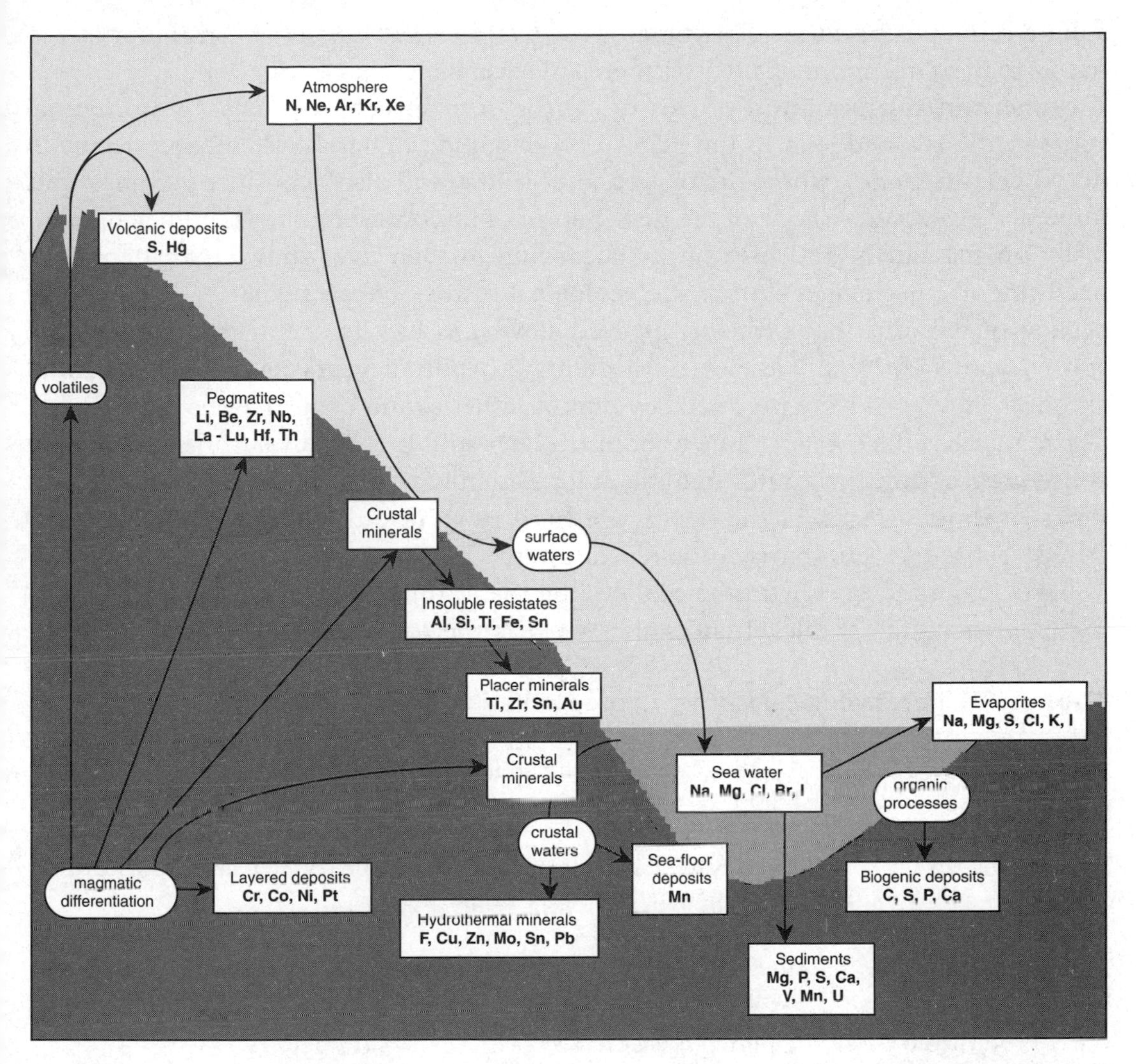

Fig. I.31 Sources of the elements. A summary of the types of geochemical process involved in forming minerals and other important sources of elements. Only a selection of elements in each category is shown.

Surface processes also contribute. As discussed above in §4.4 (see Fig. I.28), weathering separates soluble and insoluble elements, the former passing into the oceans and forming sediments or evaporites. Solid grains of resistates may be mechanically sorted by running water according to their size and density, resulting in local concentrations of *placer minerals*. The involvement of redox chemistry in sedimentation was emphasised in §4.4. Oxidising conditions precipitate some elements, especially iron, but cause others such as copper, nickel, and uranium to become more soluble. Elements passing into solution on oxidation may form sedimentary oxide minerals such as carbonates, or they may be precipitated again by the action of sulfate-reducing bacteria. This type of *secondary enrichment* has been important in the formation of economically useful ores of several elements.

Some mineral formation processes can be seen in action today. Hydrothermal vents in the sea bed lead to the deposition of sulfide minerals. Reduction occurs in deep ocean trenches, where the oxygen level is low, and also deposits many insoluble minerals. However, a study of the distribution of important minerals, both geographically on the Earth, and historically according to their age, shows that there have been important changes during the geological history of our planet.[41] The tectonic cycling of the crust has probably evolved slowly, as has its chemistry: although life started quite early, it was not until about 2.5 billion years ago that substantial amounts of oxygen were produced by photosynthesis, and oxygen levels did not start to rise towards the present value until much later still. It is hardly surprising that, as a consequence, different routes to mineral formation have been most active at different times. Although many of the details are hard to work out and still sometimes disputed, Table I.11 summarises one recent interpretation.

Early magmatic and hydrothermal cycling seems to have been effective in mobilising elements such as nickel and gold, whereas the hydrothermal deposits of some

Table I.11 The geological history of ore formation (based on Brimhall, 1991)

Era (billion years before present)	Geological and chemical features	Major ores formed
Early Archaean (3.8–3.0)	Submarine trench formation; basic magma flows give primary greenstones	Fe, Ni, Cu sulfides, Au
Late Archaean (3.0–2.5)	Recycling of primary greenstone, hydrothermal processes	Cu, Zn hydrothermal sulfides
Early Proterozoic (2.5–1.7)	Uplifted crust erodes	Au placers
	O_2 produced by photosynthesis, oxidation of Fe^{2+}	Banded Fe formations
Mid–late Proterozoic (1.7–0.7)	Thick continental crust forms	Ti, Cr oxides, Fe sulfide, Pt metals
	Atmospheric O_2 increases; active sulfur redox chemistry	Co, Cu, U deposits
Phanerozoic (0.7–present)	Extensive crustal recycling	Hydrothermal Cu, Zn, Mo, Sn, Pb
	Tropical weathering conditions	Al, Fe resistates
	Secondary enrichment	Co, Ni, Cu minerals

41 Meyer (1985) and Brimhall (1991) discuss the history and geography of ore formation. Broader discussions of the geological history of the Earth can be found in Smith (1981).

other elements, especially molybdenum, tin, and lead, are much more recent. Copper is more varied in its occurrence, being deposited during all the eras shown. It seems that some elements have been concentrated only after extensive recycling of the continental crust. It is likely that changing redox conditions, allowing the oxidation and reduction of sulfur compounds, have also played a part.

Major iron oxide deposits first appeared when sufficient oxygen was available from photosynthesis; these deposits are the *banded iron formations*, so called from their appearance of alternating layers. Further redox processes developed as the atmospheric oxygen content rose, allowing the oxidation of insoluble sulfides and UO_2, and their subsequent reduction under oxygen-poor conditions.

The time at which minerals were laid down has an important influence on their geographical location. Ores formed in the Archaean and Proterozoic eras are generally located in rocks unchanged from very ancient continents. These stable areas include much of Canada, parts of central and southern Africa, central Asian regions of the CIS, and northern and western Australia. They are major sources of many valuable and relatively uncommon elements, such as chromium, nickel, gold, and the platinum metals. On the other hand, recent hydrothermal minerals are common in geologically active regions such as the 'Pacific Rim', and especially the South American Cordillera.

Fossil fuels form another crucial resource, not shown in Table I.11. They are sedimentary in origin, and relatively recent. Coal was formed in large quantities by the swamps of the Carboniferous period (around 300 million years ago). Oil is mostly 'younger' in geological terms. The major oil-fields of the world are located in sedimentary basins generally away from serious tectonic activity.

The geographical location of mineral reserves will be discussed again in Chapter 5. More information about the minerals in which individual elements are found is given under the appropriate headings in Part II.

5

The human impact on the environment

All living species have some impact on the environment, in some cases a profound one. The influence of life on redox chemistry, and on other aspects of chemistry on the Earth, were discussed in Chapter 4. The production of free oxygen by photosynthesis has changed the nature of the Earth's surface much more radically than anything humans are likely to do (short of global nuclear war). This natural evolution has however taken place very slowly, whereas the accelerating progress of technology risks producing changes at a much greater rate. The present chapter discusses some of the background to this environmental concern, insofar as it involves the use of elements and their compounds. First, there is a brief account of how we have come to use the elements, and the ways in which they are extracted from their natural ores. This is followed by a discussion of element reserves and their geographical locations. The final section is concerned with an aspect of technology which many people wrongly assume to be the only point of connection between chemistry and the environment: that is, pollution.

5.1 A short history of element usage

The fashioning of tools, weapons, and ornaments has long been regarded as a characteristic human activity. It is out of this activity—initially confined to purely natural materials such as wood, bone, or stone—that the first deliberate use of particular elements began.[42] Some kind of ancient metallurgy seems to have developed independently in America, China, India, and the Middle East, but it is in the last of these locations (in Asia Minor) that the earliest remains have been found. Copper pins dating from around 8000 BC are thought to have been rolled or beaten out of native (i.e. elemental) copper, followed by some heat-treatment below the melting temperature of the metal to anneal them. Gold was also first used several millennia BC, although archaeological finds are rarer than for copper, probably because gold was more highly prized, and generally reused rather than being left in burial mounds.

42 For an account of early metallurgy, see Tylecote (1992).

It is no accident that two elements from Group 11 in the periodic table were the first to be used in this way. Having a low affinity for oxygen (see §3.1), they can both be found as native, uncombined, elements; furthermore, they have relatively low melting temperatures and are ductile and easily worked, even without melting. The amount of native copper found naturally is small however, and its extensive use from about 3500 BC onwards required the development of smelting techniques to extract the molten metal from its ores. Carbonates such as malachite were reduced with carbon—another element important in early technology, obtained in the form of charcoal by the incomplete combustion of wood:

$$CuCO_3 \rightarrow CuO + CO_2;$$

$$2CuO + C \rightarrow 2Cu + CO_2.$$

Later developments used the commoner sulfide ores, which required cycles of heating in air (to convert to oxide) and then reduction with charcoal. Ores with a significant concentration of arsenic seem to have been particularly favoured. Some of this ends up in the copper and improves its working properties, although some volatile and toxic As_2O_3 is also produced. Another additive which makes copper easier to work is tin, made by reduction of cassiterite (SnO_2) with charcoal. The alloying of copper with tin to make *bronze* was widespread in the Middle East by 2000 BC, and eliminated the need to use arsenic-rich ores with their associated toxic hazard. Brass, an alloy of copper with zinc, was first made in China around 200 BC, and spread, first to Egypt, and then to Roman Europe. Two other metals used widely by Roman times were lead and silver, the latter not occurring in significant native deposits like copper and gold, but often found as a minor constituent of the lead ore *galena* (PbS).

All these metals can be found either in metallic form or as easily reduced ores. The early use of iron is more surprising, as it is more reactive and therefore harder to extract from its compounds. It also has a higher melting point and is harder to work. Native iron occurs only as a constituent of meteorites, and it is from this source that the earliest known uses derived. The greater difficulty in working was compensated by a much stronger material, which therefore became prized for weapons. Extraction from oxide and sulfide ores was developed and widely used by 1200–1000 BC. It has been suggested that the growth of ancient Greece as an important power in the first millennium BC was partly due to the use of iron derived from its own natural resources.

Other elements came to be used in ways which did not require chemical extraction from naturally occurring minerals.[43] The manufacture of glass, by fusing together silica (SiO_2), lime ($CaCO_3$ or CaO made by heating it), and soda (natural Na_2CO_3), was started in Egypt around 2000 BC. Lime was also used in mortar for building, and soda as a cleansing material before the invention of soap. Another important development was the invention of gunpowder by the Chinese around

43 Sherwood Taylor (1957).

AD 900. Nitrates such as saltpetre (KNO_3) are the strongest naturally occurring oxidising agents, and form a highly combustible mixture with charcoal and sulfur (another element that occurs naturally). Natural sources of saltpetre were supplemented by a primitive form of biotechnology: soil was deliberately fertilised by animal urine, giving a high nitrogen content which encourages the growth of nitrifying bacteria, hence producing nitrate. A few other elements, such as arsenic and antimony, were also known in the Middle Ages, and used as drugs.

These developments and subsequent ones are summarised in Table I.12. Progress has accelerated greatly with the development of modern chemistry and technology in the last two centuries. The Industrial Revolution did not originally entail the applica-

Table I.12 History of element usage. The use of natural materials (e.g. wood, stone) without processing is excluded

Application	Era of first major usage of element or its compounds			
	Prehistoric (before 2500 BC)	Pre-industrial (2500 BC–AD 1750)	Industrial (AD 1750–AD 1940)	High-technology (after AD 1940)
Metals: vessels, tools, coins, weapons, etc.	Cu, Au	Fe, Zn, Ag, Sn, Pb	Ni, Mo, W	Zr, Nb
Metals: construction, transport	–	–	Al, Cr, Mn, Fe	Be, Mg, Ti
Fuels and explosives	C	N, S	–	U, Pu
Glass, ceramics, refractories	–	Na, Ca, Si, Pb	Mg, Zr, Th	Li, B, La–Lu
Pigments and dyeing	–	Al, Fe, Co, Cu, Cd, Hg, Pb	N, Zn, As, Se	Ti
Pharmaceutical	–	S, As, Sb, Hg	C, Bi, Br, I, Ra	Li, Pt
Fertilisers and pesticides	–	–	N, K, P, Cl, As, Br, Hg	Sn
Industrial chemicals and catalysts	–	–	N, F, Na, S, Cl, K, Hg, Pt	Ar, Rh, Ba, La–Lu, Re
Electrical and electronics	–	–	Cu, In, Pb, W	Si, Ga, Ge, As, Se Ta, Ir
Household goods and chemicals	–	–	C, N, Na, Cl	B, P, Br, Sn

tion of new elements; rather there was a greatly increased use of carbon derived from fossil fuels (coal), which allowed iron to be made more easily, and in much larger quantities. This was the start of the 'energy-intensive' civilisation in which we live today. New ideas and new manufacturing processes helped to stimulate the spirit of enquiry which led to the development of modern chemistry, and to the rapid discovery of many new elements during the nineteenth century. Some of these were incorporated into new materials such as steels with improved mechanical properties and corrosion resistance; others were used as constituents of industrial chemicals (acids, alkalis, solvents, catalysts, etc.) required for other processes. The provision of food for the increasing urban populations also stimulated the use of inorganic fertilisers containing the essential elements nitrogen, phosphorus, and potassium.

The twentieth century has seen the application of many elements which can only be extracted or exploited by advanced technological methods. Aluminium—the most abundant metallic element in the Earth's crust—is highly electropositive, and its oxide cannot be reduced by carbon. Its major use as a metal depends on extraction by electrolysis, and hence on the availability of cheap electric power. Other very electropositive metals introduced even more recently include titanium and magnesium; both of these are produced by reduction of compounds with metallic sodium, itself made by electrolysis.

Developments since the middle of the century have been of a rather different character. In contrast to the major construction materials and high-volume industrial chemicals of the previous era, 'high technology' involves smaller quantities of material and much more sophisticated processing. The silicon chips of computers form the most notable example; others include additives to improve the properties of plastics, specialised glasses and materials for displays, new pharmaceuticals, and catalysts for the petrochemicals industry.

The current usage of elements, giving the order-of-magnitude quantities involved and the type of application, is shown in Fig. I.32. Only the major use of each element is shown; for example the amount of silicon made into electronic components is completely dwarfed by its use in ferro-silicon steels, and the uses of magnesium and titanium as metals are smaller than applications of the oxides, in refractory materials (e.g. blast-furnace linings), and white pigments, respectively. More details about the applications of many elements are given in Part II.

The logarithmic scale in Fig. I.32 should be noted. The amounts of elements extracted each year span a range from nearly 10^{13} kg (10 billion tonnes) for carbon, to a few thousand kilograms (a few tonnes) for rhodium and osmium. Both in absolute terms and relative to its abundance on Earth (see below), the use of carbon in fossil fuel burning is more significant than the use of any other element.

One of the elements known since ancient times has been deliberately omitted from Fig. I.32, as in spite of its production in appreciable quantities (2×10^6 kg per year) it is almost totally 'useless'. The major applications of gold have always been, and remain

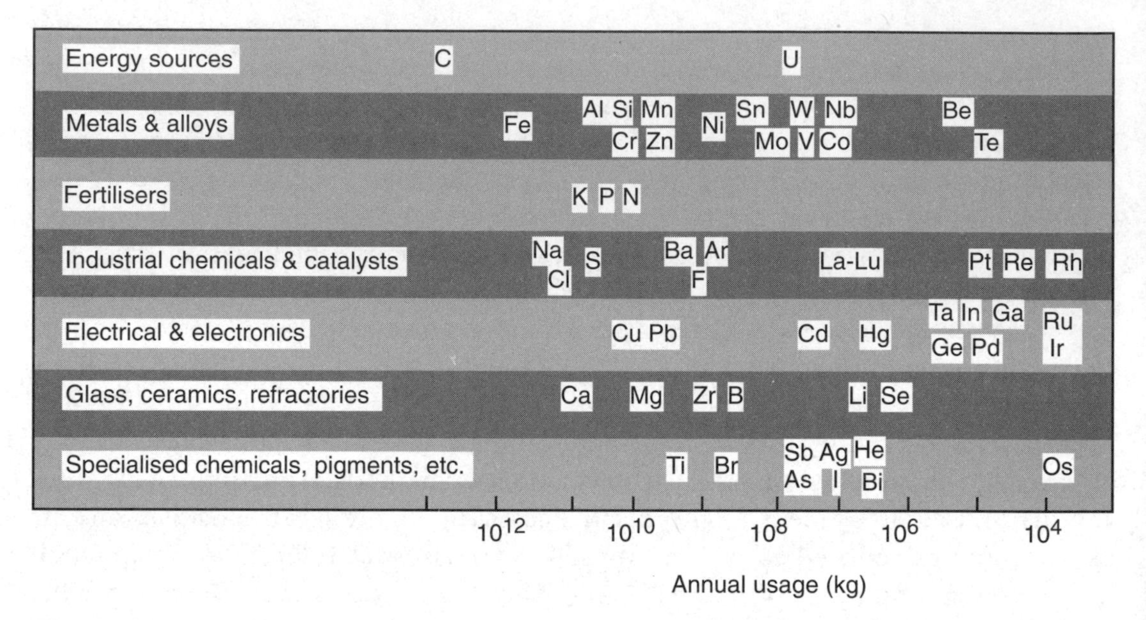

Fig. I.32 Current (1990) usage of elements, either extracted or in purified compounds. Only the major application of each element is noted. (Data mostly taken from Crowson, 1992.)

today, for ornamental and monetary purposes. There is a further reflection which is more sinister. Several elements have first been used in a major way for military purposes: this list includes iron, nitrogen (in gunpowder), and uranium (atomic weapons). Human vanity, greed, and aggression have always been more powerful motivators of change than the desire to improve the condition of mankind. They remain—together with overpopulation of the planet—the most serious threats to the environment.

5.2 The extraction of elements

For some applications such as glass, it is unnecessary to extract pure elements from their naturally occurring compounds. In many uses however, especially with metals, the elements must be chemically extracted from their ores. The first metals used in this way were ones for which this extraction is relatively easy. Reduction of an oxide by carbon has several advantages. A fairly pure form of this element can be obtained as charcoal by the incomplete combustion of wood, although this was later replaced by fossil fuel sources. The oxidation product CO_2 is a gas which escapes and does not contaminate the product. Another important factor is that a reaction such as

$$C(s) + 2MO(s) \rightarrow CO_2(g) + 2M(s)$$

becomes thermodynamically more favourable at higher temperatures. At room temperature, as Fig. I.12 (§3.1) suggests, carbon has a lesser affinity for oxygen than do iron and tin. But carbon becomes a better reducing agent at higher temperatures, and can be used for the extraction of several elements. This is not possible, however, for highly electropositive metals such as Mg or Al, or for elements such as W which form stable carbides. The extraction of these elements depends on finding more effective reducing agents such as hydrogen or sodium, or on electrolytic methods of reduction.[44]

Table I.13 summarises the chemical methods used currently for the extraction of elements. Reduction by carbon is still the preferred method for a number of elements, especially iron. For chalcophilic metals occurring as sulfide ores, the traditional route is to roast the sulfide in air to give an oxide, followed by reduction of this with carbon (see discussion of copper in §5.1). In some cases, however, this subsequent reduction step is unnecessary: as discussed under *sulfur, the sulfides of several elements, including copper, can be converted directly to the metal by controlled treatment with oxygen.

The highly electropositive metals, especially of Groups 1 and 2, were first prepared by electrolysis. This is still used for many of them, important examples being sodium, obtained together with chlorine by the electrolysis of brine (concentrated NaCl solution in water), and aluminium, made by the electrolysis of molten cryolite, Na_3AlF_6. The process is very expensive in electric power, and aluminium is often manufactured in countries, such as Canada and Norway, where cheap hydroelectric power is available. The convenience of the electrolytic process depends on the availability of suitable compounds, and is not always possible. An alternative is to use a yet more electropositive metal, itself made by electrolysis, to reduce a compound, generally a halide. Titanium for example is manufactured from $TiCl_4$ and magnesium. Metallic sodium can also used for this kind of procedure.

Table I.13 The extraction of elements from their compounds. (See Greenwood and Earnshaw (1984) for more details)

Method of extraction	Elements
Reduction of oxide with C	Si, P, Mn, Fe, Sn
Conversion of sulfide to oxide, then reduction with C	Co, Zn, Pb, Bi
Reaction of sulfide with O_2	Cu, Hg
Electrolysis of solution or molten salt	H, Li, Be, B, F, Na, Ca, Al, Cl, Ni, Cu, Ga, Sr, In, Ba, La–Lu, Tl
Reduction of halide with Na or other electropositive metal	Be, Mg, Si, K, Ti, V, Cr, Rb, Zr, Cs, La–Lu, Hf, U
Reduction of halide or oxide with H_2	B, Ni, Ge, Mo, Ru, W, Re
Oxidation of compound with Cl_2	Br, I

44 See Shriver *et al.* (1990) for a further discussion of the basic principles, and Greenwood and Earnshaw (1984) for details on the extraction of many individual elements.

In many cases, the best method depends on the ultimate use to which the element will be put. Elements intended as constituents of steels (silicon, vanadium, chromium, etc.) do not need to be prepared in pure form first, and so their oxides can be reduced along with that of iron, using carbon. Purer forms for applications such as electronics require a 'cleaner' procedure. Very pure silicon for 'chips' is made by reduction of a halide with hydrogen gas.

A few elements, such as F, Cl, Br, and I, occur naturally in reduced forms rather than oxidised forrms. These may either be extracted by electrolysis, where they are liberated at the anode rather than the cathode, or by oxidation with another element. For example, bromine and iodine are manufactured by the reaction of their salts with Cl_2.

5.3 Element reserves

The significance of current element usage appears in another light when it is compared with quantities available to us. Figure I.33 shows the annual production of elements, divided by the average concentration of each in the Earth's crust. The

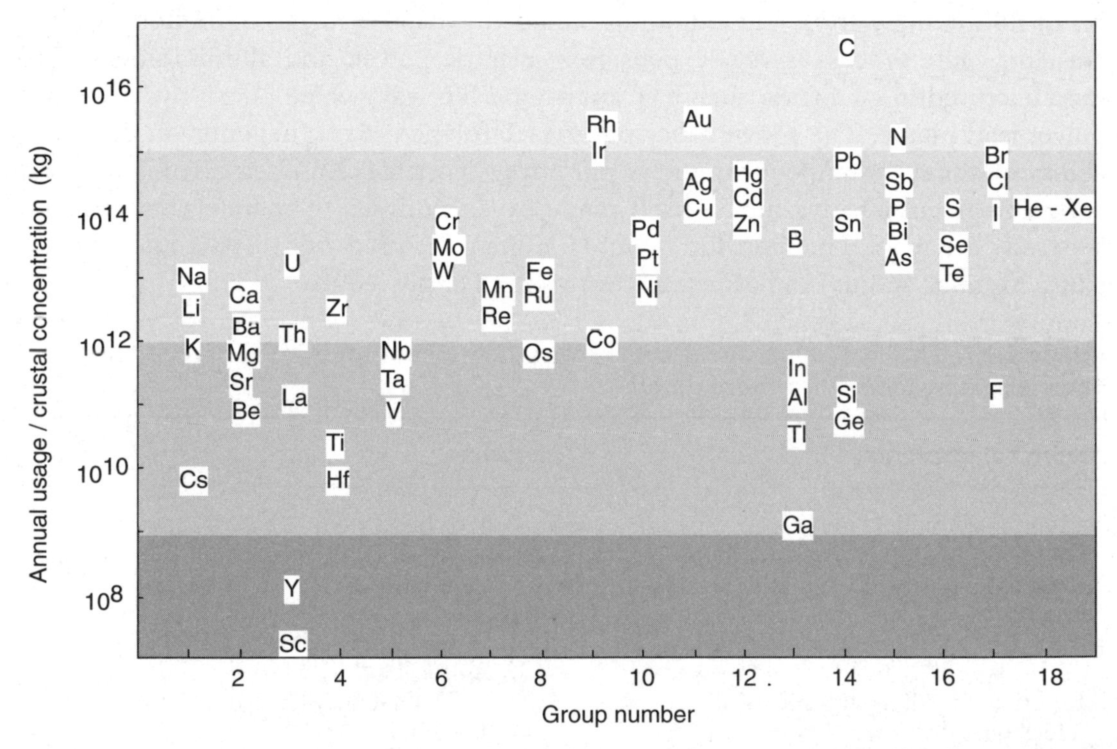

Fig. I.33 Annual element usage divided by average crustal concentration. The vertical (logarithmic) scale shows the mass of earth that would have to be processed annually to obtain each element, if it were uniformly distributed in the crust. In spite of this artificial assumption, the plot gives an idea of the intensity of human exploitation of different elements, in comparison with the amounts available. (Data mostly for 1990, from Crowson, 1992, and Emsley, 1990.)

figures in the plot span many orders of magnitude, from carbon (fossil fuels) at the top, to some little used elements such as scandium and yttrium at the bottom. Some elements used in large quantities, such as aluminium, do not appear high up in this plot because of their high crustal concentrations.

Figure I.33 gives some idea of the intensity of human usage of elements, in comparison with the total natural resources. Elements such as carbon, gold, nitrogen, and indeed many of the elements from Groups 11–17 in the periodic table, appear to be in much greater demand than elements from earlier groups. This is a reflection partly of the more diverse chemical properties of elements in later groups, which make them more desirable for industrial applications, but partly also of their lesser general abundance on Earth. Nevertheless, in some ways the data in Fig. I.33 are misleading. If elements were indeed extracted from 'average' crust as implied here, the quantities involved would be outlandish: to extract even an element in a middle position such as uranium in this way, it would be necessary to dig up some 10^{13} kg of crust annually. This is 10 billion tonnes, corresponding to well over a cubic kilometre of earth. For carbon, one would need about a thousand cubic kilometres. Apart from the amount of digging involved, the difficulty of chemical processing required to extract the relevant elements would be stupendous. In reality, nearly all elements are obtained from specific minerals and other deposits, where they have already been concentrated by the geological and other processes described in Chapter 4. Economically viable sources of most elements require ores with a sufficient concentration factor relative to the average crustal abundance. Table I.14 shows typical

Table I.14 Current reserves of some elements, showing the concentration factor (compared with average crustal concentration) for economically viable reserves, the amount of known reserves, their lifetime based on current usage, and the countries where major reserves are found. (Based on Mason and Moore, 1982, and Crowson, 1992)

Element	Concentration factor	Reserves (10^9 kg)	Lifetime (years)	Locations of major reserves
Al	4	20 000	220	Australia, Guinea, Brazil, many others
Fe	6	66 000	120	CIS,[a] Australia, Canada
Mn	350	800	100	S. Africa, CIS,[a] Gabon
Cr	3000	400	100	S. Africa, Zimbabwe, CIS[a]
Cu	160	300	36	Chile, USA, CIS,[a] Zaïre
Ni	125	47	55	Cuba, CIS,[a] Canada, New Caledonia
Zn	600	150	21	Canada, Australia, USA
Sn	5000	5	28	China, Brazil, Indonesia, Malaysia
Pb	3000	71	20	Australia, USA, CIS,[a] Canada
U	500	2.8	58	Australia, CIS,[a] S. Africa, USA

[a]No individual breakdown is available for nations constituting the Commonwealth of Independent States (formerly USSR).

values required for a number of elements: they are not large for common elements such as aluminium and iron, but they may be well over a thousand for rarer ones such as chromium, tin, or lead.

Also shown in Table I.14 are the known reserves of elements, and their lifetime estimated at the current (1990) rate of production. Some of the latter figures are quite alarming, and suggest that suitable sources of elements such as zinc, tin, and lead could be exhausted in 20 years or so. These figures are however susceptible to alteration. The rate of usage of many elements is generally increasing, but for some others such as lead, it is actually declining. The known reserves of elements can increase for two reasons. Geological exploration can lead to the discovery of previously unknown deposits. But there is also a more subtle factor. Whether a particular mineral deposit counts as a useful reserve depends on many economic and technological considerations. As the most concentrated deposits are exhausted, new methods of extraction may be developed, and/or the price of the element may rise, making it possible to exploit less favourable sources; these may now be regarded as reserves where previously they were not. There has been a historic tendency for known reserves of many elements to increase roughly in proportion to their production, but clearly this cannot happen indefinitely.

Countries where some of the major reserves are located are also listed in Table I.14. As discussed in §4.6, these different locations are a consequence of the diverse geological origins of minerals, and a function of the type of rocks making up the crust in particular areas of the world. The uneven geological distribution of desirable elements has had an important impact on world history. To give a few examples, the Spanish conquest of Central and South America in the sixteenth century was spurred on (and financed) by silver and gold in use by the indigenous peoples; the location of coal and iron ore in the British Midlands partly determined the site of the early Industrial Revolution; and in our own century, the geopolitics of oil reserves have played a role in several international conflicts, including the Second World War (as the Germans and Japanese were largely without their own supplies, and as a result were desperate to reach areas such as the Middle East). Element extraction and production is a very significant factor in world trade. Table I.15 lists some countries which are major exporters of elements or their ores. In nearly all cases, the principal importing countries are the highly industrialised ones: the USA, Japan, and countries of the European Community. In many of the countries listed in Table I.15, their exports make up a sizeable fraction of the entire economy. This is true not only with oil production in the Middle East, but also for some countries of southern Africa and South America, which depend heavily on their sources of valuable metals such as gold, copper, chromium, manganese, or molybdenum. Such resources sometimes give countries an important economic advantage over less well-endowed neighbours, but they can also create problems. The demand for elements, and their prices, are subject to variables that may have little to do with the resources themselves. Economic recessions in developed countries, wars, technological changes, and stock-

Table I.15 Major exporting countries for elements and their ores (based on data from Crowson, 1992)

Countries	Elements
Albania, Zimbabwe	Cr
Australia	Li, Al, Ti, V, Mn, Fe, Zn, Se, Zr, Ag, La–Lu, Ta
Bolivia	Sb
Brazil	Be, Mn, Fe, Nb, Sn, La–Lu, Ta
Canada	K, Ti, Ni, Cu, Zn, Se, Mo, Ta, U
Chile	Li, Cu, As, Se, Mo, Ba, Re
China	Ba, Sb, La–Lu, W, Hg, Bi
Czech and Slovak Republics, Namibia, Niger	U
Gabon	Mn
Germany	K, U
Guinea, Jamaica	Al
Indonesia	C,[a] Ni, Sn
Iran, Iraq, Kuwait, Libya, Nigeria, Saudi Arabia, United Arab Emirates, Venezuela	C[a]
Israel	P, K
Jordan	P
Malaysia	Ti, Sn, La–Lu
Mexico	C,[a] Cu, As, Mo, Ag, Pb, Bi
Morocco	P, Ba
New Caledonia	Ni
Norway	Al, Ti
Peru	Cu, Zn, Mo, Ag, Pb, Bi
Philippines	As
Poland	S, Cu
South Africa	Ti, V, Cr, Mn, Zr, Pt metals, Au
Sweden	As, Pb
Turkey	B, Cr, Cu
Zaïre, Zambia	Co, Cu

[a]Fossil fuel (chiefly oil) production of C.

market manipulations are among the factors that can have a disproportionate economic effect on countries which are highly dependent on a single type of resource.

5.4 Environmental pollution

'Pollution' is an emotive term, meaning different things to different people: a reasonable general definition might be 'too much of something in the wrong place'.[45] This can include entirely natural substances. For example, one of the most serious pollutants in many countries is sewage—mostly human excreta—which contaminates

45 Quoted from Carthorne and Dobbs, in Harrison (1990).

rivers with pathogenic bacteria and depletes them of dissolved oxygen. This section discusses briefly the forms of pollution that arise specifically from the technological uses of elements, and is not concerned with the wider forms of pollution resulting from human activity. Table I.16 summarises the elements chiefly implicated.[46] More specific details about individual elements can be found in Part II.

Previous sections have discussed several issues which are relevant to pollution by specific elements. Figures I.32 and I.33 give information about the amounts of different elements used, both in absolute magnitudes (I.32) and relative to their natural abundance (I.33). Elements used intensively according to either of these criteria may have some potential for disturbing the natural environment. The general mobility of an element in the environment is also important (see §4.1). Ones with compounds that are involatile and very insoluble in water are less likely to cause problems. Other factors include the toxicity of an element, and the chemical forms in which it is used: compounds of a different kind from those which appear in nature are more likely to be toxic or otherwise harmful.

It is notable that nearly all the elements appearing in Table I.16 come from the later groups in the periodic table. They are in general more intensively used (Fig. I.33) and more mobile in the environment (Fig. I.25). Their more covalent chemistry makes them versatile as constituents of varied types of compound. This is one reason for their greater usage, but it also makes it more likely that the compounds produced and used by industry may be serious pollutants.

Carbon is the most heavily used element, both in absolute terms and relative to its general abundance. Most is used in the form of fossil fuels, for industry, transport, and domestic purposes. This gives rise to several forms of pollution. CO_2 produced by complete combustion may cause global warming. Incomplete combustion produces CO, organic compounds, and particulate matter, which may be toxic and cause air pollution of a short-term and more localised nature. Some other elements are mobilised which can also cause problems: oxides of sulfur and nitrogen which are toxic and give rise to acid rain, and less abundant elements such as arsenic which are constituents of coal.

Another general source of pollution is the processing and use of metals. Those from Groups 10–14 quite heavily used in proportion to their abundance, and are toxic to varying degrees. These elements may be released into the environment in soluble form, during mining and other processing operations, and into the atmosphere as dusts. As with carbon, the use of one element may mobilise others. Pollution by the very toxic cadmium can arise during the extraction of zinc, as the two elements are often found in association.

---

46 See for example Harrison (1990, 1992), Clark (1989), Bridgman (1990), Elsom (1987), and Freedman (1989). Books on environmental chemistry, e.g. Manahan (1991), are often concerned largely with pollution.

Table I.16 Elements forming significant environmental pollutants

Element	Chemical or other form	Principal anthropogenic sources	Reason for concern
C	CO_2	fossil fuel burning	global warming
	CH_4	agriculture	global warming
	CO, organics	automobiles, industry	air pollution
	liquid hydrocarbons	petroleum extraction and transport	harmful to wildlife
	organic waste	untreated sewage	water pollution, biological oxygen demand
N	nitrates, nitrites	fertilisers	water pollution, toxic to infants
	NO, NO_2	automobiles	air pollution, acid rain
F	fluorides	industrial	toxic
Al	soluble Al^{3+}	leaching by acid rain	toxic to plants and animals
P	soluble phosphates	fertilisers, detergents	lake eutrophication
	organic compounds	pesticides, poison gas	highly toxic
S	SO_2	fossil fuel burning, industry	air pollution, acid rain
Cl	organic compounds	industrial chemicals, pesticides	toxic
	CFCs	refrigerants, aerosols	damage to ozone layer
Ni	any compounds	mining and extraction of metals	toxic dust and soils
Cu	any compounds	mining and extraction of metals	toxic to many species
As	any compounds	industry, pesticides	toxic
Br	organic compounds	fire-extinguishing agents	damage to ozone layer
Kr	^{86}Kr	nuclear power	radioactive
Cd	any compounds	industry	very toxic
Sn	organic compounds	marine anti-fouling paints	toxic to marine life
Sb	any compounds	industry	toxic
I	^{131}I	nuclear power	radioactive
Cs	^{137}Cs	nuclear power	radioactive
Hg	any compounds	industry, pesticides	very toxic
Pb	any compounds	industry, paints, automobile exhaust, water supply	very toxic
Pu	any compounds	nuclear power	radioactive

Agriculture accounts for a high proportion of the usage of some elements, nitrogen and phosphorus especially (see Fig. I.32). The intensive use of fertilisers and pesticides can give rise to many problems, especially as these compounds are spread deliberately in the environment. Excess fertilisers can contaminate water supplies, but changes in agricultural practice also have other consequences. A recent rise in the atmospheric concentrations of some trace gases, such as the 'greenhouse gases' methane (CH_4) and nitrous oxide (N_2O), may be one effect of increased fertilisation.

Biomass burning, especially of tropical rain-forest, is another source of atmospheric perturbations.

The examples of chlorine and bromine in Table I.16 illustrate the problems arising with compounds that do not occur naturally. These elements are constituents of synthetic organic molecules unknown in nature, but often very persistent in the environment. Many of these are toxic, and indeed are manufactured as insecticides or herbicides. Others, the chlorofluorocarbons and related bromine compounds, are very inert chemically and have applications depending on this property; their problem however is that they persist so long in the lower atmosphere that they can reach the stratosphere, where they may damage the ozone layer.

Artificial radioactive elements are made in much smaller quantities, and are unlikely to give rise to chemical perturbations of the environment. Their potential for harm is of a different nature, which is discussed in the following section.

5.5 Radioactive elements and ionising radiation

A selection of radioactive elements and isotopes, including artificial ones, was listed in Table I.6 (see §2.3), and a few of the latter are shown in Table I.16 as possible environmental pollutants. Radioactive substances are damaging to life because of the very high energy of the radiation emitted.[47] So-called *ionising radiation* knocks electrons out of atoms, and breaks chemical bonds. Large numbers of very reactive radicals may be produced, which initiate chemical reactions that can destroy or modify the biochemical constituents of living cells. Most compounds in the cell are being continually synthesised, with a sufficiently high turnover that they can be replaced fairly quickly. The most damaging effect of ionising radiation is on the genetic apparatus, and especially the DNA itself. The DNA in each cell is responsible for controlling the chemical reactions in the cell, and if it is damaged or modified then the cell may run partly out of control. One cell which dies has little effect, but damage which causes a cell to proliferate uncontrollably, as in cancer, is much more serious. Rapidly dividing cells are most susceptible, especially those in the bone marrow where the cells responsible for the body's immune system are manufactured. Damage to these areas causes leukaemia, one of the commonest types of cancer attributed to radiation exposure. When DNA in the sperm or ova is damaged, it may result in mutations which can be passed on to later generations.

The damaging effect of ionising radiation depends on the energy deposited in body tissues, modified by a so-called *quality factor* which depends on the ionising power of the particular type of radiation. X-rays, γ radiation, and electrons have a quality factor of one, α particles of 20. Absorbed doses are measured in units of *Grays* (Gy), equal to 1 joule deposited per kilogram of matter. When multiplied by the quality

47 See Martin and Harbison (1986), and NRPB (1986, 1988).

factor, the result is the *dose equivalent* measured in *sieverts* (Sv).[48] An acute radiation dose of 8 Gy is likely to cause irreparable damage to the bone marrow, and death within 2 months caused by failure of the immune system; larger doses may destroy the cells of the intestines and lead to even more rapid death. These doses are exceptional, and are only likely in the immediate vicinity of nuclear explosions or serious leakage from power stations such as that at Chernobyl in Ukraine in 1986. Smaller doses can increase the likelihood of developing cancer, and it is generally considered that no dose of radiation is totally 'safe'.

The *radiotoxicity* of an element depends not only on its half-life and the type of radiation produced, but also on chemical and biological factors. Elements that are excreted rapidly by the body give rise to less exposure than ones which are retained. The particular organs where an element may be concentrated are also important. Exposures to the testes and ovaries are most serious in terms of long-term genetic effects, but no elements have a particular affinity for these sites. Iodine is concentrated in the thyroid gland, and so radioactive isotopes such as ^{131}I can cause cancer there. The elements with the highest radiotoxicity rating are strontium (as ^{90}Sr), radium, and plutonium. They are taken up in bones, and not only retained for many years in the body, but can irradiate the sensitive cells of the bone marrow.

Table I.17 shows the some natural and anthropogenic sources of ionising radiation, with the average dose (in millisieverts, mSv), received annually by people living in Britain. These numbers vary considerably according to location. For example, people living at high altitude receive a greater exposure from cosmic rays, as this source of radiation comes from outer space, and is attenuated by the atmosphere. The largest source for most people is the radioactive element *radon, a gas coming from the decay of natural thorium and uranium. Especially high levels may be found in neighbourhoods where these elements are abundant in rocks; this is particularly true of acid igneous rocks such as granite.

The largest artificial source of radiation is from medical X-rays. There is also a smaller exposure from radioactive isotopes used for medical purposes. In comparison with these, and especially relative to the radiation from natural sources, the additional exposure from radioactive elements generated by nuclear weapons, and nuclear power generation, appears almost negligible. This is no reason to be unconcerned, however. Substantial amounts of many radioactive isotopes were generated by the atmospheric testing of nuclear weapons in the 1950s, before this activity was banned by international agreement. Elements of concern included fission products such as *strontium, and radioactive isotopes produced by neutron bombardment, such as those of *hydrogen and *carbon. A nuclear war, even of 'limited' extent,

---

48 These SI units have replaced the older ones, the *rad* and the *rem*: 1 Gy = 100 rad; 1 Sv = 100 rem.

Table I.17 Sources of ionising radiation, showing the doses received by an average UK resident. Both natural and anthropogenic doses may vary greatly from place to place. (Data quoted by Sutton, 1992, from NRPB, 1988)

Source	Dose (mSv yr^{-1})
Cosmic rays	0.25
Natural radioactive elements:	
ground and buildings	0.35
food and drink	0.3
radon in air	1.3
Total natural sources	2.2
Medical X-rays	0.3
Radioactive isotopes used in medicine	0.02
Fallout from nuclear weapons testing	0.01
Nuclear power production	0.001 (0.04)[a]
Misc. artificial sources (TV, air travel, etc.)	0.01
Total anthropogenic sources	0.34 (0.38)[a]

[a]The figure in parentheses includes the exceptional doses received in the first year after the Chernobyl accident in April 1986.

could be much more serious in this respect. Nuclear power production and reprocessing, described under *uranium, also attracts considerable attention. Under normal operation, the largest releases of radioactivity probably come in the form of radioactive *krypton and *xenon. The uncontrolled release of fission products in a fire, as happened at Chernobyl, gives rise to a different mix of elements than does fallout from explosions. Radioactive *iodine (in the short term) and *caesium (in the longer term) have been the main pollutants from this accident.

Apart from the possibility of future accidents, the most serious issue relating to nuclear power production is how to dispose of high-level radioactive material coming from spent fuel elements. These contain artificial transuranium elements such as *plutonium, and fission products from the uranium. Significant radioactivity will remain for thousands of years. Large amounts of plutonium have been separated in reprocessing operations for the manufacture of nuclear weapons, and the disposal of this element poses special problems. The currently favoured plan for the remaining waste is to convert it into a highly insoluble glass-like form, and deposit it deep underground, in sites with suitable geological features so that it should remain undisturbed for the foreseeable future.

Part II

Part II gives an individual discussion of the elements up to fermium (atomic number 100), the heaviest one ever detected outside the laboratory. Elements are listed in alphabetical order of their English names, with the atomic number and chemical symbol following in parentheses; for example $_{89}$*Ac indicates an atomic number of 89. Alternative US spellings are also shown.*

*Cross-references such as '§1.1' refer to the appropriate sections in Part I; those of the form '*oxygen' to the corresponding element in Part II.*

Concentrations of elements shown in the tables are given in parts per hundred (%), per 10^6 *(ppm), or per* 10^9 *(ppb). They refer to the mass fraction of the element, except in the atmosphere, where the mixing ratio, defined as the volume fraction of the appropriate compound (and equal to the relative numbers of molecules present), is normally used. (Note that to a good approximation in water, 1 ppm = 1 mg per* dm^3 *and 1 ppb = 1 μg per* dm^3*.)*

Data in the tables have been obtained from the following sources unless specified otherwise: Emsley (1991) for general abundance; Henderson (1982) average concentrations in rivers; Fergusson (1990) for environmental concentrations of As, Se, Cd, Sb, Hg, and Pb. Some useful references on the chemistry of the elements are: Greenwood and Earnshaw (1984) for general chemistry; Wayne (1991) for atmospheric chemistry; Stryer (1988) for general biochemistry; and da Silva and Williams (1991) for the biological chemistry of specific elements. References to other sources are given explicitly at the appropriate point.

Actinium ($_{89}$Ac)

Actinium is a radioactive element formed in the decay series of *uranium and *thorium. ^{227}Ac is formed from ^{235}U, and has a half-life of 22 years; ^{228}Ac, with a half-life of 6 hours, comes from ^{232}Th. These isotopes occur naturally in the minerals of thorium and uranium, and contribute to the natural radioactivity in the environment. As both isotopes are relatively short-lived, however, their environmental significance is small.

Aluminium ($_{13}$Al; US aluminum)

Aluminium is the commonest metallic element in the Earth's crust, following oxygen and silicon in order of abundance. It is a highly electropositive element and a strong lithophile, being found exclusively as Al^{3+} in combination with oxygen. Some of its important minerals are shown in Table II.1. The aluminosilicate feldspars are the most abundant minerals of the crust, being major components of igneous rocks such as granite. At the Earth's surface they are broken down by weathering reactions through the action of water and carbon dioxide, to form clay minerals. These are important constituents of most soils, and play an essential role in retaining water and nutrient elements such as *potassium and *calcium. Further weathering under tropical conditions leads to the removal of most elements and the formation of oxide and hydroxide minerals which make up bauxite, the major mineral source of aluminium.

Aluminium and its compounds are widely used. The metallic element is potentially highly reactive towards air and water, but forms a very stable film of the oxide Al_2O_3 on its surface, which protects it from further oxidation. This natural protective film may be deliberately enhanced by an electrochemical anodising process.

Table II.1 Some important Al-containing minerals. Compositions are approximate, as most minerals contain smaller proportions of many other elements. See under *silicon for more details concerning silicates

Mineral	Typical composition
Feldspars:	
plagioclase	$(K,Ca)[(Al,Si)_4O_8]$
alkali	$(Na,K)[AlSi_3O_8]$
Clay minerals:	
kaolinite	$Al_4[Si_4O_{10}](OH)_8$
montmorillonite	$(Ca,Na)_{0.7}(Al,Mg,Fe)_4[(Si,Al)_8O_{20}](OH)_4.nH_2O$
Constituents of bauxite:	
gibbsite	$Al(OH)_3$
boehmite and diaspore	$AlO(OH)$

Aluminium is used in cooking utensils and foil, soft-drinks cans, and in lightweight alloys for aircraft and many other construction purposes. Because it is so electropositive, the element itself must be extracted from compounds by electrolysis, usually of molten cryolite, Na_3AlF_6. The extraction process is very demanding in energy, often making use of hydroelectric power. Some *fluorine compounds are also released into the environment as a result. These environmental problems have stimulated the recycling of aluminium from soft-drinks cans, etc. Compounds of aluminium find various uses. An important one is for the treatment of drinking water, where addition of aluminium sulfate is used to assist the coagulation and precipitation of suspended matter. Because it is precipitated as $Al(OH)_3$, the amount of aluminium remaining in the water is normally very low.

Aluminium hydroxides are very insoluble in water at neutral pH, and as shown in Table II.2 the dissolved concentration of Al^{3+} in most natural waters is very low compared with the crustal abundance. However, the solubility depends strongly on pH, and, as shown in Fig. II.1, increases in both acid and alkaline conditions. The occurrence of acid rain therefore leads to increased Al^{3+} concentrations in streams and lakes, which may account for some of the ecological damage. One consequence of high aluminium concentrations is to limit the availability of *phosphorus, because of the extreme insolubility of aluminium phosphate, $AlPO_4$ (see Fig. II.30, below).

Aluminium is not thought to be essential for life, and is now widely regarded as a rather toxic element. Under normal conditions, very little is absorbed by the body. A normal daily diet may contain 3–5 mg Al, of which only about 10 μg is absorbed, and the same amount excreted in urine. Inhalation of dust may also contribute to the intake. The normally insoluble Al^{3+} is taken into the body by *transferrin*, a special molecule designed for the uptake of *iron in the insoluble Fe^{3+} form. Excess aluminium may interfere with phosphate metabolism, as $AlPO_4$ is very insoluble, but it may also be confused with iron in some reactions. Symptoms of aluminium toxicity include anaemia, bone disease, and problems with brain function. They have arisen with patients undergoing renal dialysis, where the blood is circulated directly through the dialysis machine and can take up aluminium contamination much more easily than is possible through the gut. Abnormal concentrations of aluminium have also been reported in the brains of

Table II.2 Aluminium concentrations in the environment

Location	Concentration
Crust	8.2 %
Sea water	0.5 ppb (surface), 0.2 ppb (deep)
Fresh waters	1 ppb ?
Human body	0.9 ppm

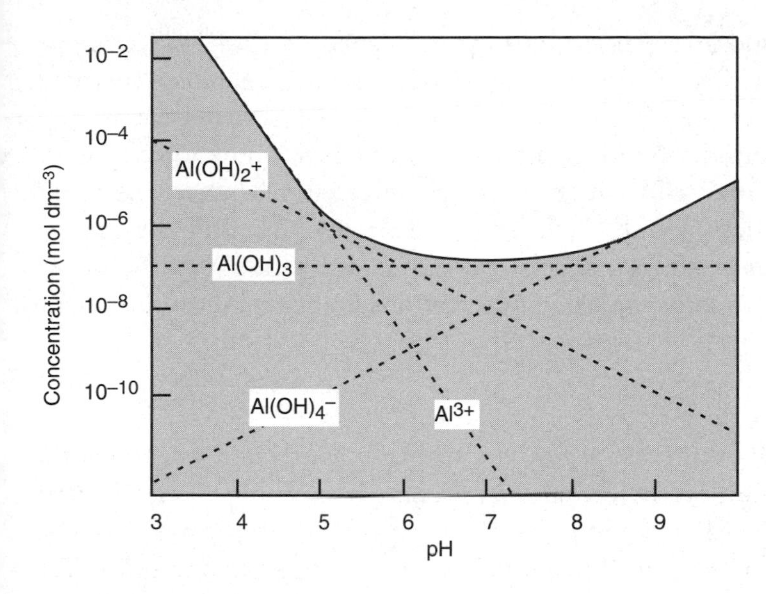

Fig. II.1 Solubility of aluminium hydroxide species as a function of pH. The dashed lines show the solubility of different complexes, and the solid line the overall Al^{3+} solubility. (Based on Lindsay, 1979.)

people suffering from Alzheimer's disease, a form of senile dementia. The significance of this is still disputed, however: on the one hand, some correlation has been claimed between the aluminium content of drinking water and the incidence of Alzheimer's disease; on the other hand, it is thought that the accumulation of aluminium in the brain may occur as a result of the disease, rather than being its cause.

Americium ($_{95}$Am)

One of the radioactive transuranium elements, americium does not occur naturally in the environment. It is formed by neutron bombardment of *plutonium, and is produced as a minor by-product during the fission of *uranium or plutonium in nuclear explosions and nuclear power stations. The most important isotopes are ^{241}Am and ^{243}Am, which are α emitters with half-lives of 432 and 7370 years, respectively. Small quantities are separated during the reprocessing of nuclear waste, and used for specialised purposes, e.g. as a tracer in diagnostic medicine. Americium will provide a significant source of radiation in nuclear waste or fallout after a lapse of a few hundred years, when the shorter-lived fission products have decayed.

Antimony ($_{51}$Sb)

A minor element in the natural environment, antimony is a strong chalcophile and occurs as the sulfide ore stibnite, Sb_2S_3, as well as in combination with many other chalcophilic elements. Its major use is in the production of alloys, for example with *lead for use in batteries. It is also used in some medicines. Antimony is found in small concentrations in the oceans, probably as the complex ion $[Sb(OH)_4]^-$. Its compounds are rather volatile and are released into the atmosphere during the incineration of waste and in the smelting of some metals. The element is therefore found in atmospheric dust, where its concentration may be several hundred times the average crustal value, as well as in soils surrounding industrial areas. It has no known biological function, and like *arsenic, with which it has chemical similarities, it is toxic. Owing to its lower abundance (about 10 times less than arsenic), and the relative insolubility of most compounds, it is not normally regarded as a serious environmental contaminant. Table II.3 shows some typical concentrations found in the environment.

Argon ($_{18}Ar$)

One of the noble gas elements, argon is chemically inert, and occurs as the uncombined gaseous element Ar. It forms 0.9 per cent by volume of the Earth's atmosphere, having a much higher abundance than the other noble gases, because ^{40}Ar is formed in the decay of ^{40}K, the natural radioactive isotope of potassium. The amount of ^{40}Ar found in potassium-containing rocks depends on the time elapsed since their formation, and so measurements of argon in rocks can be used for geological dating purposes. Argon is extracted from liquid air, and has some important applications in industry and in chemical research. Its chemical inertness makes it ideal as an environment in which to handle very reactive substances. The major industrial uses are in metallurgy (e.g. argon-arc welding) and in gas-filled lamps. Argon has no known biological function.

Arsenic ($_{33}As$)

A chalcophilic element of minor occurrence in the natural environment, arsenic is often found in combination with *sulfur in minerals such as arsenopyrite, FeAsS. It also has chemical similarities with phosphorus, so that arsenic may occur in minerals containing this element. Natural waters contain low levels of arsenic, in the As(III) form as $As(OH)_3$ and as As(V) in $H_2AsO_4^-$. Table II.4 shows typical concentrations in the environment.

Table II.3 Antimony concentrations in the environment

Location	Concentration
Crust	0.2 ppm
Sea water	0.3 ppb
Fresh waters	0.3–5 ppb
Soils	0.2–10 ppm
Human body	0.1 ppm

Arsenic has been used in pigments, medicines, and poisons. Although many of these applications have been abandoned owing to its toxicity, some have continued into the twentieth century.[49] The arsenic-containing drug salvarsan, discovered in 1909 by the chemotherapy pioneer Paul Ehrlich, was widely used to treat syphilis until it was superseded by penicillin. Some arsenical drugs are still used in the treatment of tropical diseases. A major present-day use is in organoarsenic compounds for pesticides and wood preservatives. It is also a constituent of gallium arsenide, GaAs, an important semiconductor that may increasingly replace silicon in electronic devices. The manufacture of GaAs involves the very poisonous gas arsine (AsH_3), which must be rigorously contained. Apart from agricultural uses, the major anthropogenic sources of arsenic in the environment are 'accidental' ones arising from the occurrence of the element as an impurity in various minerals, such as phosphate fertilisers. Compounds such as As_2O_3 and the sulfides are also relatively volatile, and are released into the atmosphere through the burning of fossil fuels (especially coal) and in the extraction of some elements from sulfide minerals (see Table II.5). Most arsenic used in industry is obtained as a by-

Table II.4 Arsenic concentrations in the environment

Location	Concentration
Crust	1.5 ppm
Sea water	1.5 ppb
Fresh waters:	
normal	1–10 ppb
polluted	10–1000 ppb
Soils:	
normal	1–10 ppm
contaminated	up to 200 ppm
Atmosphere	trace
Human body:	
average	0.25 ppm
hair	1 ppm

Table II.5 Arsenic emissions into the atmosphere. (Based on Fergusson, 1990)

Source	Emission (10^6 kg year^{-1})
Volcanic and soil dust	3
Biogenic gases	21
Anthropogenic dusts	78

49 See Lenihan (1988).

product from the extraction of *copper and *lead. Airborne dusts from industrial regions contain arsenic at concentrations up to 1000 times that of the crustal average, and this, together with agricultural sources, can lead to high concentrations in soils.

Arsenic is well known as a highly toxic element. Surprisingly, it is thought to be essential to life in trace amounts, although its function is unknown. It is most toxic in the As(III) form, which, like other chalcophilic elements, combines with –SH groups and interferes with the function of enzymes. The oxidised As(V) form $H_2AsO_4^-$ is chemically similar to phosphate and may interfere with phosphate metabolism, although it is less toxic than As(III). The interconversion of different chemical forms of arsenic is known to be performed by some micro-organisms, as shown in Fig. II.2. The methylated forms, especial trimethyl arsine, $(CH_3)_3As$, are volatile, and are present in trace amounts in the atmosphere.

Volatile methyl arsenic compounds may have been responsible for some accidental poisonings. Two green copper arsenate pigments, *Scheele's green* and *Paris green*, were widely used for wallpaper in the nineteenth century. Symptoms of arsenic poisoning appeared in Germany from 1815 onwards, and in 1839 the famous chemist Gmelin commented on the 'garlic odour' associated with moulds growing on wallpapers. The odour, and the source of toxicity, was identified as trimethyl arsine, $(CH_3)_3As$, in 1932, and provided the first evidence for the biological methylation of toxic ele-

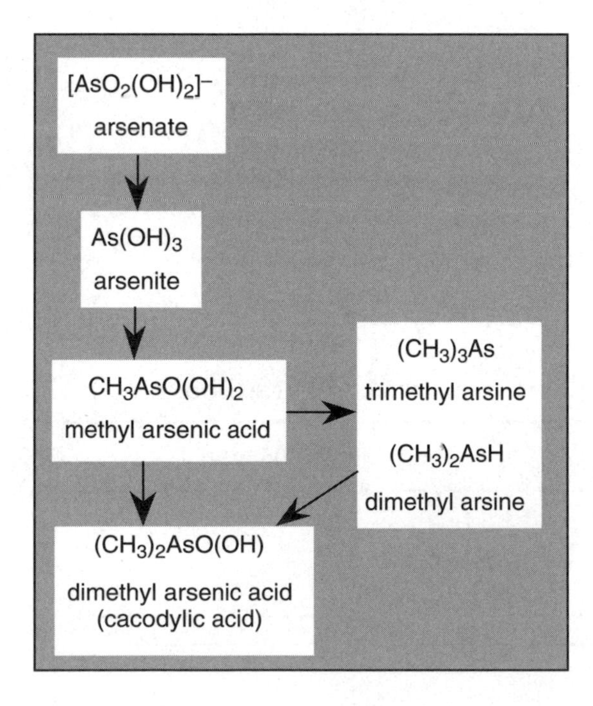

Fig. II.2 Biological interconversions of arsenic

ments. This process is now well established for many other elements, including *mercury (see also under *carbon).

Arsenic in the form of As_2O_3 has frequently been used as a deliberate poison since its discovery, attributed to the alchemist Geber in the eighth century AD. It has also been responsible for some accidental poisonings. The affinity for sulfur makes arsenic concentrate in hair, which contains proteins with a high concentration of –SH groups; analysis of hair has thus been used to diagnose arsenic poisoning. The strangest story in this context is that of Napoleon Bonaparte, who died in exile on the island of St. Helena in 1821. Doctors could not agree on the cause of his death. Samples of his hair were later analysed, and found to contain arsenic concentrations of 10–20 ppm, considerably larger than the normal at that time. There is still disagreement about whether Napoleon died of poisoning. It seems likely that he had been given arsenic-containing medicines—widely popular in the eighteenth and nineteenth centuries—for some time before his death. It is possible however that volatile arsenic emissions from wallpaper pigments might have played a role.

Acute poisoning by arsenic causes severe gastro-enteritis, and the long-term effects of small doses include loss of hair and skin lesions. Whether the overall arsenic concentration in the environment poses a health hazard is difficult to say. Widespread pollution, in the sense of concentrations larger than found in the natural environment, does certainly exist. But most of this is associated with mining and other industries that have been current for some centuries, and there is evidence from the analysis of hair that general arsenic levels in human populations have been declining over the past century.

Astatine ($_{85}At$)

Astatine is a radioactive element for which the longest-lived isotopes have half-lives of only a few hours. It is formed as part of the decay series of ^{235}U (see *uranium), but in concentrations so small as to be virtually undetectable. It is estimated that the whole of the Earth's crust contains only a few milligrams of astatine, making it one of the rarest naturally occurring elements.

Barium ($_{56}Ba$)

An element of moderate abundance in the crust (comparable to sulfur) and in the oceans, barium is a strong lithophile and is widespread as a minor component of silicate and other oxide minerals. It occurs particularly as the ore *barite*, $BaSO_4$, which is one of the most insoluble sulfate salts. Very little elemental barium is extracted, but $BaSO_4$ is used in large quantities to make a 'mud' slurry for oil and gas drilling. Barium compounds have a high density and absorb X-rays strongly. $BaSO_4$

is given as 'barium meals' to patients prior to X-ray investigations of the digestive tract. Barium is not essential to most life, although crystals of barite are found in some aquatic organisms, which exploit the high density of the mineral, either as a weight to cause them to sink into the mud, or as a gravitational sensor to orient them. Soluble barium compounds are rather toxic, but $BaSO_4$ is so insoluble as to be essentially harmless.

Berkelium ($_{97}Bk$)

Berkelium is a short-lived transuranium element made in minute amounts in nuclear reactors and explosions by neutron bombardment of ^{239}Pu. The longest-lived isotope is ^{247}Bk, with a half-life of 1400 years; the others have half-lives of only hours or days.

Beryllium ($_{4}Be$)

Beryllium is a rare element, occurring in trace concentrations as the Be^{2+} ion in silicate minerals, and as the mineral *beryl*, $Be_2Al_2Si_6O_{18}$. Its solubility in natural waters is extremely low. Beryllium is one of the lightest metals, and finds applications in alloys (especially with copper) and in nuclear and space technology. Very small amounts of beryllium oxide dust are produced by burning fossil fuels. Beryllium is one of the most toxic of all elements, and although the general levels in the environment are very low and cause no concern, stringent precautions are needed to avoid exposure in industrial workplaces using it.

Bismuth ($_{83}Bi$)

Bismuth is an extremely rare element, both in the crust and in natural waters. It occurs as oxide (Bi_2O_3) and sulfide (Bi_2S_3) ores, but is mostly obtained as a by-product of the extraction of lead, zinc, or copper. Bismuth is used in some alloys, in catalysts for the chemical industry, in pigments and cosmetics, and in drugs used to combat stomach infections by bacteria which cause indigestion. Although it is a toxic element, the insolubility of most forms renders it relatively harmless.

Boron ($_{5}B$)

A non-metallic element of low general abundance, boron is chemically rather similar to *silicon, and occurs in low concentrations in silicate minerals. Borates are however more soluble in water than silicates, and extensive deposits of borax,

$Na_2B_2O_5(OH)_4.8H_2O$, and other borates have been produced by a process of concentration and fractional crystallisation associated with hot springs in volcanic neighbourhoods. The borate ion $[B(OH)_4]^-$ is found in the ocean. The principal use of boron is in the manufacture of heat-resistant borosilicate glass such as Pyrex. Boron is essential for plant life, although its biochemical function is not understood; it is not thought to be essential for animals.

Bromine ($_{35}$Br)

Bromine is a rather rare element in the crust, occurring exclusively as the bromide ion Br^-. It is chemically similar to *chlorine, and bromide is found as a minor constituent of all natural chloride salts such as sodium chloride (NaCl, common salt). Bromides are very soluble, and Br^- is relatively abundant in the ocean. Most bromine used in industry is extracted from sea water and other natural brines. It is rarely used as the element itself, being principally converted into organic bromine compounds. Ethylene dibromide, CH_2BrCH_2Br, is used as a fuel additive, acting as a 'scavenger' for *lead from tetraethyl lead. With the phasing out of lead additives, this use is declining. Organic bromides are used as pesticides and as fire retardants for synthetic fibres. Another significant use is as a constituent of 'Halon' fire-extinguishing gases such as CF_3Br, which resemble CFCs in their ability to damage the ozone layer.

As the bromide ion, the element is universally present in life along with the similar chloride. It is not thought to be essential, although several brominated compounds are found in marine organisms. The molecule shown in Fig II.3 can be extracted from the snail *Murex brandaris*, and forms the purple dye known to the Romans as Tyrian purple. The elemental form Br_2 is (like chlorine) highly toxic. Elevated concentrations of Br^- appear to have a depressive effect on the nervous system, and bromide has been used as a sedative and as an anti-convulsant for treating epilepsy.

Traces of bromine compounds occur in the atmosphere, at concentrations of a few parts in 10^{12}. Volcanic emissions contain a little HBr, but this very soluble gas is quickly washed out and contributes to the Br^- concentration in the oceans. Methyl

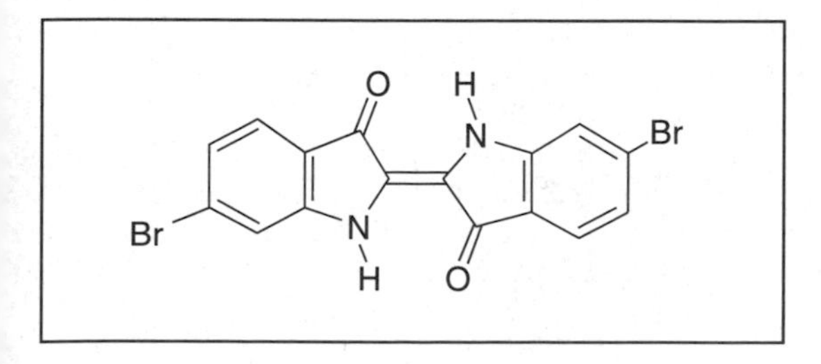

Fig. II.3 The bromine-containing dye Tyrian purple.

bromide, CH_3Br, is the dominant natural species, being emitted by biological action in the ocean. To this must be added minor amounts of industrially produced organobromine compounds, including the CFC-like Halons such as CF_3Br. Like the analogous *chlorine compounds, these unreactive molecules have very long lifetimes in the lower atmosphere, and so reach the stratosphere. Bromine atoms produced there by the action of UV radiation can potentially interfere with the ozone layer. The present concentrations are generally much lower than those of chlorine compounds, but atom for atom it is predicted that bromine may have a more damaging effect than chlorine, so that any increase is a matter for concern.

Cadmium ($_{28}$Cd)

A relatively uncommon element in the natural environment, cadmium is a strongly chalcophilic metal, found predominantly in combination with *sulfur. The mineral *greenockite*, CdS, is rare, and most cadmium is found in low concentrations in zinc ores, and obtained as a by-product of their processing. The element is used as an anticorrosion coating for metals, and in batteries, but most anthropogenic cadmium in the environment comes not from these applications, but from the mining and processing of zinc and other chalcophilic metals. Table II.6 lists some data on its abundance.

Table II.6 Cadmium concentrations in the environment

Location	Concentration
Crust	0.1 ppm
Sea water:	
surface	0.001 ppb
deep ocean	0.1 ppb
polluted coasts	up to 15 ppb
Fresh waters:	
clean	< 1 ppb
polluted	1–10 ppm
mining areas	100–700 ppb
Drinking water (limit)	5 ppb
Soils:	
clean	0.35 ppm
polluted	up to 1000 ppm
Edible plants	0.05–2 ppm
Human body:	
average	0.7 ppm
kidney	10–30 ppm
liver	2–3 ppm

Cadmium is not believed to be essential for life, and is very toxic. It is strongly scavenged by marine organisms, which accounts for its much lower concentrations in surface waters, where life is common, than in the deep ocean. Abnormally high concentrations can be found in rivers and coastal estuaries near mining and industrial centres, and in soils. Much of this pollution arises from airborne dust, although direct leaching from waste deposits also occurs. The particularly high levels of contamination found in some sewage sludges pose a serious problem of disposal.

Most cadmium in humans comes from food, an average daily consumption being around 35 μg per day, of which about 2 μg is absorbed. Tobacco also contains cadmium, and heavy smokers probably receive about the same amount again. Intake of the element stimulates the production of *metallothionein* in the liver. This enzyme contains an unusually high proportion of *sulfur-containing cysteine residues, which bind Cd^{2+} and form a complex containing up to seven metal atoms per molecule. The complex is then transported to the kidneys, which therefore concentrate a significant amount of the body content of cadmium. This process is designed as a defence against cadmium and other heavy metals, and the toxic effects presumably arise from its incomplete take-up by metallothionein, especially when excess doses are received. Cadmium competes with several essential metals, including *calcium, *zinc, and *copper, and interferes with their metabolism. Symptoms of poisoning include damage to the function of lungs and kidneys, and a softening of the bones leading to intense pain in the joints. Outbreaks of cadmium poisoning in Japan, where it is known as Itai-Itai ('ouch-ouch') disease, have been associated with exceptionally high intakes as a result of pollution from zinc refineries. The toxic effects are exacerbated by a diet deficient in calcium or in vitamin D. There is some evidence that hard water with a high calcium content gives a degree of protection, as does adequate dietary *selenium; the latter element has an even higher affinity for cadmium than does sulfur, and so may combine with it preferentially.

Caesium ($_{55}$Cs; US cesium)

A rare element of the alkali metal group, caesium has little use, no known biological role, and is non-toxic. However, radioactive isotopes of caesium are formed as fission products from *uranium, and are of environmental concern. Because of the low charge and large radius of the Cs^+ ion, compounds such as CsI are more volatile than most fission products, and so are relatively easily released during overheating of fuel elements in a reactor core. ^{137}Cs, with a half-life of 30 years, has formed a source of radioactive pollution in Europe following the fire in a nuclear reactor at Chernobyl in Ukraine in 1986. The element was spread by prevailing winds, and deposited largely in rain. Cs^+ is strongly adsorbed in soils and retained by plants, and is very persistent in the environment.

Calcium ($_{20}Ca$)

A dominant element in the natural environment, calcium is fifth in order of abundance (between iron and magnesium) in the Earth's crust, common in all natural waters, and essential for all life. Calcium is a strongly lithophilic metal, being present in all natural compounds as Ca^{2+}.

Most calcium on Earth is present in silicate minerals, usually in combination with other elements such as sodium, magnesium, and aluminium: the examples in Table II.7 show idealised formulae for two of these. Along with Na^+, Ca^{2+} is the ion most easily released in the weathering of silicates; unlike sodium however, calcium forms insoluble salts with many important anions, including carbonate (CO_3^{2-}), sulfate (SO_4^{2-}), phosphate (PO_4^{3-}), and fluoride (F^-). Calcium salts are therefore the dominant minerals containing these ions, and are common in sedimentary rocks. Apatite and fluorite are the major sources of *phosphorus and *fluorine, respectively, and the insolubility of apatite limits the availability of phosphorus for many plants. Vast deposits of calcium occur as carbonates, $CaCO_3$ being the major constituent of limestone and chalk, and dolomite, $CaMg(CO_3)_2$, being another major mineral of the crust.

Because of its easy release in weathering, Ca^{2+} is the commonest metal cation in most fresh waters, although its concentration varies widely according to the type of rocks around. In sea water, its concentration is limited by the formation of insoluble $CaCO_3$, and is less than that of Na^+ or Mg^{2+}. As discussed below, reactions involving calcium and carbonate are involved in the cycling of carbon in the environment, and in controlling the pH of natural waters.

Metallic calcium has a minor use in alloys, but calcium carbonate and compounds manufactured from it, especially the oxide CaO ('quicklime') and hydroxide

Table II.7 Some important minerals containing calcium. The silicate formulae are idealised: in reality, these minerals contain small proportions of many other elements

Name	Chemical formula
Feldspars: anorthite	$Ca[Al_2Si_2O_8]$
Pyroxenes: augite	$CaMg[Si_2O_6]$
Calcite	$CaCO_3$
Dolomite	$CaMg(CO_3)_2$
Gypsum	$CaSO_4.2H_2O$
Apatite	$Ca_5(PO_4)_3(OH,F,Cl)$
Fluorite	CaF_2

$Ca(OH)_2$ ('slaked lime') are used on a scale exceeding nearly all other elements. Lime mortar was used by the Romans. The major present-day applications are in the manufacture of concrete, in metallurgical processes such as iron and steel making, and in the treatment of water and soils to remove acidity.

Calcium is present in all living things, and is an essential element. As shown by the figures for blood cells and plasma in Table II.8, Ca^{2+} is maintained at a much lower concentration within cells than in their surroundings. One reason for this is that a high concentration of Ca^{2+} would lead to precipitation of calcium phosphate, and so limit the availability of phosphate, which is essential for cell metabolism. Concentration gradients are maintained across cell membranes by special 'pumps' similar to those for *sodium. Ca^{2+} is admitted to cells through selective membrane channels, and is widely used in biochemical control mechanisms. The ion binds strongly to appropriately negatively charged groups in proteins (such as carboxylates $-COO^-$) and thus may cause the molecules to change conformation. This can be used to switch the activity of enzymes 'on' and 'off'. The influx of calcium is used in this way to trigger many important cell changes, such as the release of neurotransmitters at synapses (nerve-cell junctions) and the contraction of muscles.

Another important function for calcium is to make skeletons from insoluble salts. The external shells of many marine organisms are made from calcium carbonate, and the consequent precipitation of this mineral in the sea forms part of the calcium cycle discussed below. Internal skeletons and teeth, such as in humans, are largely made of apatite (calcium hydroxy-phosphate). The metabolism of calcium (and other metal ions) in the human body is influenced by vitamin D, and an adequate supply both of this vitamin and of calcium is essential in the diet. Pregnant and nursing women and growing children are particularly at risk of deficiency, which affects the bones and teeth.

Table II.8 Calcium concentrations in the environment

Location	Concentration
Crust:	
average	4.1 %
sedimentary rocks	2–40 %
Ocean	400 ppm
Fresh waters	15 ppm (*very variable*)
Human body:	
average	1.4 %
bone	17 %
blood plasma	80 ppm
red blood cells	4 ppb

The calcium cycle

Through its general abundance and the insolubility of many salts, calcium exerts a significant control on the mobility and availability of several other elements in the environment. It is especially important in the context of the *carbon cycle, as over 99.9 per cent of all the CO_2 on Earth is effectively locked up in carbonate minerals, chiefly those of calcium. Things are very different on the planet Venus, where the higher surface temperature and the absence of liquid water prevent the formation of carbonate minerals: there most CO_2 is in the atmosphere, at a pressure 300 000 times the partial pressure of CO_2 in the Earth's atmosphere. On Earth, the cycling of calcium and carbon are closely linked. Three major types of reactions are important:

1. The weathering of rocks, especially silicates, through the action of water and CO_2 liberates soluble calcium and bicarbonate ions, leaving insoluble minerals such as quartz:

$$CaSiO_3 + 2CO_2 + H_2O \rightarrow Ca^{2+} + 2HCO_3^- + SiO_2. \quad (1)$$

   Here '$CaSiO_3$' represents calcium silicates, which are in reality much more complex than the simple formula suggests.

2. The soluble ions are washed into the sea and concentrated, and insoluble $CaCO_3$ is precipitated:

$$Ca^{2+} + 2HCO_3^- \rightarrow CaCO_3 + CO_2 + H_2O. \quad (2)$$

   Although the net reaction is in the direction shown, these species are nearly in equilibrium in the sea, as discussed below.

3. Tectonic processes lead to uplift and exposure of some carbonate minerals on land. Other processes however can bury the sediments deep in the crust, where they are heated, and decomposed:

$$CaCO_3 + SiO_2 \rightarrow CaSiO_3 + CO_2. \quad (3)$$

   As in (1), this is written in a very simplified form. But the overall effect is to regenerate silicates, which may be subsequently uplifted and re-exposed at the surface. The CO_2 is emitted into the atmosphere through volcanoes.

The cycle formed by these reactions is illustrated in Fig. II.4, which shows the approximate magnitudes involved; the annual transfer of around 5×10^{11} kg (that is, 500 million tonnes) of calcium is larger than for any other involatile element (see Table I.10 in §4.1).

The ultimate driving force for the inorganic reactions (1)–(3) is the physical cycling of the atmosphere, water, and the crust. Although they would continue in the absence

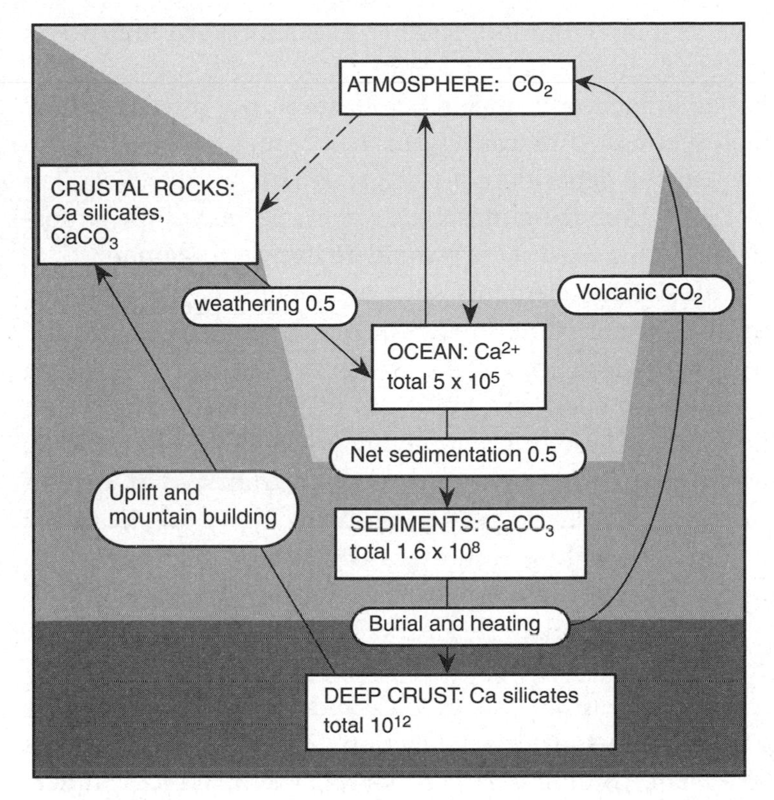

Fig. II.4 The calcium cycle, emphasising its connection with the cycling of carbon. Reservoirs (square-cornered boxes) and annual fluxes (round-cornered boxes) are given in units of 10^{12} kg Ca.

of life, living organisms influence them considerably. The weathering of rocks, reaction (1), depends on the CO_2 content and general acidity of the water. Organic matter in soils is acidic, because CO_2 is produced by bacteria which decompose plant litter, and because humus contains many organic acids. It has been estimated that plants and bacteria may accelerate the weathering of rocks by as much as a hundred times. Reaction (2), leading to the formation of $CaCO_3$ deposits, is also effectively under organic control. The surface layers of the sea are supersaturated in $CaCO_3$, but the spontaneous precipitation of this mineral is very slow under inorganic conditions, especially in the presence of Mg^{2+}. Most solid $CaCO_3$ is produced by marine life, mainly in the form of external skeletons and shells. They may be formed on the bottom in shallow waters (e.g. coral reefs), or else they sink after the death of the organism. The solubility of $CaCO_3$ increases in deep water, and below a point known as the *carbonate compensation depth* (CCD) the equilibrium in reaction (2) above moves to the left. Falling shells may therefore start to redissolve before they reach the bottom. The average CCD is about 3700 m in the Atlantic and 1000 m in the Pacific, the difference being due to the different circulation patterns which allow a greater build-up of dissolved CO_2 in the Pacific. Even when the sediment has formed, complex reactions with sea water may continue, leading to chemical modification; magnesium may be introduced to form dolomite, $CaMg(CO_3)_2$, in this way.

The presence of life also has an influence through other reactions not directly involving calcium. For example, photosynthesis uses up CO_2. In the ocean, this helps to drive reaction (2) to the right and assists the deposition of $CaCO_3$. It also has the effect that the oceans are net importers of CO_2 from the atmosphere; otherwise half of the atmospheric CO_2 consumed in the weathering reaction (1) would be liberated again in (2).

One feature of the cycling of metallic elements such as calcium is the immense time-scale involved. The mean residence time of Ca^{2+} in the ocean, defined as the total amount present divided by the flux (see §4.1), is about a million years. The times involved in cycling sedimentary rocks are hundreds of millions of years, and comparable to the age of the Earth. These times are many powers of 10 larger than those appropriate to cycles involving volatile elements. For transfers of carbon between the oceans, the atmosphere, and living matter, residence times are measured in years. The cycling of calcium carbonate is important because of the huge reservoirs involved, but it can only influence the carbon cycle in the very long run.

The equilibrium between calcium carbonate rocks and soluble forms is important in fresh waters as well as in the ocean. Limestone is a very common sedimentary rock, and does itself weather under the influence of CO_2 and water. The high CO_2 contents of waters draining from some soils drives reaction (2) to the left and gives solutions of Ca^{2+} and HCO_3^-. The solubility of carbonate rocks produces underground rivers and caves, which are a common feature of limestone country, and leads to hard waters with high Ca^{2+} contents. Cave formations such as stalactites and stalagmites arise from the liberation of CO_2 and the precipitation of $CaCO_3$; the same effect happens when water is heated in kettles and shower heads, where hard water forms carbonate scales.

The presence of $CaCO_3$ also has an influence on the pH of natural waters (see under *hydrogen and *carbon). It maintains an alkaline pH in soils, and only where it is absent can organic acids (or acid rain) give rise to acid waters. Adding calcium carbonate has been used to counteract the effects of acid rain on some lakes; a preferable method seems to be to treat the surrounding land, from which the water drains.

Californium ($_{98}$Cf)

Californium is a short-lived transuranium element made in minute amounts in nuclear reactors and explosions, by neutron bombardment of ^{239}Pu. The longest-lived isotope is ^{251}Cf, with a half-life of 890 years. ^{252}Cf, which has a half-life of 2.6 years, is extracted in very small quantities and used in the treatment of cancer by radiation therapy.

Carbon ($_6C$)

Carbon is the third or fourth most abundant element in the Universe. It is relatively much less abundant on Earth, making only 480 ppm of the crust, but it is a crucial element in the environment. Carbon is the major element of life; it occurs (mainly as CO_2) in the atmosphere, and is abundant as the bicarbonate ion, HCO_3^-, in all natural waters, and as carbonate minerals in sedimentary rocks. Around 20 per cent of the carbon in sedimentary rocks is in a reduced chemical state, mostly hydrocarbons. Sometimes these compounds occur in concentrated deposits, as coal, oil, or natural gas, which form the fossil fuels on which twentieth-century civilisation so much depends.

Carbon is an extremely mobile element in the environment (see §4.1), and its cycling involves biochemical processes such as photosynthesis and respiration, as well as inorganic reactions such as weathering and sedimentation. Bicarbonate plays a major role in regulating the pH of natural waters. In the atmosphere, CO_2 absorbs infrared radiation and is thus an important 'greenhouse' gas which raises the temperature of the Earth's surface and hence influences the entire global environment (see §1.4).

The major use of carbon is in the form of fossil fuels, for energy production and for the manufacture of synthetic organic chemicals such as plastics and pharmaceuticals. Environmental problems arise in the organic chemicals industry as many essential intermediates are volatile and toxic. The major current concern, however, is the increase in atmospheric CO_2 concentration arising from fossil fuel burning, and the possible changes to the Earth's climate that may result.

The radioactive isotope ^{14}C is produced by cosmic-ray bombardment of the upper atmosphere, and forms a small proportion of all carbon in the atmosphere and in living organisms. ^{14}C is also produced by nuclear explosions in the atmosphere, and would contribute a significant source of radioactivity following a nuclear war.

Most of the issues summarised above are discussed in more detail in the sections which follow.

Carbon compounds in space and on the early Earth

The origin of life is one of the great unsolved problems of modern science. Although around 25 other elements are necessary for life today, and may have been involved in the earliest stages, the unique chemistry of carbon is crucial. It is very hard to see how life could have started without a supply of complex carbon compounds, such as fatty acids, amino acids, and other *nitrogen compounds such as the purine and pyrimidine bases which form essential constituents of DNA. For this reason, there has been intense interest among chemists about the chemical forms of carbon originally present on Earth, and about the reactions that may have taken place in the 'prebiotic' environment before life began.

Carbon is common in space, being with oxygen the third or fourth most abundant element in the Universe. These elements are synthesised from hydrogen and helium by nuclear reactions inside stars, and the fact that they are both made in similar proportions seems to depend on some peculiar 'accidents' of nuclear physics: if the constants of nature were only very slightly different from their actual values in the Universe, either very little carbon would be made at all, or there would be hardly any heavier elements, including oxygen. Either way, life as we know it could not exist.[50]

The elements synthesised in stars are blown out into space, either slowly from the surface of a star, or more dramatically in gigantic explosions—*supernovae*—when a star ends its active life. As the gases from a star expand and cool, chemical reactions take place leading to the formation of molecules and solids. Radio-astronomers have detected the spectra of many carbon-containing compounds in deep interstellar space, the commonest being CO. Other observations suggest that carbonaceous dust particles—mostly graphite, but possibly also the rarer and less stable form diamond—are present in interstellar space. Another, recently discovered form of carbon, the C_{60} molecule *buckminsterfullerene*, is also thought to exist in space. Carbon compounds are known from spectroscopic observations on the atmospheres of other planets: CO_2 is the dominant form on the inner planets Venus and Mars, whereas CH_4 predominates on the outer planets such as Jupiter and Saturn, and on their satellites such as Titan.

Carbon compounds are also constituents of smaller bodies in the solar system, meteorites and comets. Many meteorites shown signs of having been heated strongly, and have lost most of their carbon. But there is a rare type, known as *carbonaceous chondrites*, that contain a reasonable proportion of carbon, some in the form of graphite and involatile compounds of high molecular weight, but also a range of very interesting complex molecules, including amino acids and some other compounds necessary for life. Comets are often thought of as 'dirty snowballs' made of frozen water, methane, and ammonia, but they also contain other compounds. One of the surprises of the space missions to Halley's comet in 1986 was the finding that the solid nucleus is almost black, apparently because of solid carbon and high-molecular-weight compounds. Both comets and carbonaceous chondrite meteorites are thought to be remnants of an early stage in the formation of the planets, and their chemical composition is particularly interesting, as it may be representative of the material from which the Earth was formed.

---

50 The relationship between the laws of nature and the existence of life (and ourselves) in the Universe is the subject of Barrow and Tipler (1986).

A selection of the carbon compounds found in various regions in space is shown in Table II.9. It is notable that the predominant form of carbon differs from place to place: CO and solid C are commonest in interstellar space, CH_4 on the outer planets, and CO_2 on the inner planets. This variability is partly a consequence of the thermodynamics of carbon compounds, displayed in Fig. II.5. The dominant, thermodynamically most stable form of carbon in combination with hydrogen or oxygen, is shown according to the degree of oxidation (horizontal scale) and the temperature (vertical scale). The borderlines between different regions are not rigid, but are meant to indicate shifting equilibria.

In the absence of oxygen, on the left-hand side in Fig. II.5, solid carbon and methane are stable. Gases usually predominate at higher temperatures, but the reverse is true here. In the exothermic reaction

Table II.9 Carbon compounds in space, illustrating the variety of chemical forms identified. Bold type indicates the dominant carbon-containing species in each location. (See Duley and Williams (1984) for more details on interstellar chemistry, and Wayne (1991) for planetary atmospheres)

Location	Type of compound	Examples
Interstellar space	solid C	graphite, diamond
	fullerenes	C_{60}
	hydrocarbons	CH, C_2H, CH_4, C_2H_2
	oxides	CO
	organic O compounds	H_2CO, H_2CCO, HCO_2H, CH_3OH, C_2H_5OH, CH_3CHO
	organic N compounds	CN, HCN, C_3N, CH_3CN, HC_3N, HC_5N, $HC_{11}N$
	organic S compounds	H_2CS, CH_3SH
Comets and meteorites	solids	graphite, diamond, high-MW compounds
	small molecules	CO_2, CH_4
	carboxylic acids, amino acids, purine bases	HCO_2H, CH_3CO_2H, $H_2NCH_2CO_2H$
Atmospheres of Jupiter, Saturn, and Titan	hydrocarbons	CH_4, C_2H_6, C_2H_4, C_2H_2, CH_3CCH, C_3H_8
	oxides	CO
	organic N compounds	HCN, $(CN)_2$, HC_3N
Atmospheres of Venus and Mars	small molecules	CO_2, CO

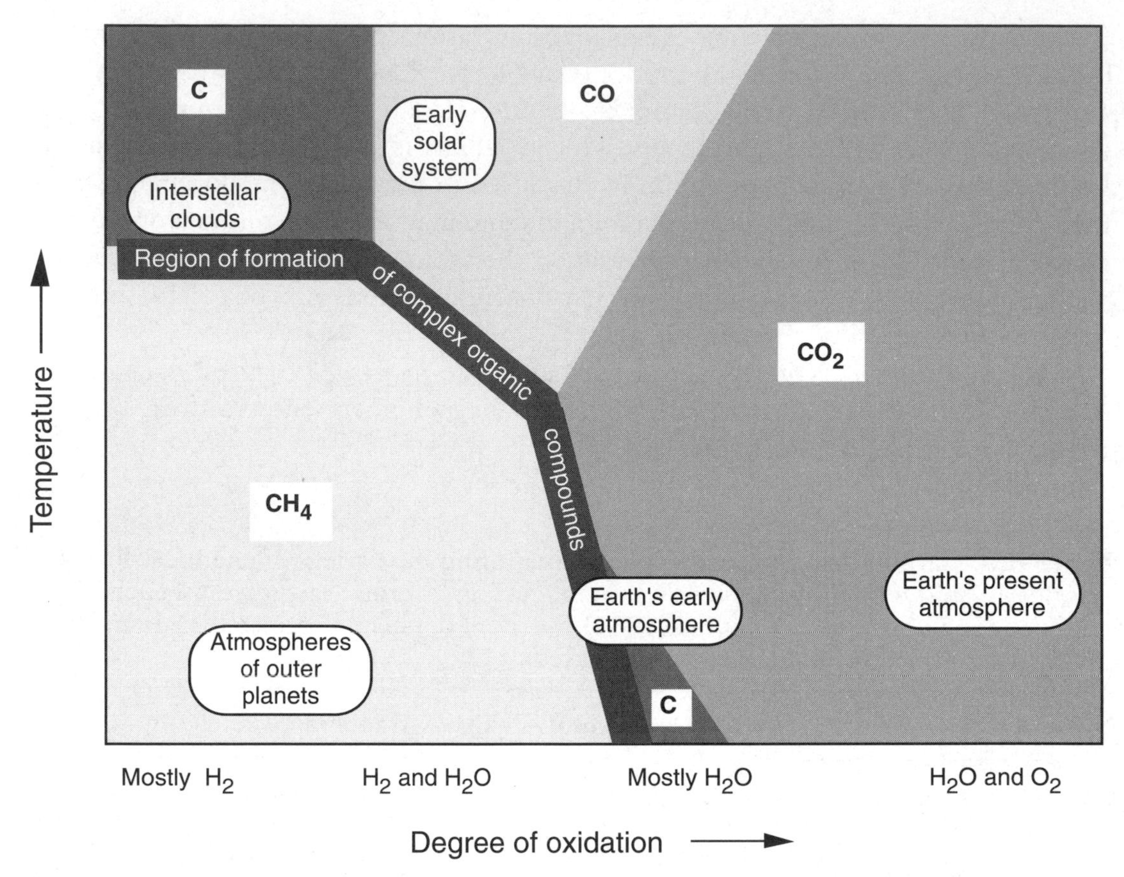

Fig. II.5 The thermodynamic stability of carbon compounds (oxides and hydrides). Each region is labelled with the form most stable under the given conditions. The horizontal axis represents the availability of oxygen (mostly in the form of water vapour) relative to hydrogen. Strongly reducing conditions are on the left, and strongly oxidising ones on the right. The vertical scale is the temperature, ranging from below zero centigrade, to a thousand or so degrees. The boundaries between different areas are not rigid, but show the changing balance between different forms. Complex organic molecules are not thermodynamically stable under any conditions; the circumstances under which they are most likely to be produced are however marked. Also shown are the conditions appropriate to five important environments: interstellar clouds, the solar nebula from which the planets formed, the atmospheres of outer planets (Jupiter, Saturn, etc.), and the Earth's early and present atmosphere.

$$C(s) + 2H_2 \rightarrow CH_4$$

two gaseous H_2 molecules are converted into only one of methane. Thus CH_4 is a low-entropy form of carbon under these conditions, and is more stable at lower temperatures, whereas solid carbon predominates at higher temperatures. In the presence of oxygen, carbon is first oxidised to CO, which predominates at high temperatures if there is enough oxygen present. Such was probably the situation in the *solar nebula*, the cloud of gas and dust which surrounded the sun as it formed, and from which the planets were made (see §§1.1 and 3.1). There may have been regions where oxygen was less common, and some solid carbon present. This certainly appears to be the situation in parts of interstellar space, where observations show that there are clouds containing graphite particles and reduced carbon compounds. The conditions appropriate to these interstellar clouds, and to the early solar system, are labelled in Fig. II.5.

As the solar nebula cooled below about 300°C, the equilibrium in the exothermic reaction

$$CO + 3H_2 \rightarrow CH_4 + H_2O$$

shifted to the right, and methane became the thermodynamically most stable form of carbon in the early solar system. In the absence of a catalyst, however, this gas-phase reaction is extremely slow. Dust particles containing metallic elements and oxides would have been able to act as catalysts. The catalytic reaction between CO and H_2 is known as the *Fischer–Tropsch* reaction, and has been extensively studied in the laboratory. As well as the favoured product methane, many more complex carbon compounds can be formed. If nitrogen is present as ammonia, NH_3, it is also possible to make small quantities of the amino acids and other compounds. A wide variety of these have been found in carbonaceous meteorites, and may also be constituents of comets. Complex carbon compounds such as these should not really appear in Fig. II.5, because they are never thermodynamically stable. Their formation illustrates one of the most important aspects of carbon chemistry: its reactions are often dominated by *kinetics*. The conditions of thermodynamic stability shown in the diagram are nevertheless very important, as the opportunity to make complex molecules depends on the possibility of shifting the equilibria between the different types of compounds. If either oxidised (CO or CO_2) or reduced (C or CH_4) carbon were always the stable form, there might be be no interesting compounds formed at all.

Whereas carbon and oxygen are about equally common in space, there is an enormous difference in their abundance on Earth, carbon being relatively several thousand times less common. Most carbon was present as volatile compounds (CH_4, CO, or CO_2) when the Earth formed, and largely escaped condensation into solid materials. The giant outer planets, Jupiter and Saturn, were large enough and cold enough to capture part of the gaseous nebula before it dispersed, and their present atmospheres probably reflect the original composition of this quite closely. During the formation of the inner planets such as the Earth, however, very little gas was picked up directly, and the volatile components that they now contain (including water) were originally trapped in minerals, and subsequently released by heating. The Earth's atmosphere, therefore, is of secondary origin. About 80 per cent of carbon on Earth is in the form of CO_2 or carbonate rocks, but this has not necessarily always been the case. Some scientists believe that the early atmosphere may have been dominated by methane; it has also been proposed that large amounts of methane may still be trapped deep below the crust in the Earth's mantle, although the evidence for this is not strong. It is generally thought that, even if CH_4 or CO were dominant in the early atmosphere, they would not have lasted long. As there was no ozone layer to absorb energetic UV radiation, complex photochemical reactions would have occurred, splitting up methane and generating hydroxyl (OH) radicals from water. Although there was very little O_2 present, the oxidation of CH_4 and CO could have occurred just as easily as it does today, with hydrogen being lost to space, and carbon ending up as CO_2 and carbonate. The main difficulty with this scenario is that there is no reasonable source for the oxygen required to oxidise nearly all the carbon on Earth in this way. It is most likely therefore that the oxidised forms of carbon were dominant from the beginning, and that CO_2 was the major carbon compound in the atmosphere. There may well have been also some solid carbon, and complex compounds similar to those found in meteorites.

Some of these complex carbon compounds from space may have formed the basis on which life began. But there have been many other suggestions about early carbon chemistry on Earth.[51] Best known are the experiments performed by Miller and Urey in 1953, which attempted to reproduce the conditions that may have been present in the early atmosphere. Complex molecules, including amino acids, are produced when a mixture of methane, ammonia, and water vapour is exposed to a prolonged electric discharge, intended to simulate the effect of lightning. The original experiments have been criticised on the grounds that carbon dioxide was probably present rather than methane. Similar results can however be obtained by starting with CO_2, although the yield of complex compounds is lower, and it is essential that there is no O_2. Many refinements and other suggestions have been made, and there seems no doubt that many of the basic molecules required for life can be made in this way. On the other hand, molecules formed in space before the Earth was made could have been equally important.

Complex carbon compounds could therefore have come from a variety of plausible sources. The origin of life itself remains a mystery because it is not sufficient to have complex molecules: they must somehow get together and organise themselves to produce something that can reproduce. The origin of nucleic acids such as RNA and DNA is still very hard to understand, and it may be that the earliest forms of 'life' had a different chemistry, possibly involving inorganic substances such as silicates or

51 See Mason (1991) for a detailed discussion of these issues.

iron sulfides. Even more speculative—and discounted by most scientists—is the idea that life did not originate on Earth, but arrived from space.[52]

Carbon dioxide in natural waters and biology

CO_2 undergoes a series of reactions which are important in controlling the pH of many natural waters. The first of these is

$$CO_2\ (g) \rightleftharpoons CO_2\ (aq),$$

and gives rise to dissolved CO_2 molecules, at a concentration of about 10^{-5} mol per dm^3 for water in equilibrium with the normal atmosphere. The next reaction,

$$CO_2\ (aq) + H_2O \rightleftharpoons H_2CO_3,$$

produces carbonic acid, but it has a very small equilibrium constant, and the carbonic acid concentration is typically only 2×10^{-7} mol per dm^3. The acid dissociates, first to give bicarbonate,

$$H_2CO_3 \rightleftharpoons H^+ + HCO_3^-,$$

and then carbonate ions,

$$HCO_3^- \rightleftharpoons H^+ + CO_3^{2-}.$$

Figure II.6 shows the pH dependence of the concentrations of these species, in equilibrium with the normal atmosphere.

Pure water in equilibrium with atmospheric CO_2 has a pH controlled largely by the first dissociation of carbonic acid, giving a value of 5.7 where the H^+ and HCO_3^- concentrations are about equal. This is the 'natural' pH of rainfall, away from other sources of acidity coming from the oxidation of atmospheric *nitrogen and *sulfur compounds. For ground waters, however, the presence of carbonate minerals is important. They provide an additional source of carbonate ions, and lead to an increase in pH. Figure II.6 shows that a significant concentration of CO_3^{2-} can only be present at fairly high pH values; normally the reaction

$$CO_2 + H_2O + CO_3^{2-} \rightarrow 2\ HCO_3^-$$

acts to increase the bicarbonate concentration relative to that of CO_2. On the other hand, many soils may have a CO_2 concentration significantly higher than that of the normal atmosphere, derived from the aerobic decomposition of organic matter. In

---

52 Cairns-Smith (1982) argues for early life forms based on silicate minerals, rather than organic compounds. Hoyle and Wickramasinghe (1978) suggest that life originated in space.

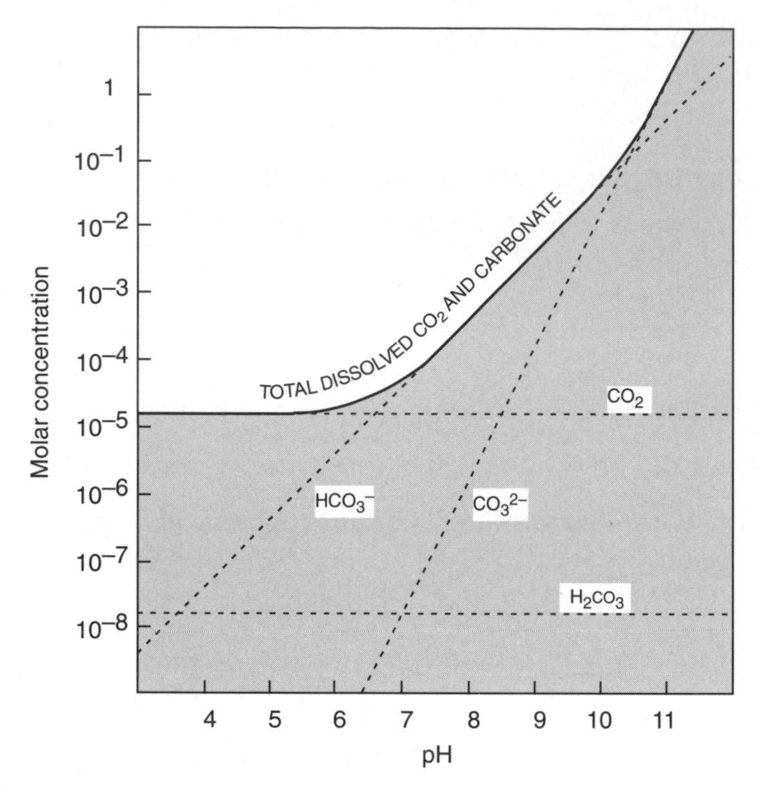

Fig. II.6 Natural carbonate equilibria (I), showing the concentrations of dissolved CO_2, carbonic acid, bicarbonate, and carbonate ions in equilibrium with atmospheric CO_2 as a function of pH. The heavy line shows the total dissolved concentration of CO_2 and carbonates.

the absence of carbonate minerals, elevated CO_2 will lead to more acid conditions with a lower pH.

Figure II.7 shows the result of calculations on the effects of both gaseous CO_2 concentration and dissolved carbonate minerals. It can be seen that a pH as low as 4 could be found in water in equilibrium with an atmosphere of pure CO_2, whereas values of 11 or above can result from high dissolved concentrations of carbonate minerals. Very alkaline waters can come from saline deposits of evaporite minerals, which may contain some soluble Na_2CO_3. Such extreme values are rare, partly because of the effect of other mineral equilibria, but also because the commonest carbonate found naturally is $CaCO_3$, which has a rather low solubility. The heavy line in the figure shows this solubility limit, and its effect on pH. Water with very little atmospheric CO_2 can dissolve about 10^{-4} moles (10 mg) of $CaCO_3$ per dm^3, giving a pH around 10. In the presence of atmospheric CO_2 at its normal concentration the pH is predicted to be 8.4, with about three times as much dissolved $CaCO_3$. The pH of sea water is 8.1, and is largely controlled by this combination of atmospheric CO_2 and solid $CaCO_3$. (See *hydrogen for a further discussion of pH.)

$CO_2 - HCO_3^-$ reactions also play a role in biology. The pH of body fluids is maintained fairly close to 7 by the buffering effect of several species, including phosphate.

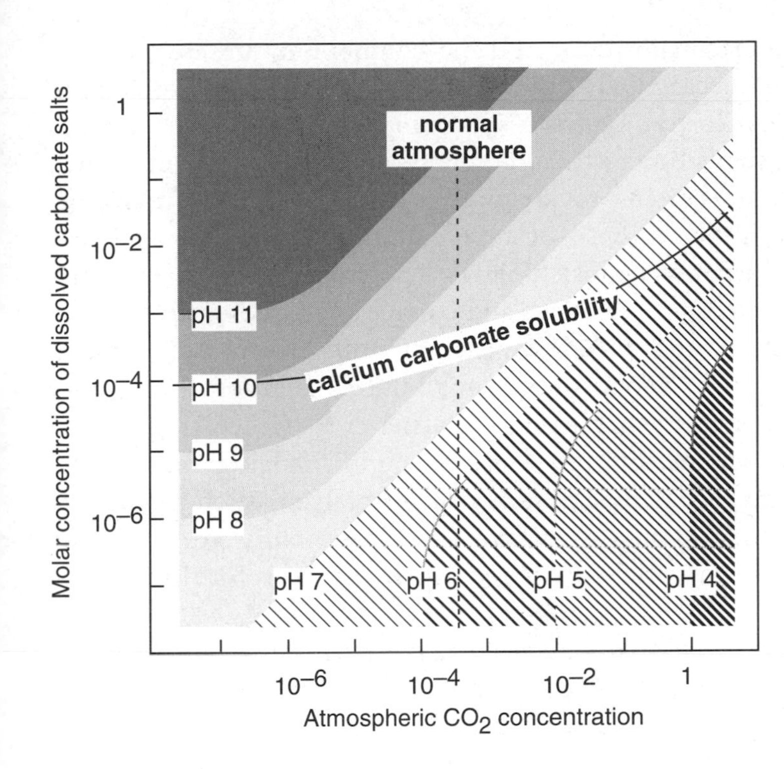

Fig. II.7 Natural carbonate equilibria (II), showing the effect of varying atmospheric CO_2 concentrations (horizontal axis), and dissolved carbonate salts (concentrations given by the vertical axis). The shading in the diagram shows regions of different pH. The heavy line represents the maximum equilibrium solubility of calcium carbonate, $CaCO_3$. The vertical dashed line shows the normal atmospheric CO_2 concentration.

But as CO_2 is produced in respiration the pH falls, and the increased acidity stimulates the release of O_2 from haemoglobin. 'Overbreathing', as a result of stress for example, leads to a decrease in the CO_2 content of blood, and an increase in pH, which can have various physiological consequences including unconsciousness. Another interesting aspect of respiration is the final release of CO_2 in the lungs, through the reactions

$$HCO_3^- + H^+ \rightarrow H_2CO_3 \rightarrow CO_2 + H_2O.$$

The dehydration of carbonic acid is slow in the absence of a catalyst, and there is a specific *zinc-containing enzyme, *carbonic anhydrase*, which is used to catalyse it. The reverse reaction is also catalysed, and is important in the uptake of CO_2 by green plants for photosynthesis.

Carbon chemistry and life

The unusual diversity of carbon chemistry stems partly from the ability of a carbon atom to form strong covalent bonds which are resistant to attack. This is particularly true of bonds between two carbon atoms, and although the element is not unique in forming the chains and rings familiar in organic chemistry, these structures are much more stable with carbon than with any other element. Strong bonds to other atoms are equally important however, and the compounds present in living systems have H, N, O, S, and occasionally other elements directly linked to carbon. A small selection of molecules illustrating this is shown in Fig. I.22 (in §3.6).

The presence of other elements bound to carbon has several consequences which are crucial in life. Different atoms often provide more reactive centres in a molecule, which can be used to aid both the synthesis of more complex compounds and their subsequent breakdown. For example, most biological polymers, such as carbohydrates (starch and glycogen), proteins (enzymes and connective tissues such as collagen), and nucleic acids (RNA and DNA) consist of units connected through elements such as nitrogen, oxygen, or phosphorus in addition to carbon. They differ in this from most synthetic polymers such as polyethylene or PVC, which are composed entirely of long carbon chains with other elements only 'hanging' on the side. The consequence of this difference is that decomposing organisms such as bacteria can destroy biological polymers easily, but have greater difficulty attacking the unreactive carbon chains of synthetic polymers, which are therefore not easily 'biodegradable'.

Some important functions of elements such as oxygen and nitrogen bound to carbon were explained in §3.6: they provide polar and ionisable groups which control the water solubility of molecules and the structure of biological systems; they also provide sites where metal ions can bind.

One very important feature of carbon chemistry is the kinetic stability of many of its compounds. All carbon compounds except CO_2 are thermodynamically unstable in the presence of atmospheric O_2. The complex molecules of life are also potentially unstable even in the absence of O_2, as the decomposition to simpler compounds, especially CO_2 and CH_4, is a thermodynamically favoured process (§4.2). But these oxidation or decomposition reactions are very slow in the absence of any catalyst. Because of the high strength of bonds involving carbon, high-energy intermediates must be formed during the transformations between different molecules. In biology, catalysis of carbon chemistry is performed by enzymes, specialised protein molecules which because of their varied shapes and functional groups provide extremely specific catalytic centres for biochemical reactions. A crucial function of the genetic instructions contained in DNA is to direct the synthesis of the appropriate enzymes, which in turn control the biochemical reactions that build up and break down the desired organic molecules.

Although kinetics is crucial, the thermodynamics of carbon chemistry is also important in understanding how living things operate. All organisms require a continual supply of free energy, to synthesise required molecules, to maintain their internal conditions of temperature and chemical composition, to move around, and so on. The immediate source of free energy for most biochemical reactions is ATP, an 'energy-rich' poly-phosphate molecule (see *phosphorus). ATP in its turn is synthesised as a result of energy-giving reactions, most of which are of the oxidation/reduction (*redox*) type. As discussed in §4.2, the aerobic oxidation, e.g. of glucose,

$$\underset{\text{glucose}}{C_6H_{12}O_6} + 6O_2 \rightarrow 6H_2O + 6CO_2,$$

is the type of reaction that yields the most energy, giving about 475 kJ (114 kcal) per mole of carbon oxidised. This reaction forms the principal energy source for all air-breathing organisms, including humans. The ultimate source of the organic matter (i.e. food) and of the atmospheric oxygen required is from photosynthesis by green plants. This is essentially the reverse reaction to that written above, and the energy input of 475 kJ per mole needed to perform this comes from sunlight. (The reactions producing and using O_2 are described in more detail under *oxygen.)

The reactions of photosynthesis on the one hand, and respiration and aerobic decay on the other, are the most important biological components of the carbon cycle described below. But as discussed in §4.2, there are many other possible redox reactions. In *anaerobic* conditions, where free oxygen is absent, organic matter may be oxidised by other compounds, such as sulfates and nitrates. These reactions, known respectively as *dissimilatory sulfate reduction* and *denitrification*, are important in the cycles of *nitrogen and *sulfur. In the very early environment on Earth, even these sources of oxidising power were probably not present, and the earliest life forms may have started by utilising *fermentation* reactions in which less stable carbon compounds decompose to more stable ones without the intervention of an external oxidising agent. The familiar fermentation of glucose to give ethanol and CO_2 is only one of a variety of possible reactions, many of them very complex. Some liberate dihydrogen, as in

$$\underset{\text{glucose}}{C_6H_{12}O_6} + 4H_2O \rightarrow 2CH_3COO^- + 2HCO_3^- + 4H^+ + 4H_2$$

which has an energy yield of about 30 kJ per mole of carbon. Other reactions give methane, CH_4, either by reduction of CO_2 using H_2 produced by other micro-organisms, or directly from organic carbon, for example by

$$CH_3COO^- + H^+ \rightarrow CH_4 + CO_2$$

which gives 18 kJ per mole of carbon.[53]

The development of aerobic biochemistry can only have started in earnest when there was sufficient oxygen in the atmosphere, probably not more than a billion years ago. Oxygen itself comes from photosynthesis, a process that must have evolved when the complex carbon compounds originally fairly abundant in the environment became depleted through their use by life. The advent of oxygen produced some

53 See Levett (1990).

dramatic changes, as the O_2 molecule is very toxic towards most primitive anaerobic organisms. With the evolution of new, aerobic species, anaerobic chemistry persisted, and continues today, in a variety of suitable oxygen-free environments. These include water-logged soils and marshes, lake and ocean sediments, and the digestive tracts of animals, especially grazing species such as cattle. Anaerobic decomposition of organic matter in these environments gives rise to the major source of methane in the Earth's atmosphere.

Carbon therefore has a dual role in life, as an element required in the organic molecules of life, and as a component of the redox reactions which supply free energy. The production and transfer of organic carbon forms the 'backbone' of the food chain in land and marine ecosystems. Figure II.8 shows a very simple model of such a food chain, operating in an environment such as a temperate forest.

The chain begins with *primary producers*, principally green plants which perform photosynthesis to 'fix' atmospheric CO_2 into organic matter. Such species are also known as *autotrophs*, as opposed to *heterotrophic* organisms that require a source of organic carbon as food. Photosynthesising plants reoxidise about 50 per cent of the fixed carbon, so that the *net photosynthetic yield* in an ecosystem is only about half that of the gross photosynthesis occurring. Some of the fixed organic matter passes to the next *trophic level* through grazing, and a fraction of that to the highest level, that of predators at the end of the food chain. The amount of organic carbon passed along the chain at each stage is only a small fraction of that received (less than 1 per cent).

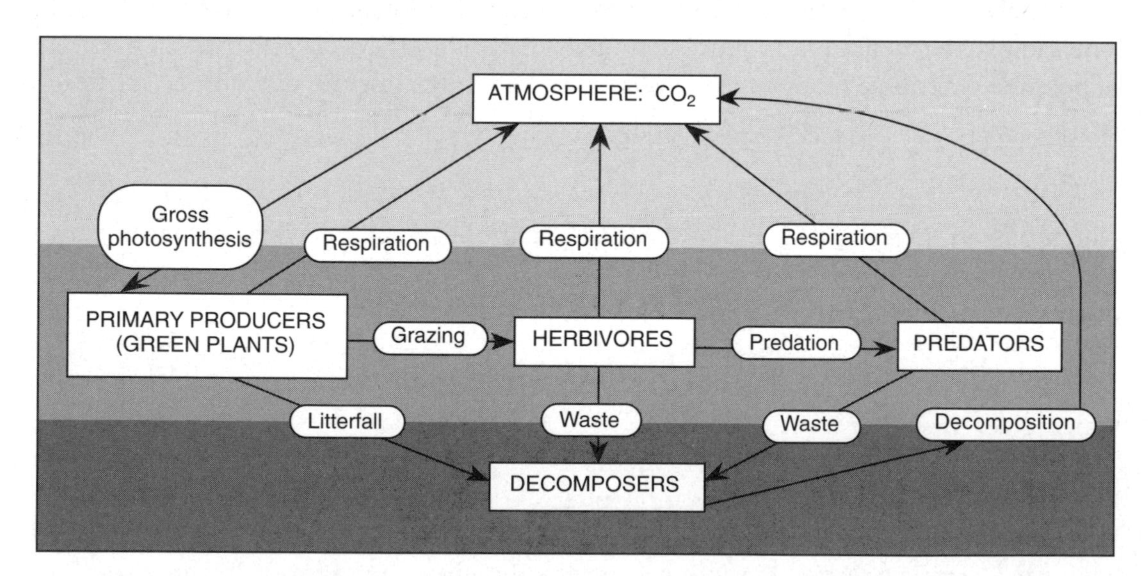

Fig. II.8 Carbon transfers in a terrestrial ecosystem. A very simplified model, showing *primary producers* (green plants), *herbivores*, *predators*, and *decomposers* only. In reality the 'food chain' is much more complex, and is better described as a 'food web'. (See for example White *et al.*, 1984.)

A major part is returned to the atmosphere as CO_2 in respiration, and a significant fraction also goes into the soil as waste (dead leaves, animal faeces, etc.) and is used by a great variety of *decomposers*, micro-organisms such as fungi and bacteria which operate under both aerobic and anaerobic conditions. Most carbon eventually returns to the atmosphere as CO_2, and a relatively minor amount as methane.

Fossil fuels

Fossil fuels—natural gas, oil, and coal—constitute a reserve which is more heavily used than that of any other element (see §§5.1 and 5.3). Although various theories of their origin have been put forward, it is generally agreed that fossil fuels are of biological origin, coming from organic carbon that has escaped decomposition or reoxidation. Only a small fraction of such carbon is available as extractable material, the majority being widely dispersed in the crust, and known as *kerogen*. As with some other types of element deposit, the details of fossil fuel formation are not easy to reconstruct. Geological conditions now may be different from those in the past; and the very slow chemical reactions involved in the conversion of plant and animal remains into fuels are not easy to study in the laboratory. Nevertheless, some broad outline ideas can be agreed.[54]

Kerogen can be derived from a wide variety of plant and animals remains, but the type with the best potential for generating oil comes from marine algae. Its initial composition is high in lipid (fatty) molecules from cell membranes; these have long hydrocarbon chains so that the oxygen content is relatively low. Figure II.9 shows schematically how the composition evolves after burial. Both microbial processes and the effects of heat may be involved, as oxygen and hydrogen (with other minor

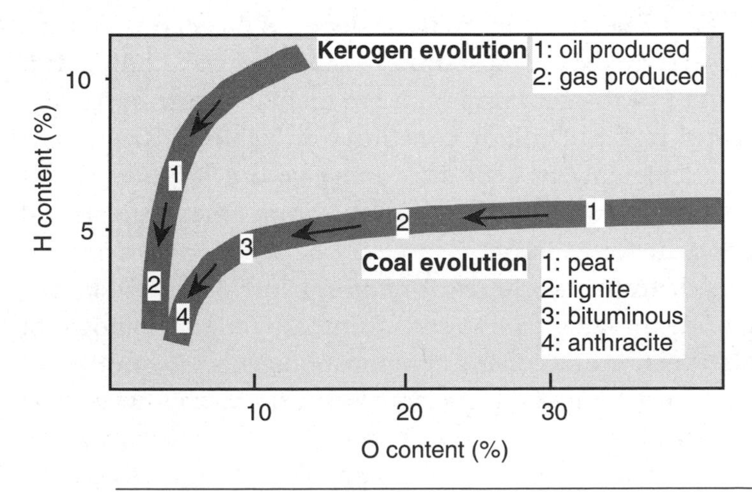

Fig. II.9 Fossil fuel formation, showing schematically the evolution in composition of kerogen (upper line, with regions of oil and gas formation shown) and coal (lower line, with classification). The scales give the weight per cent of oxygen (horizontal) and hydrogen (vertical), most of the remainder being carbon.

54 Mason and Moore (1982) and Hutchison (1983) give a discussion.

elements) are progressively lost. Kerogen itself is an amorphous material composed of complex organic molecules of high molecular weight. Aromatic (benzene-like) structures predominate. With prolonged burial, it evolves into forms more closely resembling graphite, with functional groups (e.g. *CO and –COOH) and hydrogen being progressively lost. This slow breakdown also yields a range of smaller molecules, including water, CO_2, and hydrocarbons of successively smaller size. These migrate, and hydrocarbons may be trapped by impermeable rocks, and collect as deposits of oil or gas.

Coal is derived from more concentrated organic deposits, mostly from the remains of land plants in river deltas and other swampy conditions. The evolution in its composition shows a similar pattern to that of kerogen, although the original organic material normally contains much more carbohydrate (cellulose), and so has a higher oxygen content. The progression towards higher grades of fuel, from peat, to lignite, bituminous coal, and eventually anthracite, is shown in Fig. II.9. Lighter organic compounds are also formed. Bituminous coal is rich in aromatic compounds, and eventually methane is produced. High temperatures may be involved in coal formation, whereas oil can only be generated between 70 and 175°C, because methane is the main product of the breakdown of kerogen at higher temperatures.

Coal contains numerous fossil plant remains. Obviously this cannot be the case with oil, which is liquid and has migrated from its original source. Nevertheless, oil does contain 'geochemical fossils', molecules clearly derived with little modification from their biological source. They include nitrogen-containing porphyrin molecules found in life, for example in combination with *magnesium and *iron as chlorophyll and haem, respectively. The original metallic elements seem to be easily lost, but the remaining organic compound has strong complexing properties and may combine with other metals, especially vanadium and nickel. The vanadium content is particularly high in oil from Venezuela, and comes from mineral deposits through which the oil has migrated. It is possible that *nickel is picked up in the same way, although it may be the direct remnant of the F-430 coenzyme found in methanogenic bacteria.

Coal contains a much greater range of minor constituents. Table II.10 lists the typical concentrations of some elements in coal ash, many with a high degree of enrichment compared with average crustal concentrations. Sometimes exceptional concentrations have been reported; for example up to 7.5 per cent for germanium. These concentrations must result from the chemical nature of the organic deposits giving rise to the coal. Some of the organic compounds formed by the breakdown of plant remains, such as humic acids, are excellent complexing agents. A number of elements (e.g. Be, B, Sc, Ge) may be picked up from solution in this way. The reducing conditions (which must be present if organic deposits are to survive) also convert sulfates into sulfides, and chalcophilic elements (e.g. Co, Ni, Mo, Bi) can be

Table II.10 Rare elements in coal ash. The average concentrations shown are subject to wide variation. The final column shows the enrichment factor, relative to average crustal composition. (From Mason and Moore, 1991)

Element	Concentration in coal ash (ppm)	Enrichment factor
Be	45	16
B	500	60
Sc	60	3
Co	300	12
Ni	700	9
Ga	100	7
Ge	500	330
As	500	250
Mo	50	30
Ag	2	20
Cd	5	25
Pb	100	8
Bi	20	100
U	2.7	150

precipitated (see *sulfur). *Uranium may be precipitated as insoluble UO_2 in the same circumstances.

Elements forming volatile compounds when fossil fuels are burned have a particular environmental significance. *Arsenic is one such, as its oxide As_2O_3 is volatile at combustion temperatures. Coal burning therefore contributes to the spread of arsenic in the environment. *Sulfur, however, is the cause of most concern. Its presence in fossil fuels comes partly from the sulfur content of living organisms, and partly from sulfide minerals deposited along with the organic matter. Sulfur in oil is present in organic molecules such as thiophen, C_4H_4S; coal may contain these in addition to inorganic sulfide minerals. On combustion, sulfur compounds are converted to SO_2, which is a serious air pollutant: in combination with organic compounds and soot particles it causes respiratory problems; and it is one of the principal sources of acid rain (see also *hydrogen). Modern petroleum refining aims to remove a high proportion of the sulfur content. This is done by *dehydrosulfurisation*, using catalysts containing cobalt and molybdenum to reduce organic sulfur compounds to H_2S. Removing sulfur from coal is much harder, and coal-burning power stations contribute a large proportion of the SO_2 emissions to the atmosphere. SO_2 can be removed from the flue gases after combustion, although this is quite expensive. A promising alternative is to burn the coal in powdered form using 'fluidised-bed' technology. Much of the SO_2 can be removed at the time of combustion by adding calcium carbonate, which reacts to form calcium sulfate.

Atmospheric carbon chemistry

Many carbon compounds are volatile and are present in the atmosphere. Table II.11 shows a selection of the important ones, giving an estimate of their present-day average concentration, the mean residence time for molecules in the atmosphere, and the rate of proportionate increase in their concentration. These increases are the cause of much current concern, in the case of *chlorine compounds because of their potentiality to damage the ozone layer as described elsewhere, and for CO_2 and CH_4 because of their contribution to the 'greenhouse' effect, as discussed later. Some other organic molecules containing other elements and present only in trace amounts in the atmosphere are discussed later (see Table II.12).

CO_2 is by far the major carbon-containing constituent of the atmosphere. Its input comes about equally from two chemically different types of source: volatilisation of dissolved CO_2 from the ocean and (to a much smaller extent) from weathering of carbonate rocks on land; and from oxidation of organic carbon by respiration, decay and burning of biomass, and from fossil fuel burning. It is primarily this latter source which is giving rise to the present increase in atmospheric concentration. CO_2 is removed from the atmosphere by two routes which are also of roughly equal importance: by dissolving in water, and by photosynthesis to give organic carbon.

Methane in the atmosphere comes mostly from the decomposition of living matter by bacteria under anaerobic conditions; as mentioned above, the conversion of

Table II.11 Carbon compounds in the atmosphere, giving their average concentration (mixing ratio), and estimate of the residence time, and of the annual rate of increase, as a proportion of the present concentration. (Data from Ramanathan *et al.*, 1987; there are indications that the increase for methane may be slowing)

Compound	Concentration (ppm by volume)	Residence time (years)	Annual rate of increase (%)
CO_2	315	3	0.4
CH_4	1.6	10	1
Other hydrocarbons	< 0.01	< 0.5	?
CO	0.25	0.2	1
COS	0.0005	1	0
Formaldehyde	0.0002	0.001	0
Chlorine compounds	0.001	10–200	4

organic carbon into CO_2 and CH_4 can provide metabolic energy when no oxidising agents or light are present. The main sources are natural wetlands (swamps and marshes), rice paddies, and the digestive tracts of ruminant animals (cattle, etc.) and termites. An increase in rice growing and other farming practices may be leading to some increase in methane input to the atmosphere, but there are also other anthropogenic sources, such as biomass burning, waste disposal in landfills, and fossil fuel (including natural gas) extraction. Heavier hydrocarbons come in much smaller quantities from a variety of sources, both anthropogenic ones such as internal combustion engines, and natural ones such as volatile terpenoid compounds from certain plants. Hydrocarbons are principally removed from the atmosphere after oxidation by OH radicals as discussed below. They may also form intermediates which react with *nitrogen oxides, to give products which are noxious and contribute to urban air pollution.

Carbon monoxide in the atmosphere comes partly from incomplete combustion in the burning of fossil fuels and biomass; the atmospheric oxidation of methane and other hydrocarbons also makes an important contribution. Although the global contribution of CO to the carbon cycle is low, local concentrations from traffic fumes in urban areas may reach toxic levels of 50 ppm or more. CO combines with haemoglobin in red blood cells, and so inhibits the transport of oxygen from the lungs to body tissues. Incomplete fuel burning can give a variety of other pollutants, including heavy hydrocarbons and particulate carbon (soot). Increasing controls specifying the types of fuel burnt, and requiring the cleaning of exhaust gases, e.g. by catalytic converters, have limited the increase of some of these pollutants, but the atmospheric CO concentration is nevertheless increasing.

The atmospheric reactions of reduced carbon compounds mostly start with attack by hydroxyl radicals (see §4.3 and *oxygen):

$$OH + CH_4 \rightarrow CH_3 + H_2O;$$

$$OH + CO \rightarrow CO_2 + H.$$

The latter reaction is much the faster one, and in spite of the higher concentration of CH_4, about 70 per cent of OH radicals react with CO, and 30 per cent with methane, in the unpolluted atmosphere. The subsequent reactions of methyl radicals are complicated, and depend on the concentrations of other species such as *nitrogen oxides. A typical sequence is:

$$CH_3 + O_2 \rightarrow CH_3O_2;$$

$$CH_3O_2 + NO \rightarrow CH_3O + NO_2;$$

$$CH_3O + O_2 \rightarrow HCHO + HO_2.$$

The formaldehyde (HCHO) formed in the last step here can be decomposed photochemically:

$$HCHO + h\nu \rightarrow CHO + H,$$

$$CHO + O_2 \rightarrow CO + HO_2,$$

with the CO being finally oxidised to CO_2. Other products are possible, and heavier hydrocarbons can give a variety of molecules, including the molecule peroxyacetyl nitrate (PAN, $CH_3CO.O_2.NO_2$), which is a powerful lachrymator.

The atmospheric lifetime of methane is long enough for some of it to reach the stratosphere. It is the main hydrogen-containing molecule to do so, as the very low temperature at the tropopause (where the troposphere and stratosphere meet) acts as a 'cold trap', preventing water vapour from rising further. Methane then reacts with oxygen atoms,

$$O + CH_4 \rightarrow CH_3 + OH,$$

and is the principal source of stratospheric H, OH, and H_2O, which are important in reactions involving the ozone layer (see *oxygen). However, the greatest current worry about the stratosphere concerns halogenated carbon compounds such as CFCs, which are discussed under *chlorine.

Organoelement compounds in the environment

Organoelement compounds are ones with organic carbon-containing groups such as methyl (CH_3), attached to another element. Such compounds can be made for nearly all the elements in the laboratory, but mostly they are highly reactive, especially to oxygen and water, and are not found under natural conditions. Exceptions occur when the carbon–element bond is sufficiently strong to prevent rapid decomposition. This is the case with non-metallic elements, and with metals in the lower right-hand region of the periodic table (Sn, Hg, and Pb in particular).

Organic compounds of N, O, and S are of course major components of living systems. A number of other organoelement compounds are produced in small quantities by natural processes, especially those of life. To these must be added industrially produced compounds which are released into the environment and sometimes persist for some time.[55] A selection of both types of compound is shown in Table II.12.

Organoelement compounds make a negligible contribution to the amount of carbon in the natural environment, but for the other elements concerned they may be significant. Frequently they are the only naturally occurring volatile forms of that

55 See Fergusson (1990) and Ashby and Craig (1990).

Table II.12 Organoelement compounds found in the environment. More details are given under the element concerned. Note the abbreviations: Me = CH_3; Et = C_2H_5; Bu = C_4H_9. (From Fergusson, 1990, and Harrison, 1990)

Element	Biogenic compounds	Synthetic compounds
F	–	CFCs, e.g. CF_3Cl
S	Me_2S, Me_2S_2	many
Cl	MeCl	many
P	– ?	many
Ge	Me_4Ge	–
As	Me_3As, $Me_2AsO(OH)$	many
Sn	– ?	$Bu_3SnOCO.Me$, many others
Se	Me_2Se, Me_2Se_2, Me_2SeO	–
Br	MeBr	CH_2BrCH_2Br, CF_3Br, others
Te	Me_2Te	–
I	MeI	–
Hg	Me_2Hg, $[MeHg]^+$, MeHgSMe	–
Pb	Me_4Pb	Et_4Pb

element, and are found as traces in the atmosphere, where they may contribute to the global cycling of the element. A few of them are significant in atmospheric chemistry. For example, Me_2S (dimethyl sulfide, DMS) is one of the most important naturally produced *sulfur compounds in the atmosphere, and its oxidation contributes to thc acidity of rainfall, and possibly to cloud formation processes. Among the methyl halides, MeCl and MeBr form the only natural sources of *chlorine and *bromine atoms in the stratosphere (although now outweighed by anthropogenic compounds).

Another reason for the interest in organoelement compounds is that many of them are toxic; some are deliberately manufactured as pesticides or fungicides. For elements such as *arsenic, *mercury, and *lead, which are toxic in any chemical form, this toxicity is enhanced by the presence of organic groups, as they promote solubility in fatty tissues such as cell membranes, and so increase the ease with which the element can be absorbed by living organisms.

Most naturally produced organoelement compounds contain the simplest organic group, methyl. Although the mechanisms by which they are made are not always known, some routes have been established. Micro-organisms in lake or ocean sediments are often involved, and in some cases methylcobalamin (a *cobalt-containing vitamin B_{12} derivative) is the source of the methyl group. This molecule is unusual in containing a bond between a methyl carbon and the metallic element cobalt. It is involved in some bacterial reactions such as those which produce methane; schematically,

$$MeCoB_{12} + [H] \rightarrow CH_4 + CoB_{12},$$

methylcobalamin

where [H] represents a source of reducing *hydrogen such as NADH. It is thought that metals such as mercury can intervene in this reaction, for example,

$$MeCoB_{12} + Hg^{2+} \rightarrow MeHg^{+}.$$

In some other cases sulfur compounds are involved. The decomposition of dimethyl sulfonium propionate (DMSP) to give Me_2S is mentioned under *sulfur; other elements may be methylated by S-adenosyl methionine, which like DMSP contains a sulfonium ion, R_2S^+Me. Sometimes the biological methylation may be an 'accidental' by-product of cell chemistry; but it may be that some mechanisms have evolved in micro-organisms as a way of ridding themselves of toxic elements. The irony of this strategy is that in doing so, they convert elements into forms which are much more toxic for other species, including ourselves!

All the above reactions involve the transfer of a methyl group from one element to another. Such *transmethylation* must ultimately start with a biogenic methyl compound, but it can proceed under abiotic conditions. Methyl iodide is thought to be an effective transmethylating agent for some elements.

The carbon cycle

Attempts to understand the carbon cycle have taken on a new urgency in recent years, largely because of concern with the increasing concentration of atmospheric CO_2 and its influence on the 'greenhouse effect'. The major chemical transformations of the cycle are illustrated in Fig. II.10. Inorganic carbon—CO_2 and bicarbonate—is converted to organic form by photosynthesis. The reverse transformation occurs naturally via the respiration, decay, and burning of living matter. A small amount of organic carbon is buried forming kerogen and fossil fuels; this is subject to natural weathering, but a much greater flux these days comes from the burning of fossil fuels. Human intervention also contributes to the burning of living vegetation. Inorganic carbon is also subject to cycling, between the atmosphere and ocean by precipitation and evaporation, and through carbonate rocks by sedimentation and weathering (see *calcium).

Estimates of the reservoirs and fluxes in the carbon cycle are shown in Fig. II.11. In spite of some uncertainties, it is clear that the reservoirs of carbon in the environment follow the order of size:

atmosphere < biosphere < ocean bicarbonate < sediments.

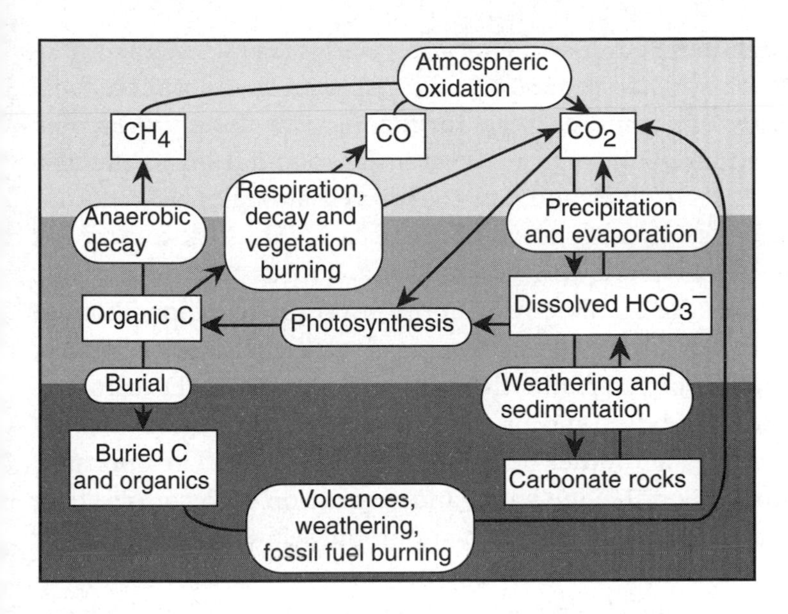

Fig. II.10 Chemical transformations in the carbon cycle.

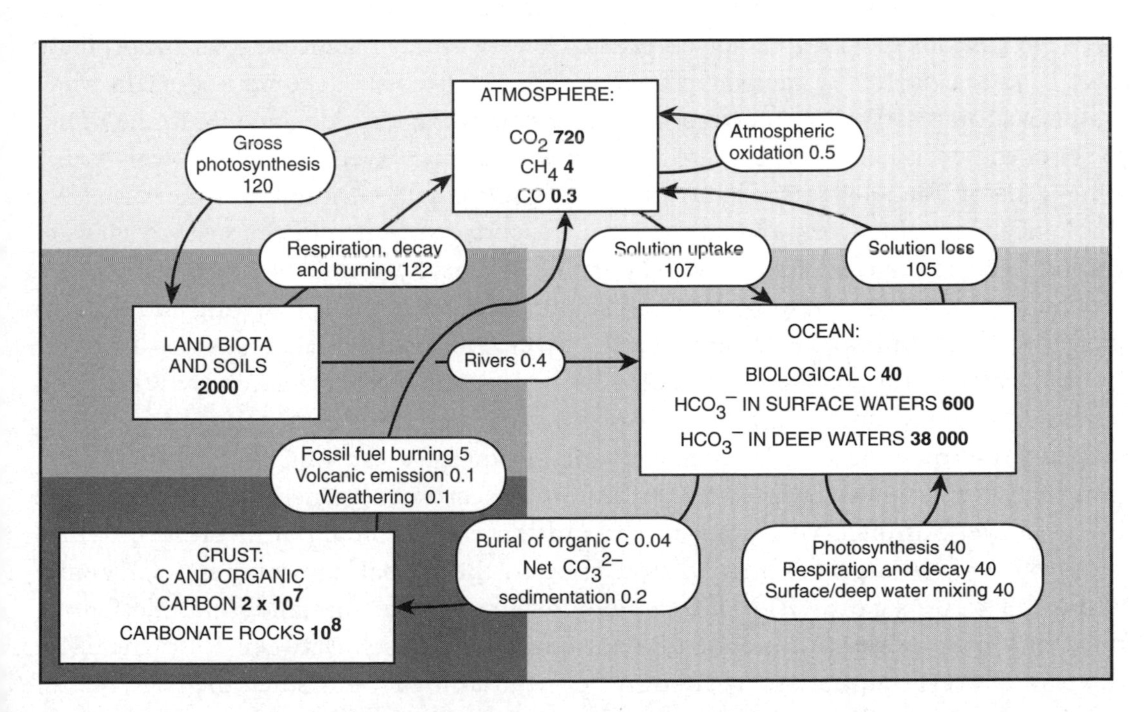

Fig. II.11 The carbon cycle. Reservoirs (square-cornered boxes) and annual fluxes (round-cornered boxes) are given in units of 10^{12} kg carbon. (Based on Schlesinger, 1991, Houghton *et al.*, 1990, and Schneider and Boston, 1991.)

The natural cycling through these compartments also takes place on several different time-scales: a few years for transfer between the atmosphere, the biosphere, and the ocean surface; of the order of a hundred years for mixing with the much greater reservoir of the deep ocean; and millions of years for transfer to and from sediments.

Over 2×10^{14} kg, that is 200 Gt of carbon, are transferred annually between the atmosphere and either land biota (mostly plants) or the oceans. Thus the mean residence time of CO_2 in the atmosphere is 3–4 years. About 80 per cent of the CO_2 turnover is taken up by photosynthesis, directly from the atmosphere by land plants, or from dissolved bicarbonate by marine plankton. Half of the photosynthesised carbon is returned by respiration of the plants themselves. The remainder is passed along the food chain as discussed above, and provides fuel to be oxidised in respiration by animals, or by soil bacteria and other decomposing organisms. The opposing fluxes provided by photosynthesis on the one hand, and respiration or decomposition on the other, are nearly in balance both on the land and in the sea. Only a very small amount of reduced carbon ends up buried as fossil fuels.

In addition to the photosynthesis/respiration transfers, the physicochemical exchange of carbon between the atmosphere (CO_2) and surface waters (HCO_3^-) provides a flux which is much larger than for almost any other element. The equilibrium between gaseous CO_2 and dissolved HCO_3^- is very evenly balanced; thus atmospheric CO_2 readily dissolves in rainwater, and bicarbonate is easily decomposed again when water evaporates. This exchange process gives an average residence time in the upper parts of the ocean of about 6 years. Mixing with deeper layers is much slower, with a time-scale of 50–200 years. Rising and sinking currents contribute, but there is also a 'biological pump' at work. Organisms near the surface produce skeletons and shells of insoluble *calcium carbonate, which later sink and partly dissolve in deep waters.

The formation of carbonate sediments, and their subsequent recycling and decomposition in tectonic process, are much slower still, the overall residence time for 'mobile carbon' (atmosphere + biosphere + ocean) being around a hundred thousand years. The burial and weathering of organic carbon involves a similar time-scale, although the magnitudes are somewhat smaller than for carbonates. However, these natural fluxes are now dwarfed by the amount of carbon returned to the atmosphere by fossil fuel burning. Measured in terms of the mass of material involved, it represents by far the largest human disturbance to the global environment. Currently about 1.8×10^{13} kg (18 Gt) of CO_2 enters the atmosphere annually from this source, most of it from power production and automobile exhausts. Between 1900 and 1970 the rate of CO_2 output rose at about 4 per cent annually, but since then the rate of increase has slowed down somewhat, although it is still above 2 per cent a year. Another type of anthropogenic perturbation to the cycle comes from deforestation and other changes in land use. Direct inputs of CO_2 into the atmosphere come from vegetation burning; loss of green plants also reduce the outputs from photosynthesis. These two factors may contribute as much as 8 Gt CO_2, or 2 Gt C per year.

As a result of these disturbances, the atmospheric CO_2 concentration has increased from around 288 ppm in 1850 to over 350 ppm in 1990. The increase represents about 50 per cent of the total carbon input during the period concerned. Most of the remainder must have been taken up by the ocean, but there is some uncertainty as to the fate of all the CO_2 produced by human activities. There is some 2 Gt C per year 'missing' in current estimates of the uptake from the atmosphere. Some land plants may have responded to the increased atmospheric CO_2 by increasing their rate of photosynthesis.

The different magnitudes and time-scales just discussed are important for understanding the future consequences of fossil fuel burning. Carbon equilibrates quite rapidly in the biosphere and in the surface layers of the ocean, but only a limited amount of excess CO_2 can be accommodated in these reservoirs. The overall capacity of the ocean is much more substantial, but this can only operate on a longer time span, determined by surface/deep water mixing. Furthermore, the 'biological pump' which transfers carbon to deep waters cannot respond easily to increased CO_2 content, as the biological productivity of the ocean is limited by the availability of other elements such as nitrogen and phosphorus.

Over 99.9 per cent of all the carbon on Earth is in the form of rocks, as insoluble carbonates such as $CaCO_3$, and in reduced form derived from the burial of organic matter. Ultimately, the extra CO_2 derived from continued fossil fuel burning must return to the crust. Although this will happen much too slowly to be significant on a human time-scale, the processes of sedimentation and weathering of crustal carbon are important in the control of atmospheric CO_2 over geological times. It is thought that the sun's output has increased by some 25 per cent during the lifetime of the Earth, but that surface temperatures here have not increased as would be expected from this. A likely explanation is that the CO_2 content of the atmosphere has decreased over this period, thus reducing the 'greenhouse effect' (see §1.4). The rate of removal of carbon from the oceans, and ultimately from the atmosphere, depends on the weathering of crustal rocks to release metal ions such as Ca^{2+}, which form insoluble carbonates. Weathering rates should respond to global temperature changes, as most chemical reactions speed up under warmer conditions. The presence of life can also accelerate weathering, by increasing the acidity of soils through extra CO_2 content and through humic acids produced in the decay of plant remains. Thus global temperatures may be linked to the carbon cycle through a variety of feedback loops. Proponents of the Gaia hypothesis have taken these ideas further, by suggesting that life on Earth exerts a control over the composition of the atmosphere which is in a sense 'deliberate', and designed to maintain a suitable temperature.[56]

56 See Lovelock (1982) for a non-technical exposition, and Schneider and Boston (1991) for several detailed articles.

Global warming

The background to discussions on the greenhouse effect was given in §1.4. Molecules in the atmosphere trap out-going infrared emission from the Earth, with the result that the mean surface temperature is about 33°C higher than predicted without any atmosphere. The contributions made by different atmospheric constituents depend in a very complicated way on their concentration and other properties, such as the wavelength at which they absorb radiation, and the strength of that absorption. (The major components N_2 and O_2 do not absorb infrared radiation at all, and so make no contribution.) The effects of different molecules are non-additive, especially when they absorb radiation in the same region of the spectrum. For example, in the spectral ranges where CO_2 and H_2O vapour absorb strongly, all out-going infrared radiation is absorbed up to an altitude of several kilometres in the atmosphere. Increasing their concentration does increase the greenhouse effect, but to a much smaller extent than a similar increase of the CFC molecule CF_2C_2, which absorbs in a different region of the spectrum where no other absorption currently takes place. In spite of these complications, it is possible to calculate the relative contributions to an increased greenhouse effect owing to the current emissions of different gases. Table II.13 shows an estimate of such effects over a period of 100 years from 1990, and based on the rates of emission in that year.[57]

The major greenhouse gas in the atmosphere is water vapour. This is omitted from Table II.13 because its atmospheric concentration is outside human control. The indirect effects of some gases, such as the contribution of CH_4 to stratospheric water vapour, and of NO_x species, where their effect arises from the production of other molecules such as ozone, are however included. It is apparent from the table that carbon dioxide is expected to have by far the largest effect on global warming, followed by methane.

The predictions in Table II.13 are tentative, because of uncertainties in the atmospheric lifetimes of some constituents, and the complexities of indirect effects. But even if these relative effects could be taken as a reasonable guideline, there is much greater doubt over the absolute magnitude expected for global warming. Although it is possible to calculate the extra radiation trapping caused by a given molecule such as CO_2, the Earth's climate responds in very complicated non-linear way to this effect. There are numerous feedbacks, both physical and chemical in nature. The most important is the atmospheric water-vapour concentration (see §1.4), but there may also be changes in cloud cover, in the extent of the ice-caps, in the rate of uptake of CO_2 by the ocean, and in the rate of photosynthesis. Other perturbations caused

57 Houghton *et al.* (1990) give a detailed but relatively non-technical review aimed at policy-makers. See also Mitchell (1989) and Ramanathan *et al.* (1987); the latter is fairly technical.

Table II.13 The relative contributions of different compounds to the *additional* greenhouse warming expected over a 100-year period from 1990. Estimates are based on current emission rates. (Based on Houghton *et al.*, 1990)

Molecule	Current anthropogenic emission (10^9 kg year^{-1})	Contribution to additional warming effect (%)
CO_2	26 000	61
CH_4	300	15
Other hydrocarbons	20	0.5
N_2O	6	4
$CFCl_3$ (CFC-11)	0.3	2
CF_2Cl_2 (CFC-12)	0.4	7
Other CFCs	0.3	2
CH_3CCl_3	0.8	0.2
CO	200	1
NO_x	66	6

by human activity will also contribute. A reduction in ozone concentration will lead to a cooling of the upper atmosphere. *Sulfur emissions may have a cooling effect, by forming sulfate aerosols which increase the reflection of sunlight. Current estimates suggest that a 'business as usual' scenario, with emissions of greenhouse gases continuing to rise at the present rate, could lead to an increase in average temperature of 1°C by the year 2025, and of 3°C by the end of the next century. There would be widespread local variations, and changes in the pattern of rainfall, as well as an appreciable rise in sea level. In spite of all the uncertainties, these predictions are worrying enough to have convinced many people that something must be done to reduce fossil fuel burning, the major cause of increased CO_2.

Isotopes of carbon

Carbon has two stable isotopes, ^{12}C, which makes up 98.9 per cent of normal carbon, and the heavier ^{13}C, which constitutes the remaining 1.1 per cent. In addition to these, the radioactive isotope ^{14}C, which has a half-life of 5730 years, is produced by nuclear reactions of nitrogen with cosmic rays in the upper atmosphere. It is present at a very low concentration in the atmosphere, and thence is incorporated into carbonates and into all living organisms.

Although the chemical properties of different isotopes are very similar, they are not absolutely identical (see §2.4). Small isotopic differences appear in equilibrium constants and in rates of reactions, and these lead to variations in the composition of carbon from various sources. A selection of data are shown in Table II.14, where the ^{13}C content of carbon from different sources is specified in terms of the deviation in

Table II.14 Isotopic composition of carbon from various sources. The ^{13}C abundance is expressed as the variation in parts per thousand, relative to a standard calcite sample. (See §2.4 for definition of the δ scale. Data from Mason and Moore, 1982)

Source	$\delta^{13}C$ (parts per thousand)
Atmospheric CO_2	−2 to −6
Yellowstone Park emissions:	
CH_4	−28
CO_2	−4
Carbonate sediments:	
range	+5 to −5
average	0
Organic sediments (kerogen):	
range	−10 to −35
average	−25
Diamonds	−2 to −5
Marine algae	−7 to −17
Land plants	−22 to −28
Coal	−22 to −28
Petroleum	−20 to −30

parts per thousand from a standard sample. Negative values correspond to depletion in the heavier isotope. Some of the differences arise from chemical equilibria. For example, when the reaction

$$4H_2 + CO_2 \rightleftharpoons CH_4 + 2H_2O$$

comes into equilibrium, the methane will have less ^{13}C than the CO_2. This is seen in emissions from volcanoes and hot springs, and the isotopic composition of gases from the Yellowstone National Park corresponds to the equilibrium constant expected at about 200°C. Similar but smaller deviations occur in the equilibration of atmospheric CO_2 and marine carbonates. Especially significant differences arise between oxidised and reduced forms of carbon, and these come largely from rates of reactions; the enzymes that reduce CO_2 during photosynthesis act slightly faster on ^{12}C than on ^{13}C, and so give rise to the negative $\delta^{13}C$ values seen with organic carbon in living organisms and sediments, including fossil fuels.

The different isotopic compositions of carbonate and organic sediments have been used to estimate the relative proportions of these types of carbon in the Earth's crust. To do this, it is necessary to know the average ^{13}C content of all the carbon on Earth. The average isotopic composition of carbon from meteorites gives an overall $\delta^{13}C$ of about −5, and a very similar estimate is obtained from diamonds, which are thought to be derived from the mantle, with an isotopic composition that has not been subjected to chemical fractionation since the formation of the Earth. Given then the average values of 0 for carbonates, and –25 for organic deposits, it follows that the latter must constitute about 20 per cent of all sedimentary carbon, in order to make up the correct overall average. Surprisingly, it appears that this proportion has been almost constant for nearly 4 billion years. This suggests that living processes must have operated very early on in the history of the Earth.[58]

The ^{14}C content of carbon on Earth varies for a different reason. It is subject to chemical fractionation effects like ^{13}C, but its radioactive decay is much more important. It is normally assumed that all living matter takes up ^{14}C in proportion to its atmospheric content. When an organism dies, ^{14}C decays with a half-life of 5730 years. The age of fossil remains, or of charcoal or textiles derived from plants or animals, can then be found by measuring the present ^{14}C content, if the original amount present is known. This is the basis for the important *radiocarbon dating* technique. The half-life is appropriate for dating specimens with ages ranging from a few hundreds to some tens of thousands of years, and is thus ideal for archaeological studies. The technique has been refined considerably over the years.[59] Modern mass spectrometers, based on high-energy particle accelerators, allow more accurate measurements to be made than was possible in the past. Much smaller samples can now be measured, as ^{14}C atoms are measured directly in the mass spectrometer, rather than through their radioactive decay as in previous methods. To give reliable results, it has also been necessary to calibrate the method properly. Originally it was assumed that the ^{14}C content of the atmosphere, and hence of living matter, had always been the same. It is now recognised that changes have occurred. Since the Industrial Revolution, natural CO_2 has been diluted somewhat with the emissions from fossil fuels; these contain almost no ^{14}C as they have been buried for millions of years. Other variations in atmospheric ^{14}C have occurred in the past, perhaps because of changes in cosmic-ray intensity.

Cerium ($_{58}Ce$)

Cerium is the most abundant of the lanthanide or 'rare earth' elements, discussed in more detail under *lanthanum. Most cerium is used in mixtures with the other lanthanides, although unlike them it forms a dioxide of formula CeO_2 which is used as a coating for self-cleaning ovens (where it acts as a catalyst for the oxidation of fats and other organic compounds) and for polishing glass. Cerium plays no role in life.

Chlorine ($_{17}Cl$)

An important element in the natural environment, chlorine occurs almost exclusively as the chloride ion Cl^-. This is very soluble in water, and is easily

58 See article by Schidlowski in Schneider and Boston (1991).

59 See articles by Delibrias and by Duplessy and Arnold, in Roth and Poty (1989).

washed out of rocks in natural weathering reactions. Although its overall abundance is not especially high (Table II.15), it accumulates in the oceans, where it has the longest residence time (4×10^8 years) and highest concentration of any dissolved non-metal. Deposits of chlorides, principally NaCl (halite, rocksalt, or 'common salt') are formed by evaporation, and are the main source of the element.

Chlorine gas (Cl_2) is made by electrolysis of NaCl solution; traditional methods used cells containing *mercury, but the toxicity of this element has led to these being phased out and replaced by diaphragm cells containing no mercury. Chlorine is used for making bleaches and in the sterilisation of drinking water, but the major applications are in making chlorinated organic compounds such as vinyl chloride (CH_2:CHCl) for the manufacture of poly(vinyl chloride) (PVC) plastics. Hydrochloric acid (HCl) is another very widely used industrial chemical, and is mostly obtained as a by-product from the manufacture of organic chlorine compounds.

Chlorine is an essential element of life, ubiquitous as Cl^- in cells and body fluids. It does not take part directly in many biochemical reactions, but fulfils the essential role of ionic balance, providing an ion of opposite charge to that of metal cations. The element Cl_2 is highly toxic to all forms of life, and was used as a poison gas in the First World War. The major environmental problems are however associated with synthetic compounds containing chlorine, especially organochlorine compounds. Some of these are highly toxic, and indeed are used as pesticides. Others, such as chlorinated fluorocarbons used in refrigeration and as aerosol propellants, and a variety of chlorinated solvents used widely in industry, are volatile and accumulate in the atmosphere. Although the total concentration of chlorine compounds in the atmosphere is currently only about one part per billion, these compounds have caused much concern because of their ability to damage the ozone layer.

Table II.15 Chlorine concentrations in the environment

Location	Concentration
Crust	130 ppm
Sea water	1.8 %
Fresh waters (average)	8 ppm
Atmosphere	*volatile compounds* 1 ppb; *also dust particles from sea salt*
Volcanic gases (typical)	50 ppm
Human body :	
average	0.12 %
blood	0.3 %

Organochlorine compounds

Apart from small amounts of methyl chloride (chloromethane, CH_3Cl) produced by marine algae, organic compounds containing carbon–chlorine bonds hardly exist in the natural environment. Large quantities are manufactured industrially however, and used for a wide variety of purposes: solvents and degreasing agents; the manufacture of plastics such as PVC; gases for refrigerators; propellants and blowing agents for aerosols and for making plastic foams; fluids for hydraulic equipment and for filling electrical transformers; and diverse pesticides such as insecticides and weed-killers. A selection of these compounds is shown in Table II.16.

The very properties which make organochlorine compounds useful can give rise to environmental problems.[60] Chemical inertness is important for many applications, and some of the compounds are consequently very persistent and resistant to breakdown. This is true both with pesticides and with the CFCs discussed in the following section. Many of the compounds are toxic, and their relatively non-polar character leads to high fat solubility, which also causes them to accumulate in food chains. For these reasons, the use of some organochlorine compounds has been banned or partially limited.

DDT was first used in the 1940s as a highly successful insecticide: its very persistence was advantageous in this respect, as repeated application is unnecessary. Through the eradication of malaria-carrying mosquitoes it must have saved very many lives, but it has now been banned in many countries. It is quite toxic to fish, and has been held responsible for failures in the breeding of some species of predatory birds. Some other organochlorine insecticides such as aldrin have also been severely controlled because of their toxicity, although many safer ones remain in use.

The compounds 2,4-D and 2,4,5-T are examples of herbicides used as weed-killers and military defoliants; the latter became notorious for its use under the name 'Agent Orange' by the United States in the Vietnam War during the 1960s and 1970s. Although both compounds may be quite toxic themselves, the major problems with their use have arisen from the production of dioxins such as TCDD (see Table II.16) as highly toxic side-products. Following a manufacturing accident at Seveso in Italy in 1976, many thousands of people were exposed to TCDD. Many suffered from a skin disease known as *chloracne*, although the long-term effects may be less than originally feared.

Aromatic chlorine compounds such as polychlorinated biphenyls (PCBs) have attracted especial concern. Many of these are carcinogenic, and the use of PCBs has now been banned in many countries. The disposal of old equipment containing PCBs, such as electrical transformers, poses problems. The normal method is high-temperature incineration, which should destroy PCBs and convert all the chlorine content to relatively harmless HCl.

---

60 Manahan (1991) discusses the environmental occurrence and toxicity of organochlorine compounds.

Table II.16 A selection of anthropogenic organochlorine compounds

Source or use	Examples	Chemical formula or structure
Solvents	Dichloromethane	CH_2Cl_2
	Tetrachloromethane (carbontetrachloride)	CCl_4
	Trichloroethane	CH_3CCl_3
	Trichloroethene	$CHClCCl_2$
Plastics manufacture	Chloroethene (vinyl chloride)	CH_2CHCl
Refrigerants, propellants for aerosols and foams	CFC-11[a]	$CFCl_3$
	CFC-12	CF_2Cl_2
	CFC-115	$CFCl_2CCl_3$
	HCFC-123	$CHCl_2CF_3$
Fluids for hydraulics and dielectrics (transformers)	Polychlorinated biphenyls (PCBs), e.g.	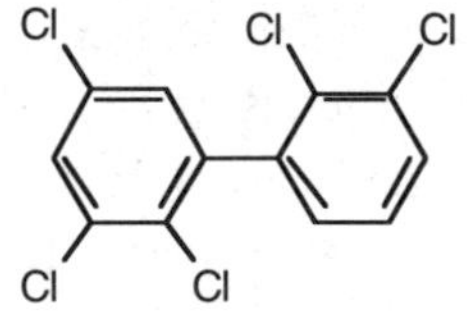
Insecticides and herbicides	DDT	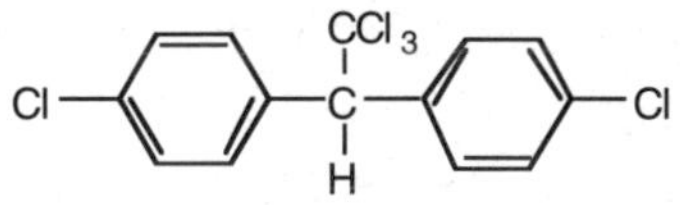
	Aldrin	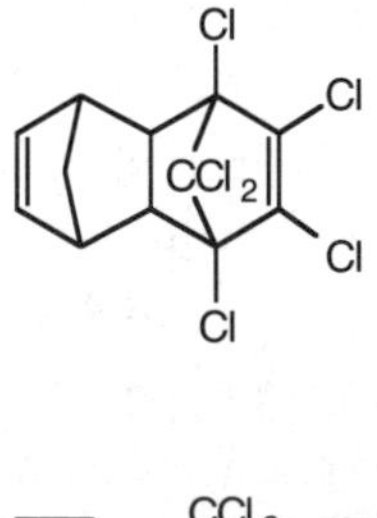
	2,4-Dichlorophenoxyace (2,4-D)	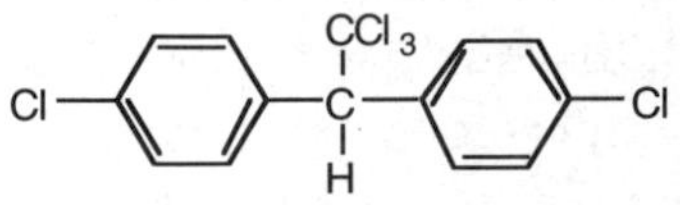

Table II.16 Continued

Source or use	Examples	Chemical formula or structure
	2,4,5-Trichlorophenoxyacetate (2,4,5-T)	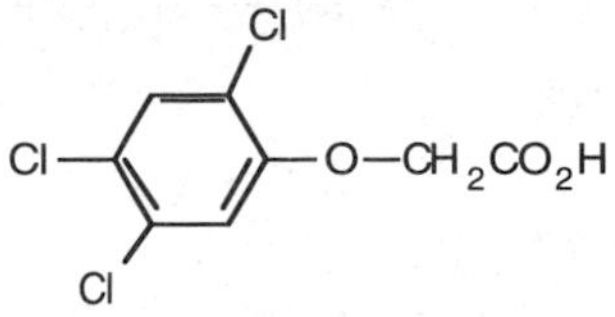 Cl, Cl, Cl, $O-CH_2CO_2H$
Toxic intermediates in manufacture and disposal	Dioxins, e.g. 2,3,7,8-tetrachlorodibenzo-*p*-dioxin (TCDD)	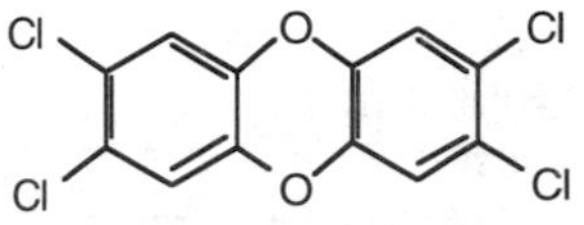 Cl, Cl, O, O, Cl, Cl

[a]The nomenclature for CFCs and HCFCs is confusing. The last digit gives the number of fluorine atoms, and the one before that the number of hydrogens plus one. A three digit number indicates more than more carbon, with the first digit equal to the number of carbons minus one; all other atoms are assumed to be chlorine.

Chlorine compounds in the atmosphere

The most stable volatile chlorine compounds are HCl and organic compounds containing C–Cl bonds. Significant amounts of HCl are produced when minerals are heated deep in the Earth, and it forms a proportion (typically around 50 ppm) of all gases liberated by volcanoes. HCl is extremely soluble in water, and is quickly washed out of the atmosphere. Over geological time-scales, a major proportion of the chloride in sea water has probably come from this source, but because of its very short residence time, the mean atmospheric HCl concentration is very low. Organic chlorine compounds are much less soluble in water and remain in the atmosphere for a long time. The major compounds present are shown in Table II.17.

Only methyl chloride is of natural origin; this molecule is apparently produced as a minor by-product of the methylation reactions which occur for many elements in the ocean (see *carbon). The other atmospheric chlorine compounds are anthropogenic, and include CFCs such as $CClF_3$, and industrial chemicals and solvents such as CCl_4. Although their concentrations are measured in fractions of a part per billion, the important point about them is that they have very long residence times. Continued use means that their concentrations are continually increasing, and the last column of Table II.17 shows an estimate of their possible concentrations by the year 2030.

The longest atmospheric residence times are associated with compounds with no C–H bonds, of which the CFCs are the most important. Molecules with C–H bonds

Table II.17 Chlorine compounds in the atmosphere. (Based on Ramanathan *et al.*, 1987, and Wayne, 1991)

Compound	Concentration (ppb volume)	Residence time (years)	Possible 2030 conc. (ppb)
CH_3Cl	0.6	1.5	0.6
$CClF_3$	0.007	400	0.06
CCl_2F_2	0.28	110	1.8
$CHClF_2$	0.06	20	0.9
CCl_3F	0.18	65	1.1
Other CFCs	<0.02	100–200	<0.2
CCl_4	0.13	50	0.3
CH_3CCl_3	0.14	8	1.5
Other organochlorine compounds	<0.1	<1	<0.1

are attacked in the lower atmosphere (the troposphere) by reactive OH radicals (see §4.3). The reaction

$$XC{-}H + OH \rightarrow XC + H_2O$$

is favourable because the new O–H bond formed is stronger (about 464 kJ per mole) than the C–H bond that must be broken (416 kJ per mole). Bonds to chlorine cannot be attacked in the same way, because a C–Cl bond (325 kJ per mole) is much stronger than an O–Cl bond (214 kJ per mole), so that the reaction

$$XC{-}Cl + OH \rightarrow XC + ClOH$$

is energetically unfavourable. Thus CFCs without any hydrogen are not broken down in the lower atmosphere. The main sink for these molecules is the upper atmosphere (the stratosphere), where the presence of UV radiation provides enough energy to break the C–Cl bonds:

$$XC{-}Cl + h\nu \rightarrow XC + Cl.$$

Chlorine atoms then react with ozone:

$$Cl + O_3 \rightarrow ClO + O_2:$$

$$ClO + O \rightarrow Cl + O_2.$$

The net reaction is

$$O_3 + O \rightarrow 2\ O_2,$$

with the chlorine atom being regenerated. The importance of these atoms, therefore, is that like some other pollutants such as oxides of *nitrogen they provide a catalytic route for the decomposition of ozone (see under *oxygen). In spite of the very low concentrations of chlorine compounds in the atmosphere, their contribution to the total destruction of ozone in the stratosphere is appreciable, and probably the largest of all anthropogenic compounds released to the atmosphere.

The atmospheric reactions of chlorine compounds are in reality a good deal more complicated than the above scheme suggests. Chlorine atoms also react with other species such as NO_2. Some other important reactions are

$$Cl + CH_4 \rightarrow HCl + CH_3$$

and

$$ClO + NO_2 \rightleftharpoons ClONO_2.$$

The latter is a reversible reaction providing a reservoir of temporarily inactive chlorine. It appears that on the surface of ice particles in polar regions, the reaction

$$HCl + ClONO_2 \rightarrow Cl_2 + HNO_3$$

occurs, and chlorine atoms can then be regenerated

$$Cl_2 + h\nu \rightarrow 2Cl.$$

This is part of the explanation for the formation of the *Antarctic ozone hole*, a phenomenon first suspected from observations made in 1982, and confirmed in 1984. During the early summer near the South Pole there is a major depletion in the stratospheric ozone layer. Mixing between the very cold polar air and the rest of the atmosphere can be very slow, and significant chemical perturbations build up during the (dark) winter months, caused by surface reactions on ice particles such as those mentioned above. The reduction in ozone is strongest in mid-September (spring in the Southern Hemisphere), as only then does significant sunlight become available to liberate the chlorine atoms.

Any perturbation to the ozone layer is worrying, as this gas is responsible for absorbing much of the hard UV radiation from the sun. One consequence of reduced ozone might be an increase in the incidence of skin cancer. But the absorption of radiation is also important in terms of the global stability and circulation of the atmosphere, so that perturbations might also affect the Earth's climate. Another reason for concern with CFCs is that their presence in the lower atmosphere makes some contribution to the greenhouse effect (see *carbon). At present this is a very minor one, but future projections based on uncontrolled use and emission predict significant increases in concentration, which could become important. For these reasons, international efforts have been made to limit the use of CFCs, or to replace them by compounds such as $CHClF_2$ which react with OH radicals and so have shorter residence times in the troposphere. These efforts have been partially successful, and emissions are probably declining.

Chromium ($_{24}Cr$)

An element of moderate abundance in the crust, chromium occurs principally in the silicates of very basic rocks (where it substitutes easily for magnesium) and as deposits of the mixed oxide mineral *chromite*, $FeCrO_4$. Rather high concentrations of chromium and *nickel are found in so-called 'serpentine soils', derived from the weathering of ultrabasic rocks (see under *magnesium). These soils are toxic to many species and support only a very restricted range of plants. In the more common Cr^{3+} form, chromium is very insoluble in water at neutral pH, and its dissolved concentration in all natural waters is very low. It is thought to be essential for life, and involved in glucose metabolism in humans, although this has been disputed. In the oxidised form chromate, CrO_4^{2-}, the element is normally much more soluble and is toxic and carcinogenic. Chromium is an important element in metallurgy, being used as a constituent of stainless steel and in 'chrome plating'. Chromate compounds are also used as pigments.

Cobalt ($_{27}Co$)

Cobalt is an element with strong chalcophilic and siderophilic tendencies, and is probably concentrated in the Earth's core (see Table II.18). In the crust, it occurs mainly as sulfide minerals in association with *nickel. Its concentration in natural waters is very low, partly because of the insolubility of cobalt sulfide. Cobalt appears to have a very short residence time in the ocean (a few years), being efficiently scavenged by marine life. The fact that its concentration is higher in surface than in deep sea water suggests that the main input comes from wind-blown dust derived from soil.

Most cobalt is obtained as a by-product of *nickel production, and it is used in alloys, pigments, and industrial catalysts. It is an essential trace element in life. Most cobalt in animals and humans is in the form of vitamin B_{12} (see Fig. II.12). This compound can be manufactured from dietary cobalt by bacteria in the gut, although in

Table II.18 Cobalt concentrations in the environment

Location	Concentration
Whole Earth	0.1 %
Crust	30 ppm
Sea water:	
surface	0.007 ppb
deep ocean	0.001 ppb
Fresh waters	0.2 ppb
Soils:	
average	8 ppm
range	1–100 ppm
Human body	0.5 ppm

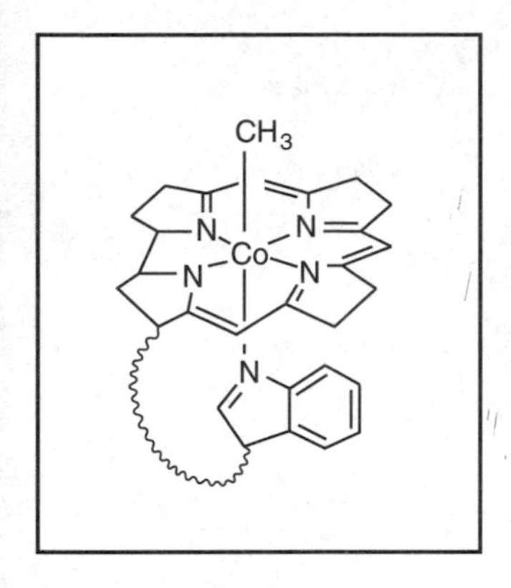

Fig. II.12 Schematic structure of methyl derivative of Vitamin B_{12}; only the important atoms surrounding cobalt are shown explicitly.

humans the process is very inefficient, as it takes place only in the colon, from where very little uptake into the blood is possible. An adequate supply of vitamin B_{12} (about 5 μg per day) is therefore essential in the diet, most coming from meat, fish, eggs, and dairy produce. Ruminant animals such as cows and sheep have a much more extensive gut population of bacteria, and can manufacture sufficient vitamin B_{12} for their own use. Even so, however, they use cobalt rather inefficiently, and problems can arise when they feed in areas where the soil concentration of cobalt is low. This can be overcome by dietary supplements, or by adding traces of cobalt sulfate to fertilisers.

One biochemical reaction accomplished by vitamin B_{12} is the transfer of methyl (CH_3) groups from one organic molecule to another. The methylated form of B_{12} is an intermediate in this, and it is interesting that cobalt is one of the few metals which forms a direct chemical bond with an organic carbon atom in biological systems. (*Nickel may be another, and has an even more limited biochemical function in life.) Methylated B_{12} can be used to make methyl derivatives of unwanted elements, especially *mercury, *arsenic, *selenium, and *tellurium. This reaction is performed by some bacteria and fungi, and may act as a defence mechanism, as the methylated products are volatile and can escape. However, they are also highly toxic, and in particular the conversion of mercury to the methyl form adds substantially to the toxicity of this element.

The radioactive isotope ^{60}Co is produced by neutron bombardment of natural ^{59}Co, and has a half-life of 5.3 years. It undergoes β decay, but produces at the same time γ radiation of unusually high energy (1.3 MeV), and therefore of high penetrating power. ^{60}Co is used as an irradiation source, in scientific and medical applications, and for the irradiation of some foods.

Copper ($_{29}Cu$)

Copper is a siderophilic element, occurring in greatest concentration in the Earth's core. It is of moderate abundance in the crust (Table II.19), where it occurs principally in combination with *sulfur in minerals such as chalcopyrite, $CuFeS_2$. Surface oxidation of these minerals leads to the release of Cu^{2+}, which

Table II.19 Copper concentrations in the environment

Location	Concentration
Crust	50 ppm
Sea water:	
surface	0.008 ppb
deep	0.02 ppb
Fresh waters	1–10 ppb
Soils:	
average	30 ppm
contaminated	100–10 000 ppm
Human body	1 ppm

Table II.20 Some enzymes and other proteins containing copper

Name	Function
Cytochrome oxidase	$O_2 \rightarrow H_2O$
Haemocyanin	O_2 transport in arthropods
Lysine oxidase	Cross-linking of collagen ($O_2 \rightarrow H_2O$)
Ascorbate oxidase	Oxidation of ascorbate ($O_2 \rightarrow H_2O$)
Amine oxidase	Removal of hormones, e.g. adrenalin
Blue proteins	Electron transfer
Superoxide dismutase	$O_2^- \rightarrow O_2 + H_2O_2$

also occurs in complex carbonates and other oxides. Some copper is also found in native (metallic) form, and this is one reason why the element played an important part in primitive metallurgy in many civilisations. Copper objects made before 2000 BC often contain a significant amount of *arsenic, but changes in metallurgical practice, especially the habit of alloying copper with tin to make bronze, reduced this. The early use of copper in coinage has continued, the other major application being for electrical wiring. The concentration of copper in natural waters is low.

Copper is an essential trace element in life, and is a component of several metalloenzymes and other proteins. A selection is shown in Table II.20. In most of these compounds, copper is coordinated by nitrogen atoms from histidine residues in the protein chain. The copper-containing protein haemocyanin is used to transport oxygen in arthropods, having a similar role therefore to iron-containing haemoglobin. There are other similarities with *iron, as many copper-containing enzymes catalyse oxidation/reduction reactions; copper is used primarily outside cells, rather than inside as is the case with iron. It has been suggested that the use of copper for this role may be relatively recent in evolutionary terms, being associated with an increase in atmospheric *oxygen, and the development of multicellular organisms.

Reactions performed by copper enzymes are important in the growth of connective tissue such as collagen. Deficiency does not normally occur with humans, but it is not uncommon in grazing animals such as sheep, as the anaerobic conditions in the gut can lead to the precipitation of insoluble sulfides. One of the symptoms, known as 'swayback disease' in young lambs, is paralysis which results from the loss of myelin, a material that forms a sheath around nerve fibres and is important for their efficient operation. Copper deficiency in sheep can be treated by dietary supplements, especially for pregnant ewes.

Myelin loss in humans is responsible for the autoimmune disease multiple sclerosis, but attempts to treat this with copper have apparently been unsuccessful.[61] It is interesting, however, that wearing copper bracelets has often been recommended as a 'folk remedy' against rheumatism, another disease associated with connective tissue. Medical opinion does not in general support their effectiveness.

Copper is toxic to many micro-organisms, and sprays containing copper sulfate are used to treat fungal infections of grapes. The element is also toxic to animals in high doses. A special protein, metallothionein, which contains many –SH groups and combines strongly with copper, is synthesised in the liver to protect the body against excess. Copper contamination of soils and water can occur near mines and in industrial areas, but the main area of environmental concern in connection with copper production is the release of other elements: *sulfur to the atmosphere from the processing of sulfide ores, and toxic elements such as *arsenic which are also found in copper minerals. A few plant species are especially tolerant of high copper levels in the soil, and some of these actually accumulate the element in large concentrations. Some of these plants have been used as indicators of sources of copper minerals in the ground.

Curium ($_{96}$Cm)

One of the radioactive transuranium elements, curium is an artificial product. The longest-lived isotope ^{247}Cm has a half-life exceeding 10^7 years, but ^{242}Cm and ^{244}Cm, with half-lives of 163 days and 18 years, respectively, are more often encountered. Small quantities are made by the neutron bombardment of *plutonium in nuclear reactors and explosions, and are therefore present in nuclear waste and in fallout. Curium is not nearly so significant a potential hazard as plutonium and *americium, which are made in larger quantities, and have longer half-lives.

---

61 Lenihan (1988).

Dysprosium ($_{66}$Dy)

An element of the lanthanide (rare earth) series, discussed under *lanthanum.

Einsteinium ($_{99}$Es)

One of the transuranium elements, einsteinium does not occur naturally, but is produced in minute amounts by neutron bombardment of *plutonium in nuclear reactors and explosions. Like fermium, it was first discovered in fallout following the testing of thermonuclear weapons. ^{252}Es has a half-life of about a year.

Erbium ($_{68}$Er) and europium ($_{63}$Eu)

Members of the lanthanide (rare earth) series of elements, discussed under *lanthanum. They mostly have very similar chemistry, but europium is unusual in that it can form Eu^{2+} under strongly reducing conditions, whereas most natural lanthanide chemistry is confined to the 3+ state. Eu^{2+} is similar is size and chemistry to Sr^{2+}, and so europium can occasionally be found in association with *strontium.

Fermium ($_{100}$Fm)

One of the transuranium elements, fermium does not occur naturally, but is produced in minute amounts by neutron bombardment of *plutonium in nuclear reactors and explosions. It was first discovered in fallout following the testing of thermonuclear weapons, and is the heaviest element to have been detected outside the laboratory. None of its isotopes has a half-life of more than a year.

Fluorine ($_{9}$F)

The lightest member of the halogen group of elements, fluorine is the most electronegative of all elements, and occurs naturally exclusively as fluoride, F^-. It is widespread in rocks, where fluoride may replace the hydroxide ion, OH^-. Its most important mineral is *fluorite*, CaF_2, although it is also found in the phosphate mineral *fluorapatite*, $Ca_5(PO_4)_3F$, and as *cryolite*, Na_3AlF_6. Hydrofluoric acid, HF, is an important industrial chemical; some elemental F_2 is manufactured by electrolysis of KF dissolved in HF. Applications include the manufacture of CFCs for refrigeration and aerosol propellants, of non-stick poly(tetrafluoroethene), (PTFE or *Teflon*), and of SnF_2 and H_2SiF_6 as a toothpaste and drinking water additives, as well as industrial applications in the manufacture of aluminium, the processing of uranium, etching glass, and in other metallurgical processes.

Although its overall abundance in the Earth's crust is higher than that of chloride, fluoride forms much more insoluble compounds, especially with the abundant cation Ca^{2+}, so that its concentration in natural waters is rather low (see Table II.21). Minute amounts of volatile fluorine compounds also occur in the atmosphere. Volcanic gases may contain a small amount of HF, coming from the decomposition of fluoride minerals at high temperatures deep underground. This gas is very soluble in water, however, and is rapidly washed out. The major atmospheric compounds are entirely anthropogenic in origin, the most important being CFCs such as CF_2Cl_2. C–F bonds are very strong, and the high chemical stability of CFCs leads to very long residence times, around 100 years, in the atmosphere. They are decomposed by hard UV radiation in the upper atmosphere, to give free chlorine atoms which have the potentiality to damage seriously the ozone layer. The fluorine itself is thought to be of less significance, and CFCs are discussed in more detail under *chlorine.

Fluoride is probably essential in life, and small quantities help to prevent tooth decay by converting some hydroxyapatite, $Ca_5(PO_4)_3(OH)$, to the more resistant fluorapatite form. For this reason it is added to toothpaste and to drinking water supplies (at a level around 1 ppm) in many areas. This latter use has caused a good deal of controversy, and the 'anti-fluoridation' lobby had considerable success in the 1970s in the prevention of water treatment. No satisfactory evidence has been produced to show that ppm levels of fluoride are harmful, and natural concentrations of this order occur in many natural waters. Significantly higher concentrations are toxic, however, and fluoride pollution associated with CaF_2 extraction and processing leads to the bone disease *fluorosis* in farm animals. HF and F_2 are very harmful, exposure to HF giving very painful and slow-healing burns. These result partly from the immobilisation of calcium as insoluble CaF_2, which upsets the ion balance of the tissues and stimulates nerve endings. HF exposure should be treated by immediate copious washing followed by the application of calcium gluconate; medical attention is essential.

Table II.21 Fluoride concentrations in the environment

Location	Concentration (ppm)
Crust	950
Sea water	1.3
Fresh waters	0.1–1
Atmosphere (CFCs)	0.0006
Human body:	
average	4
bones	2000–12 000

Francium ($_{87}Fr$)

A short-lived radioactive element, francium is formed in minute amounts in the decay series of ^{235}U (see *uranium). The longest-lived isotope, ^{223}Fr, has a half-life of only 22 minutes, and its natural abundance is minute.

Gadolinium ($_{64}Gd$)

A member of the lanthanide (rare earth) series, discussed under *lanthanum.

Gallium ($_{31}Ga$)

Gallium is borderline in its geochemical behaviour, having chemical similarities with both its vertical neighbour in the periodic table, aluminium, and its horizontal neighbour, zinc (§2.2). Gallium is widespread in both oxide and sulfide minerals, and although it is not rare, it never occurs in concentrated form. Its concentration in natural waters is very low. Gallium is obtained as a by-product from the processing of sulfide ores of zinc and copper, and is mostly used for the manufacture of semiconductor materials such as gallium arsenide, GaAs. It has no known biological role.

Germanium ($_{32}Ge$)

Germanium was one of the elements (gallium and scandium being others) that were unknown when Mendeleev devised the first satisfactory periodic table of elements in 1869. Mendeleev was able to predict its chemical properties from those of neighbouring elements, especially silicon and tin, and the confirmation of these when germanium was discovered in 1886 was influential in spreading Mendeleev's ideas. The relatively late discovery of germanium is not due to its rarity (it is more abundant than many elements known in 1870) but rather to its chemical similarity to other elements, with which it nearly always appears in combination. Like gallium, germanium is widely, though thinly distributed in both oxide and sulfide minerals, and is almost never found in high concentrations. It is extracted from flue dusts produced in the processing of zinc, and used in the semiconductor industry. Germanium compounds are very insoluble and have low concentrations in natural waters. It is not essential for life, although organic germanium compounds (like those of *arsenic) are apparently produced by micro-organisms and are found in the ocean.

Gold ($_{79}Au$)

Gold is a very rare element. The fact that it was one of the earliest known elements is due to its great chemical inertness, so that it occurs in native (metallic) form in small deposits. It is also found in combination with tellurium, and as the chloride complex $[AuCl_2]^-$ in sea-water. The attractive appearance of gold and its resistance to corrosion have led to its use since antiquity for ornaments and as currency, and these still account for the major usage of the element, although it now has some specialist applications for electrical contacts and for coating windows to cut down heat loss. Gold is not essential for life, although its compounds are used in the treatment of arthritis. It is however a very significant element in the history of human civilisation. The lust for gold has led nations to war and colonisation, and has been responsible for major environmental destruction. Current problems arising from gold prospecting include the disruption of the lives of native Amazonian populations, and the burning of rain-forest. The example of gold—an element of little technological importance—shows that human vanity and greed may have environmental consequences just as serious as those arising from industrialisation.

Hafnium ($_{72}Hf$)

A relatively rare element, hafnium has a very similar chemistry to *zirconium, and occurs in association with it in a number of oxide and silicate minerals such as *zircon*, $ZrSiO_4$. It is extremely insoluble in water, and its dissolved concentration is very low. Hafnium is a good absorber of neutrons, and is used in control rods for nuclear reactors. It is not essential for life.

Helium ($_2He$)

Helium is the second most abundant element (after *hydrogen) in the Universe, making up about 25 per cent by mass of stars and interstellar gases. It was produced from hydrogen by nuclear reactions during the first few minutes after the origin of the Universe, and its abundance in stars forms one strong piece of evidence for the Big Bang theory.[62] Helium is chemically inert, and is rare on Earth. It is found in natural gas and in oil wells, and makes up about 5 ppm of the atmosphere. This occurrence is almost entirely due to its origin in the α decay of radioactive elements such as *thorium and *uranium (see also §2.4). Helium atoms are so light that they escape eventually from the Earth's gravitational field, the atmospheric residence time being around a million years. Helium has the lowest boiling point (4.2 degrees above absolute zero) of any substance, and is used in specialised applications involving very low temperatures, such as superconducting magnets. Because it is light, chemically inert, and entirely non-toxic, it is also used for filling balloons, and in mixtures with oxygen for deep-sea diving.

---

62 See Cox (1989) for a discussion of the origin of the elements.

Holmium ($_{67}$Ho)

A lanthanide (rare earth) element, discussed under *lanthanum.

Hydrogen ($_1$H)

Hydrogen is the lightest element, and the dominant constituent of the Universe. All life on Earth depends on solar energy coming from the nuclear fusion of hydrogen to helium in the sun's interior. Hydrogen is relatively much less common on Earth, but is nevertheless a crucial element in many ways. It occurs principally as water (H_2O), which is the most mobile constituent of the environment, and is important in controlling the distribution and cycling of most other elements (see §§1.3, 3.4, and 4.4). Some minerals contain hydrogen combined either as water of crystallisation or as the hydroxide ion, OH^-. Hydrogen is also a major element of life, both as water and as organic molecules where hydrogen forms bonds with *carbon, *nitrogen, and *oxygen. The distribution of water, as rainfall, in rivers, lakes, and the ocean, has determined much of the geographical pattern of human settlement and civilisation. The presence of water vapour in the atmosphere, at concentrations up to 5 per cent, has a strong influence on the Earth's climate (§1.4).

Elemental H_2 is manufactured from hydrocarbons (fossil fuel) and by the electrolysis of water. It is an important industrial chemical, being used in the production of ammonia from *nitrogen (mainly for fertilisers), in the hydrogenation of vegetable fats to make margarine, and in the manufacture of organic chemicals. Hydrogen could be an important fuel of the future, avoiding some of the problems associated with fossil fuel burning and nuclear power. As a chemical fuel it could be burned without producing greenhouse gases such as CO_2. H_2 could in principle be made from water, using solar energy either to make electrical power to be used for electrolysis, or directly in one step by a process known as *photoelectrolysis*. Current research suggests that cells containing compounds of titanium and ruthenium might be promising for this process. Other research is investigating the nuclear fusion of hydrogen isotopes into helium, which it is claimed could provide a potentially inexhaustible and 'clean' form of nuclear energy.

The heavy isotope of hydrogen ^{2}H or *deuterium* makes up 0.0156 per cent of hydrogen on Earth. It can be concentrated from natural hydrogen by the electrolysis of water, and is used for some specialised purposes in research and nuclear power production. The radioactive isotope *tritium* (^{3}H) is produced in small amounts by cosmic-ray bombardment in the upper atmosphere and in nuclear explosions.

The hydrogen bond

In naturally occurring compounds, hydrogen invariably forms a single covalent bond to one other element.[63] The structures and properties of many compounds containing hydrogen however, suggest that a second, weaker bond to another atom can sometimes be formed. This is known as the *hydrogen bond*, and the energies involved (in the range 10–60 kJ per mole) are intermediate between those of normal covalent bonds and those appropriate to non-bonding (van der Waals) interactions between molecules. The formation of a hydrogen bond between a group A–H and another atom B requires that both A and B have a fairly high electronegativity, and the strongest bonds are formed where these atoms are N, O, or F.

Figure II.13 shows some examples of hydrogen bonding important in nature. In liquid water and ice, each hydrogen atom is attached by a normal O–H bond to one

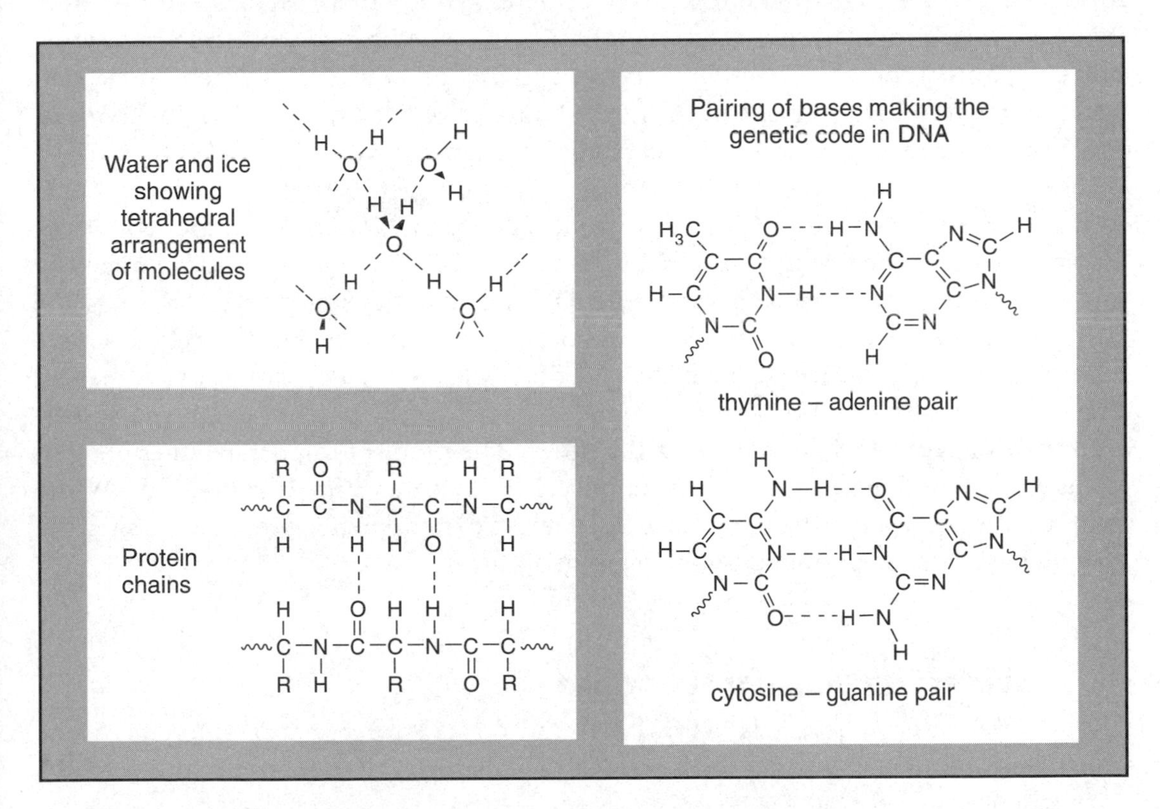

Fig. II.13 Examples of hydrogen bonding (shown by dashed lines).

63 Other forms of bonding are possible; for example hydrogen can bridge between two atoms, as in many boron hydrides (see Greenwood and Earnshaw, 1984). These compounds are not known in nature, however, as they are highly reactive in the presence of oxygen and water.

oxygen atom, and by a hydrogen bond to another oxygen. This leads to a structure where each molecule is surrounded tetrahedrally by four others. In ice the structure is regular and crystalline, except that the location of hydrogen atoms between each neighbouring oxygen pair is disordered. Water has a more disordered and fluctuating structure, based on the same principles. The existence of hydrogen bonding in water contributes to the unusual properties of this compound, discussed in the following section.

Hydrogen bonding is of crucial importance in biology. It can produce associations of molecules which form spontaneously without the need for a catalyst, and these associations can also be broken up again without an excessive expenditure of energy. Figure II.13 shows two important examples. Hydrogen bonds can form between the carbonyl (C=O) and amino (N–H) groups in protein chains, and exert a major influence on the shapes of proteins such as enzymes; these shapes are not only involved in determining the structural function of some proteins, but are also important in controlling the reactivity and selectivity of enzymes. Indeed, hydrogen bonding is often involved also in the interaction between an enzyme and its *substrate*, the molecule which it activates in a catalytic reaction.

The other example is even more familiar. The *nitrogen-containing bases, adenine (A), cytosine (C), guanine (G), and thymine (T) are the 'letters' of the genetic code. Their sequence along strands of DNA forms a set of instructions for the manufacture of the enzymes necessary to perform the chemical reactions in a cell. DNA has the famous 'double helix' structure, with two parallel strands joined by hydrogen bonds. The structure of the bases is such that A bonds with T, and C with G, as shown in Fig. II.13. The two strands are thus complementary, every base on one being associated with the corresponding base on the other. The point of this arrangement is that it allows replication of DNA, essential in cell division and in reproduction. As the two strands separate, a new strand can be built complementary to each, using the unique base pairing provided by the hydrogen bonds.

Water: an unusual substance

The water molecule is highly polar and forms strong hydrogen bonds. Some unusual properties, important for its role in the chemistry of the Earth's surface, are a consequence of these facts.

Compared with other compounds with a similar molecular mass, the freezing and boiling points of water are unusually high, and liquid water is stable over a wide range of temperatures. No other substance can exist at the Earth's surface in solid, liquid, and vapour form. The 'normal boiling point' at 100°C is the temperature at which water produces vapour at atmospheric pressure. At an average environmental temperature of 15°C the vapour pressure in equilibrium with the liquid represents about 1.7 per cent of the composition of the atmosphere, and this proportion rises rapidly with temperature. Strong hydrogen bonding in the solid and liquid accounts for the high boiling point of water, and also gives it a large latent heat of vaporisation and a large heat capacity. These latter properties are important in the physical dynamics of the atmosphere and the oceans. About 25 per cent of the flux of energy from the equatorial regions towards the poles (see §1.3) is carried by heat in ocean waters, and another 15 per cent by the latent heat of water vapour in the atmosphere. Many smaller-scale motions of the atmosphere, for example the formation of clouds and of whole tropical storm systems, are dominated by the latent heat released when water vapour condenses.[64]

Another consequence of hydrogen bonding is that it favours tetrahedral arrangement of water molecules in the crystal structure of ice (Fig. II.13), which gives a relatively open structure. When ice melts, the disordered structure of water is more closely packed, and denser. Thus ice floats on water. If this were not the case, ice would sink to the bottom of lakes and oceans in winter conditions, to be replaced by warmer water which in its turn would freeze and sink. The polar oceans and many lakes would freeze solid, and would not even melt again in the summer, as a layer of warm water formed at the top by the sun's heat would provide good thermal insulation. As it is, water is at its densest at 4°C. When a body of water has cooled to this temperature, further heat loss from the top does not produce any more convection, but merely a layer of colder water, and eventually ice, floating on top. Our planet would be very different if water were a 'normal' liquid, freezing to a denser solid.

The chemical properties of liquid water have a large influence on its ability to transport other elements. Its high polarity makes it a good solvent for many of the chemical forms of elements found on Earth, especially cations and anions with fairly low charges (e.g. Na^+, Ca^{2+}, Cl^-, and $SO_4{}^{2-}$; see §3.4). Water falling as rain produces the chemical and physical breakdown of rocks known as *weathering*, and leads to an enormous flux of elements, both in solution and as insoluble sediments, from land into the sea (see §§4.1 and 4.4).

Non-polar organic molecules are much less soluble in water, largely because of their inability to engage in hydrogen bonding. This feature is important in controlling the shapes of biological polymers, and ultimately the structures of cells and the specific functions of enzymes. Non-polar *hydrophobic* parts of the molecule avoid contact with water by associating together, whereas polar groups such as $-COO^-$and $-NH^{3+}$ substituents arrange themselves next to the aqueous medium. Lipid molecules with polar ends and long non-polar chains thus form biological membranes as shown in Fig. I.21 (in §3.5), and act as barriers between the contents of a cell and the outside environment. Many biological structures also depend on specific hydrogen bonds, as discussed in the previous section.

64 See Henderson-Sellers and Robinson (1986), for a discussion of weather systems, and the importance of atmospheric water vapour.

The pH of natural waters

Another property of water is its *self-ionisation* to give H^+ and OH^- ions:

$$H_2O \rightleftharpoons H^+ + OH^-.$$

In pure water the concentrations of these two ions are the same, and equal to 10^{-7} mol dm^{-3} at 25°C. Acid solutions have a higher H^+ and a lower OH^- concentration, the opposite being true for alkaline solutions. The normal way of specifying the acidity or alkalinity is by the pH, defined as

$$pH = -\log_{10} [H^+],$$

or equivalently

$$[H^+] = 10^{-pH},$$

where $[H^+]$ is the molar H^+ concentration. Thus neutral water has a pH of 7, acid solutions less than 7, and alkaline ones higher.

Natural waters on the Earth are rarely neutral, as there are many weakly acidic or alkaline compounds in the atmosphere and in rocks and soils. Table II.22 shows some of the reactions that can influence pH in the natural environment. The most ubiquitous acidic constituent is atmospheric CO_2, which dissolves in rainwater to give bicarbonate and hydrogen ions, at a pH of about 6. Solid carbonates such as $CaCO_3$ produce alkaline conditions, and the equilibria involving $CaCO_3$, bicarbonate, and atmospheric CO_2 are important in controlling the pH of many natural systems, including sea water. They are discussed in more detail under *carbon.

Acid soils may arise from an elevated concentration of CO_2 (from the aerobic decomposition of organic matter) and from organic plant remains known generally as *humic acids*: these are high-molecular-weight compounds of very variable composition, found in especially high concentrations in peat. Calcium carbonate is the normal cause of alkaline soils, but many other reactions contribute. For example,

Table II.22 Chemical reactions influencing the pH of natural waters

Environment	Typical pH	Reactions producing H^+ or OH^-
Unpolluted rain	6	$CO_2 + H_2O \rightarrow HCO_3^- + H^+$
Acid rain	4	$2SO_2 + O_2 + 2H_2O \rightarrow 2SO_4^{2-} + 4H^+$
		$4NO_2 + O_2 + 2H_2O \rightarrow 4NO_3^- + 4H^+$
Acid soils	5–6	$CO_2 + H_2O \rightarrow HCO_3^- + H^+$
		humic acids, e.g. $RCO_2H \rightarrow RCO_2^- + H^+$
Alkaline soils	8–9	$CaCO_3 + H_2O \rightarrow Ca^{2+} + HCO_3^- + OH^-$
Oxidising pyrites	<4	$4FeS_2 + 15O_2 + 14H_2O \rightarrow 4Fe(OH)_3 + 8SO_4^{2-} + 16H^+$
Saline deposits	>10	$Na_2CO_3 + H_2O \rightarrow 2Na^+ + HCO_3^- + OH^-$
Sea water	8	$CaCO_3/HCO_3^-/CO_2$ equilibria

clay minerals—layered aluminosilicates formed by the weathering of igneous minerals such as alkali feldspars—can take up cations of different types. Ion-exchange reactions such as

$$\text{clay–H}^+ + \text{Na}^+ \rightarrow \text{clay–Na}^+ + \text{H}^+$$

exert a 'buffering' influence which can limit pH variations owing to other effects.

pH values outside the 'normal' range of 5–9 are found in exceptional circumstances. Sodium carbonate is a constituent of many saline evaporite deposits, and is much more soluble than $CaCO_3$. Waters exposed to such deposits can have pH >10 if the normal buffering materials are absent. Strongly acid conditions can be caused by oxidation: the examples in Table II.22 show how hydrogen ions are produced by the oxidation of the atmospheric pollutants NO_2 and SO_2, and of sulfide minerals such as FeS_2.

Acid rain, resulting from the oxidation of atmospheric NO_2 and SO_2, is now recognised as a serious environmental problem. Table II.23 shows a breakdown of global contributions to the acidity of rainfall, and illustrates that anthropogenic inputs of *nitrogen and *sulfur compounds to the atmosphere outweigh natural contributions. Ammonia is included in this table, as it is the only volatile alkaline molecule occurring naturally in significant amounts, and to some extent lessens the effect of acid emissions. These emissions come principally from fossil fuel burning, especially in power generation and transport, and are concentrated in industrial areas. The major effects of acid rain are seen in upland forest regions, such as northern Britain, Norway, Canada, and the Adirondack Mountains in north-eastern USA. These regions have naturally acidic soils, lacking calcium carbonate or other constituents to neutralise the extra acidity. Both vegetation and aquatic life can be affected, although the precise causes of damage are difficult to establish.[65] One consequence of pH values of 4 or less is to increase greatly the solubility of *aluminium compounds, which may be toxic themselves, and may interfere with the availability of essential nutrients such as *phosphorus.

Table II.23 Sources of acidity in rainfall (based on data from Schlesinger, 1991)

Source	Global contribution (10^9 kg H^+ $year^{-1}$)
CO_2	1.2
NO_x from lightning	1.4
Volcanic SO_2	1.3
Biogenic sulfur	4.1
Ammonia	−3
Anthropogenic N and S	7.4
Total	12.4

65 See for example Freedman (1989).

The biological chemistry of hydrogen

As a constituent of water and of all biochemical molecules, hydrogen is ubiquitous in biological systems. The importance of hydrogen bonding and other interactions in the structure of biological molecules, including DNA, was discussed above. The present section discusses a few processes involving hydrogen: the role of hydrogen ions in biological energy production; molecules which act as carriers of reducing hydrogen in redox processes; and the role of dihydrogen in the metabolism of anaerobic bacteria.

Water both inside and outside cells contains many ionisable groups that normally act to buffer the pH close to a neutral value: the most important of these are bicarbonate (HCO_3^-) and phosphate (HPO_4^{2-} and $H_2PO_4^-$). However, the lipid membranes are fairly impermeable to ions such as H^+, and it is possible to generate concentration gradients across such membranes, which are used for specific purposes. Gradients of ⋆sodium ion concentration are used in the transport of nerve impulses, and those of ⋆calcium ions for several messenger functions. Hydrogen ion gradients have an important role in the conversion of biological energy to produce ATP (see ⋆phosphorus).

As discussed under ⋆oxygen, both photosynthesis and aerobic oxidation of food involve chains of redox reactions, in which electrons are transferred between molecules acting at different electrode potentials. These reactions are performed by molecules bound to membranes: in the case of photosynthesis, the *thylakoid membrane* in the *chloroplast* cells; and the inner membrane of *mitochondria* for oxidative metabolism. The electron transfer reactions in the redox chains also give rise to the transfer of hydrogen ions across these membranes. Thus a gradient of H^+ concentration is produced as a result of the free energy liberated by the redox chemistry. Other sites on the membrane contain molecules of *ATP-synthase*, the enzyme responsible for producing the major energy-carrying molecule in biology, ATP. Synthesis of ATP from ADP requires an input of free energy, and this is provided by passing hydrogen ions back across the membrane, from the high-concentration to the lower-concentration side.

The transfer of hydrogen ions is not itself a redox reaction. In the process described above it is essentially the charge on H^+ that is important, the membrane system being analogous to a capacitor in an electric circuit, which stores energy by accumulating a charge. Hydrogen does however have a redox role in many biochemical reactions. For example, the oxidation of organic molecules to CO_2 requires the removal of hydrogen in a redox-active form, and the reverse is true in photosynthesis. Figure II.14 shows some biological molecules which act as intermediates in this type of process. In each case, the oxidised form of the molecule (on the left) has fewer hydrogen atoms than the reduced form (on the right). Interconversion of the NAD^+/NADH pair is equivalent to adding or removing hydride ion (H^-). In the other cases shown, two neutral hydrogen atoms are transferred. These atoms or ions are far too reactive to exist as free species in biology; the molecules shown in Fig. II.14 can be thought of as 'carriers' of reducing hydrogen.

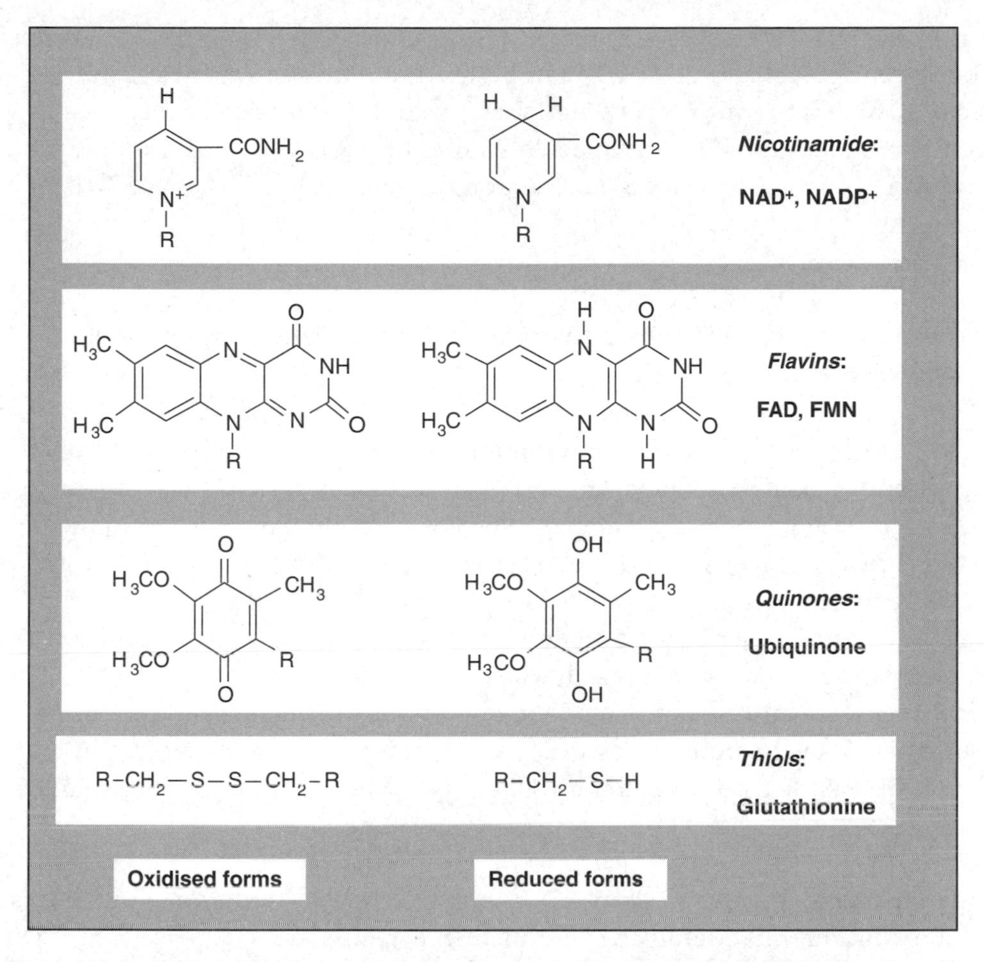

Fig. II.14 Biological molecules acting as carriers of reducing hydrogen. Only essential parts are shown; R represents a nucleoside or a protein chain. (See Stryer, 1988, for fuller descriptions of the structure and function of these molecules.)

NAD^+ is essential in oxidative metabolism, acting as a link between the reduction of O_2 to water, and the oxidation of organic molecules. Thus the overall oxidation of carbohydrate

$$O_2 + [CH_2O] \rightarrow CO_2 + H_2O$$

is essentially performed in two stages:

$$O_2 + 2NADH + 2H^+ \rightarrow 2H_2O + 2NAD^+,$$

and

$$2NAD^+ + H_2O + [CH_2O] \rightarrow 2NADH + 2H^+ + CO_2.$$

In reality each stage involves very many steps, but this separation gives a considerable flexibility to metabolism, as NAD^+ can be used to oxidise a wide range of biological molecules, without requiring any modification to the sequence of reactions in which oxygen is reduced. $NADP^+$, a molecule similar to NAD^+, is used in the reactions of photosynthesis. In this case, the net reaction of the so-called Z-scheme (see *oxygen) can be written

$$2H_2O + 2NADP^+ \rightarrow O_2 + 2NADPH + 2H^+$$

with energy being supplied from sunlight. NADPH is then used to reduce CO_2 to carbohydrates.

Of the other molecules shown in Fig. II.14, the flavin molecules FAD and FNM have similar functions to NAD^+, and ubiquinone forms a part of the electron-transfer chains in both oxidative metabolism and photosynthesis. The sulfur-containing thiols such as glutathionine perform different types of function. As discussed under *sulfur, disulfide bridges have a structural role in many proteins. Glutathionine cycles between its oxidised and reduced forms, and acts as a 'buffer' for –SH groups, as well as performing other redox functions. Among these is the decomposition of toxic hydrogen peroxide, using an enzyme which contains *selenium.

Dihydrogen is a rare constituent of our environment, being thermodynamically stable only under reducing conditions (see §3.1). There are however a number of anaerobic bacteria which both produce and metabolise H_2.[66] Fermentation reactions such as

$$C_6H_{12}O_6 + 2H_2O \rightarrow 2CH_3CO_2H + 2CO_2 + 4H_2$$

provide an energy supply of about 30 kJ per mole of carbon atoms, much less than aerobic oxidation, but useful when no oxidising agents are available. Hydrogen produced by some bacteria is then used by others, especially *methanogens* which obtain metabolic energy from the reaction

$$4H_2 + CO_2 \rightarrow CH_4 + 2H_2O,$$

which yields 130 kJ per mole. Methanogenic bacteria are thought to be among the most primitive of all surviving living organisms on Earth, and metabolic reactions involving dihydrogen may have been important in the early development of life. *Nickel is a constituent of the *hydrogenase* enzymes required.

Atmospheric hydrogen chemistry

Water vapour is the major hydrogen-containing species in the lower atmosphere; it is also the most variable of all major atmospheric constituents in its abundance. The concentration may range from almost zero in very dry air, to around 5 per

66 See Levett (1990).

cent for warm air saturated with water vapour in tropical regions. Smaller amounts of other hydrogen compounds, such as methane, ammonia, and hydrogen sulfide, also occur in the atmosphere, as does the elemental form H_2 at a concentration of 0.6 ppm.

The maximum water content varies strongly with the temperature of the air, as shown in Fig. II.15. The left-hand scale gives the *saturation vapour pressure* and the right-hand one the *saturation mixing ratio* in air at one atmosphere pressure: this is defined as the mass of water vapour per unit mass of dry air.[67] Another important quantity used to specify water content is the *relative humidity*, defined as the ratio of the actual vapour pressure to the saturation value (and normally expressed as a percentage).

Air picks up water vapour by evaporation from rivers, lakes, and the ocean, and by *transpiration* from plants. This latter source is especially important in tropical forests, which may recycle water into the atmosphere much more efficiently than could be done by evaporation alone. Warm moist air may cool by expansion as it rises, or by mixing with colder air from another source. The relative humidity rises as the temperature falls (as the saturation vapour pressure drops), until the 100 per cent point is reached, when

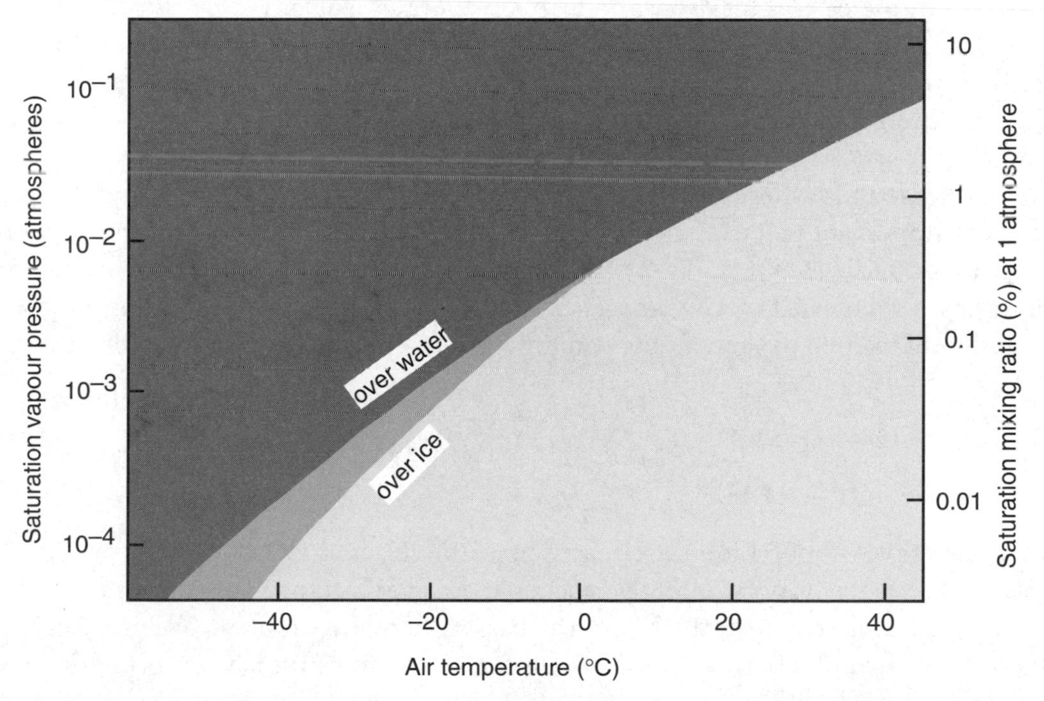

Fig. II.15 Saturation vapour pressure (left) and saturation mixing ratio in air at one atmosphere pressure (right) as a function of temperature. Note that below 0°C there are two curves, one for ice and one for supercooled liquid water.

67 See Henderson-Sellers and Robinson (1986); this definition is different from that used for most constituents, where atmospheric scientists specify parts by volume rather than by mass (Wayne, 1991).

liquid water is expected to start condensing. In fact there is a kinetic barrier to the initial formation of water droplets, because the liquid surface has a higher free energy than the bulk. Most cloud formation depends on the presence of *cloud condensation nuclei* (CCN), on which the condensation process can start. Various types of atmospheric constituent can act as CCN, including dust from soils, salt particles derived from the evaporation of sea spray, and sulfuric acid droplets coming from the oxidation of sulfur compounds. It has even been suggested that natural emissions of dimethyl sulfide from the oceans might act as a regulator of cloud cover in this way (see *sulfur).

Once water droplets have started to form, they can grow in various ways. If the humidity is sufficient, drops may continue to grow by condensation. Larger drops will start to fall faster than smaller ones, so that drops of different sizes may collide and coalesce. In cold clouds, a different mechanism is important. Once below 0°C, water may freeze to form ice, but nuclei are required to initiate freezing. A cloud at say −10°C will consist of a mixture of ice crystals and supercooled water droplets where freezing has not been nucleated. As Fig. II.15 shows, the vapour pressure of the ice is then less than that of the water. A process of distillation can therefore occur, whereby ice crystals grow and water evaporates. Silver iodide (which does not occur naturally) forms particularly effective nuclei for freezing, and has been used to 'seed' clouds and so induce precipitation. Once formed, ice crystals may grow by agglomeration and form snowflakes. Often these may melt before they reach the ground, and appear as rain rather than snow.

As discussed elsewhere (see §4.2, also *carbon, *oxygen, *nitrogen, and *sulfur), the most important reactive constituents in the lower atmosphere (the troposphere) are hydroxyl radicals (OH). These are the species mostly responsible for oxidation of minor gases such as CH_4, CO, and H_2S. Hydroxyl radicals are formed from water by the action of excited oxygen atoms coming from the photochemical decomposition of ozone:

$$O_3 + h\nu \rightarrow O_2 + O^\star;$$

$$H_2O + O^\star \rightarrow 2OH.$$

The main 'sinks' for OH are their reactions with CH_4 and CO.

Very little water vapour directly reaches the upper atmosphere (the stratosphere) because the temperature at the top of the troposphere is extremely low, and forms an effective 'cold trap'. Hydroxyl radicals are formed in the stratosphere primarily from methane:

$$CH_4 + O^\star \rightarrow CH_3 + OH.$$

They are among the species such as chlorine atoms and nitrogen oxides that can act as catalysts in the destruction of ozone (see *oxygen). Among the significant reactions are:

$$OH + O \rightarrow H + O_2;$$

$$H + O_3 \rightarrow OH + O_2;$$

$$OH + O_3 \rightarrow HO_2 + O_2;$$

$$HO_2 + O \rightarrow OH + O_2;$$

$$HO_2 + O_3 \rightarrow OH + 2O_2.$$

These have the net effect of converting O and O_3 into O_2, with OH being regenerated. Increasing methane concentrations in the atmosphere (see *carbon) may therefore have some effect on the stability of the ozone layer. Currently this problem is not thought to be nearly so serious as the effect of *chlorine atoms derived from emissions of CFCs.

Hydrogen isotopes

Hydrogen has two stable isotopes: 'normal' hydrogen ^1H, and deuterium or 'heavy' hydrogen, ^2H, often given the chemical symbol D. The average concentration of deuterium in natural hydrogen on Earth is 0.015 per cent, about one part in ten thousand. However, the chemical and physical properties of these two isotopes are slightly different (see §2.4), and this leads to variations in the isotopic composition of hydrogen in nature. The proportion of deuterium on Earth is significantly higher than that on Jupiter (which probably has a composition similar to that of the solar system as a whole), and hydrogen on Venus has even more D. In the formation of both Earth and Venus, most hydrogen was accumulated in the form of water. Reactions such as

$$H_2O\ (l) + HD\ (g) \rightleftharpoons HDO\ (l) + H_2\ (g),$$

which has an equilibrium constant of 4.5 at 0°C, may have been responsible for the higher concentration of D on the planets relative to the solar nebula. Venus has a higher abundance of D because it has lost a high proportion of its water, through photodissociation by UV radiation and the subsequent loss of hydrogen atoms into space. In this process the lighter H isotope is more reactive, and so is lost more rapidly. H_2 is also formed more rapidly than HD in electrolysis, and fairly pure 'heavy water', D_2O, can be obtained by repeated fractionation in this way. Deuterium is widely used in chemical research, and D_2O as a moderator in nuclear reactors of the Canadian 'CANDU' design (see *uranium).

On Earth, natural fractionation of H and D occurs largely through the differing physical properties of H_2O and HDO. The lighter isotope has a higher saturation vapour pressure and therefore slightly predominates in the vapour phase; this is an effect which increases at lower temperatures. Tropical rainfall has an isotopic composition close to that of the ocean, but in colder regions, and at high altitudes, rain and snow are considerably depleted in D.

The third isotope of hydrogen, ^{3}H, is known as tritium, and is often represented by the chemical symbol T. It is a radioactive β emitter, with a half-life of 12.35 years. Tritium, like the radioactive *carbon isotope ^{14}C, is produced naturally in small amounts by cosmic-ray bombardment in the upper atmosphere. It is made artificially in substantial quantities, mostly using neutron bombardment of ^{6}Li:

$$^{6}Li + n \rightarrow {}^{4}He + {}^{3}H.$$

Tritium is also generated in smaller amounts during the normal operation of nuclear reactors, and is one of the gaseous radioactive products, like *krypton, that is difficult to contain. Most anthropogenic tritium in the environment, however, has been produced by the atmospheric testing of nuclear weapons during the 1950s. Tritium is itself an important constituent of nuclear weapons, as described in the following section. It is also used for research purposes.

Hydrogen fusion

Fusion is a reaction in which the nuclei of light atoms are joined together to make heavier elements. These reactions are involved in the origin of elements, and in the energy source of stars.[68] All solar energy comes from the fusion of hydrogen atoms to make helium. The main reactions taking place near the centre of the sun are:

$$2\ {}^{1}H \rightarrow {}^{2}H + e^{+} + \nu;$$

$$^{2}H + {}^{1}H \rightarrow {}^{3}He + \gamma;$$

$$2\ {}^{3}He \rightarrow {}^{4}He + 2\ {}^{1}H;$$

where e^{+}, ν and γ are respectively a positron (the antiparticle of an electron), a neutrino, and high-energy radiation. The net reaction can be summarised as

$$4\ {}^{1}H \rightarrow {}^{4}He + \text{energy},$$

where the energy comes out in radiation and as kinetic energy. The amount liberated is tremendous, around 2.5×10^{12} J per mole of helium produced; this is several million times greater than is evolved in any chemical reaction. It is an attractive idea to try to make use of this type of energy source on Earth.

Nuclear fusion reactions involve energy barriers which are also much greater than in chemistry: it is necessary to overcome the electrostatic repulsion between two nuclei, to bring them close enough together for the nuclear 'glue', the so-called *strong interaction* between particles, to operate. Very high temperatures (several million

68 Cox (1989).

degrees) are therefore required. But the reactions starting from protons (^{1}H) are also intrinsically slow, because in the first step a proton must be simultaneously converted into a neutron, via the so-called *weak interaction*. Much faster reactions take place between heavier isotopes of hydrogen. The most favourable is the DT reaction

$$^{2}\text{H} + {}^{3}\text{H} \rightarrow {}^{4}\text{He} + \text{n},$$

where n is a neutron. This is the main reaction employed in thermonuclear devices—'hydrogen bombs'. The principal active constituent is lithium deuteride, LiD. Some tritium is included, but more is generated during the course of the explosion by the interaction of neutrons with ^{6}Li. A fission explosive (see *uranium) is used as a detonator to produce the enormous temperature required. Such devices have an explosive power measured in megatons, equivalent to millions of tons of TNT chemical explosive, and hundreds of times more powerful than the nuclear bombs which devastated Hiroshima and Nagasaki in 1945. Five countries (USA, USSR, UK, France, and China) have tested such weapons, but fortunately none has been used in war. The physical devastation would be immense, but equally serious would be the radiation produced from the explosion itself, and from the fallout of radioactive products.

Attempts at peaceful uses of fusion for an energy source have also focused mostly on the deuterium–tritium reaction. A supply of tritium would be required, from the neutron bombardment of deuterium or lithium (see previous section). As the reaction itself produces neutrons, however, this would be no problem, and it is envisaged that an operational fusion reactor could be surrounded by a 'blanket' of lithium, in order to provide tritium for future operations. The difficulty is to hold the material for long enough, and at a sufficient density, at the enormous temperatures required. Most research has concentrated on retaining a *plasma* of high-temperature, ionised gas, in a magnetic field. Among these projects is the *Joint European Torus* (JET) programme at Culham, UK.[69] The technical problems are so great that it is unlikely—short of a totally unexpected breakthrough—that fusion can realistically provide an energy supply for several decades to come.

It 1989 it seemed for a short while that such a breakthrough had been made. The electrochemists Pons and Fleischmann, working at the University of Utah, USA, announced that fusion of deuterium could be initiated without the high temperatures normally assumed to be necessary. In their apparatus, deuterium was produced by electrolysis at a palladium electrode, and reacted to form a palladium hydride compound. Unexpected amounts of energy and radiation were reported, but these results were soon questioned, and the idea of 'cold fusion' was rapidly discredited in the eyes of most scientists.[70]

69 Named after the doughnut-shaped ('torus') configuration of the plasma. Some of the difficulties and prospects of fusion research are discussed by Knief (1992).

70 Close (1990) gives a readable account of this strange story.

Indium ($_{49}In$)

Indium is a relatively uncommon element in the crust. A strong chalcophile, it occurs in low concentrations in many sulfide minerals, such as those of zinc, copper, and lead. It is obtained as a by-product of the extraction of these elements, and used in small amounts in solders and other low-melting alloys, and in the semiconductor industry. Its concentration in natural waters is very low. Indium is not essential for life, and is fairly toxic. Although it is released into the environment along with other heavy metals such as *cadmium, its low abundance and minor usage makes it a relatively insignificant pollutant.

Iodine ($_{53}I$)

An element of low abundance overall, iodine nevertheless occurs in significant quantities in the oceans (see Table II.24), and is concentrated in some seaweeds. Unlike the other halogens, fluorine, chlorine, and bromine, iodine is oxidised to the iodate ion, IO_3^-, by atmospheric oxygen (see §3.1), and small amounts are found in deposits of Chile saltpetre, $NaNO_3$. Iodine is obtained from this mineral, from brines, and from seaweeds, and is used for some specialised purposes, including dietary supplements, as an antiseptic, and as silver iodide in photographic films.

Iodine is essential for life, its only known function in the human body being as a constituent of the hormone *thyroxin* shown in Fig. I.22 (see §3.5). Iodine is thus concentrated in the thyroid gland in the neck, where this hormone is made. Deficiency of the element causes goitre, an enlargement of the thyroid. This can be prevented by dietary supplements in the form of sodium iodate, which is commonly added to table salt. The metabolism of iodine is associated with that of iron- and selenium-containing enzymes, which catalyse the incorporation and removal of iodine in organic molecules.

Regions where iodine deficiency has been a problem are somewhat remote from the sea, e.g. Switzerland and parts of China. A significant amount of iodine may be cycled through the atmosphere, both as sea-salt and as methyl iodide, CH_3I. The latter molecule is found at a concentration of around 2 parts per 10^{12} in the atmosphere, and appears to be derived from the biological methylation of iodine in the

Table II.24 Iodine concentrations in the environment

Location	Concentration
Crust	0.41 ppm
Sea water	0.05 ppm
Fresh waters	2 ppb
Human body	0.2 ppm

oceans, as with that occurring for some other elements including *chlorine and *bromine (see also *carbon). Methyl iodide is less long-lived in the atmosphere than either CH_3Cl or CH_3Br, as the weaker C–I bond can be split by UV radiation reaching the lower atmosphere (see §4.3). The IO radical which is formed as a result of this may play a minor role in the oxidation of organic *sulfur compounds.

Radioactive isotopes of iodine are produced in the fission of *uranium in nuclear reactors and explosions. Because of the relatively high volatility of iodine, they are easily released from nuclear reactors in a fire. ^{131}I, with a half-life of 8 days, formed the most important short-term source of radioactive contamination following the nuclear accident at Chernobyl in Ukraine in 1986. It is particularly dangerous as it passes efficiently through the food chain and is strongly taken up by the body, especially the thyroid. The problem can be lessened if precautions are taken promptly: large doses of non-radioactive iodine supplements dilute the radioactive iodine, and hence reduce its uptake.

Iridium ($_{77}Ir$)

One of the rarest elements in the Earth's crust, iridium is a strong siderophile and is present in somewhat higher (though never large) concentrations in metallic meteorites and probably also in the Earth's core. It occurs in conjunction with chemically similar elements such as *platinum, either in native metallic form or combined with tellurium. Its main application is in alloys with very high melting points, used for example in automobile spark plugs.

Although its direct environmental significance is negligible, iridium has played an interesting role in a long-standing controversy. Abnormally high concentrations have been found in sedimentary rocks dating from the boundary between the Cretaceous and Tertiary geological periods.[71] This is the point, about 70 million years ago, marking the extinction of many biological species, including the dinosaurs. The iridium in these rocks may have come from a major meteorite impact, which could have raised enough debris to absorb solar radiation and thus to perturb the Earth's climate for many years.

Iron ($_{26}Fe$)

Iron is the most abundant element on the Earth as a whole, making up 30 per cent of its total mass. It constitutes over 80 per cent of the core, and is also the fourth most abundant element of the crust. It has borderline lithophilic/chalcophilic properties, being common both in sulfide minerals such as *pyrites* (FeS_2), and in oxide form; indeed nearly all silicate minerals contain significant amounts of iron,

---

71 Alvarez (1987).

which may replace both magnesium and aluminium. Table II.25 shows some iron-containing minerals. Some of these are important sources of other elements: chalcopyrite for copper, ilmenite for titanium, and chromite for chromium.

A majority of the iron in the crust is present as Fe^{2+}, but this is quickly oxidised at the Earth's surface to Fe^{3+} (see §3.2). In the latter form, iron is very insoluble in water. The weathering of igneous rocks to give Fe^{2+} and its subsequent oxidation to Fe^{3+} has been very significant in the chemical evolution of the Earth's surface. In the primitive environment very little O_2 was present in the atmosphere, and sedimentary rocks containing Fe^{2+} were formed. With the development of photosynthesis, a great deal of iron was oxidised to Fe^{3+}, and a large fraction of all the O_2 ever liberated by plants is now present in this form (see §4.2 and *oxygen).

Extensive deposits of the oxide minerals *haematite* (Fe_2O_3) and *magnetite* (Fe_3O_4) form the major source of iron, which is used in larger quantities than almost any other element. 'Native' iron from meteorites was known to ancient civilisations, but it was around 1200 BC that the extraction of iron from its oxide became widespread. Originally used for weapons, it became by the nineteenth century the principal metal for structural applications. It is used alone, but to a greater extent in steels which contain carbon and many other elements. The production is demanding in energy, thus making iron a desirable element for recycling.

Magnetite is the commonest natural mineral with sufficiently strong magnetic properties to undergo orientation in the Earth's field. Known as *lodestone*, it was used as a primitive compass for navigation. The same properties are used in biology by some species.

Table II.25 Some important iron-containing minerals

Type	Examples	Formula
Sulfides	pyrites	FeS_2
	chalcopyrite	$FeCuS_2$
	arsenopyrite	$FeAsS$
Oxides and hydroxides	haematite	Fe_2O_3
	magnetite	Fe_3O_4
	ilmenite	$FeTiO_3$
	chromite	$FeCr_2O_4$
	goethite	$FeO(OH)$
Carbonates	siderite	$FeCO_3$
Silicates	olivine	$(Fe,Mg)SiO_4$
	chlorite	$(Mg,Al,Fe)_{12}[(Si,Al)_8O_{20}](OH)_{16}$

Oxide minerals containing Fe^{3+} are coloured, ranging from yellow to brown and red, depending on the structure and the iron concentration. Mixtures of Fe^{2+} and Fe^{3+} can give a dark, almost black appearance. Thus iron is an important source of colour in rocks, e.g. red sandstone, and in materials such as bricks. Iron-containing minerals have been extensively used as pigments and in pottery glazes. For example, Greek potters used iron oxide to produce a combination of red (Fe_2O_3, formed under oxidising conditions) and black (Fe_3O_4, formed by firing under reducing conditions).

Because of the insolubility of Fe^{3+} oxides, the concentration of iron in natural waters is generally low (see Table II.26). It has a very short residence time (100 years) in the ocean, its low concentration in surface waters showing that it is efficiently taken up by marine life. In coastal regions iron may be obtained from sediments, but far out to sea the main source may be wind-blown dusts. It is possible that the supply of iron may then be a limiting factor in the amount of life that can be supported in the sea. Iron is one of the most important trace elements in biological systems. An average adult human has about 4 g of iron, mostly combined in proteins of various kinds. These play many essential roles, especially in the transport of oxygen in the blood, and in the redox reaction used to generate metabolic energy by the body. A normal diet provides several milligrams of iron per day; deficiency results in anaemia, a condition causing a general lack of energy, and in serious conditions, death.

The biological chemistry of iron

Biology makes ample use of the chemical versatility of iron: its variable oxidation states, its borderline chemical character enabling it coordinate to oxygen, nitrogen, or sulfur atoms, and its ability to bind additional small molecules. Figure II.16 shows the binding of iron in some proteins. The haem group, with iron bound to four nitrogen atoms in a plane, is particularly common; the side view shows how iron in haemoglobin is coordinated to another *nitrogen atom from a histidine residue in the

Table II.26 Iron concentrations in the environment

Location	Concentration
Crust	4.1 %
Sea water:	
surface	0.01–0.1 ppb
deep	0.1–0.4 ppb
Fresh waters	0.1–1 ppm *also suspended solids*
Human body:	
average	50 ppm
blood	450 ppm

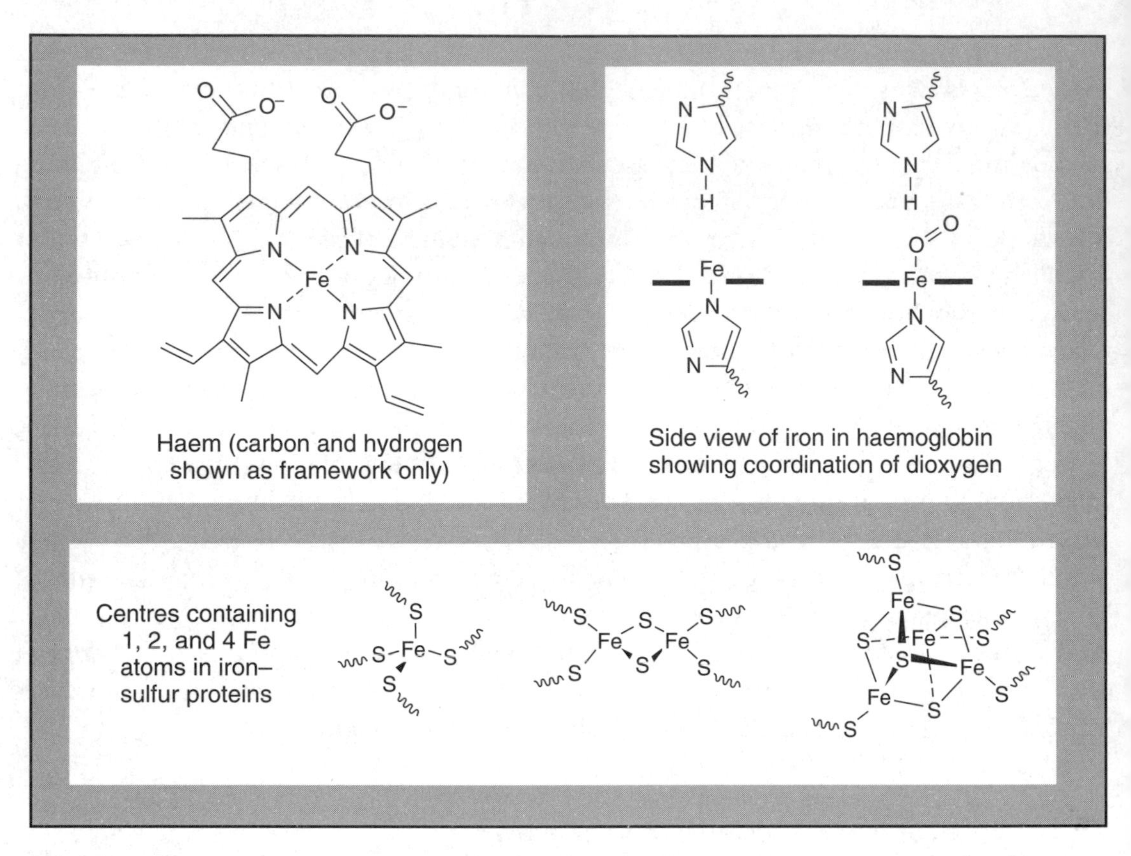

Fig. II.16 The coordination of iron in some proteins. 'Wiggly' lines indicate connections to the protein, via histidine in haemoglobin, and cysteine in iron–sulfur proteins.

protein chain. In some haem proteins a sixth ligand atom, either iron or sulfur, is present on the other side of the haem. Also illustrated in Fig. II.16 are the centres found in iron–sulfur proteins. In these molecules, which may contain one, two, four, or occasionally other numbers of Fe–S centres, iron is tetrahedrally coordinated, and some of the ligands are S atoms from cysteine residues in the protein chain.

The ubiquity of haem and iron–sulfur proteins in all life forms, including the most primitive, suggests that iron compounds must have been present in life from its origins. Many of these proteins are involved with the energy-producing reactions of cells, and it has been suggested that life could even have originated on the surface of iron sulfide particles in the early environment. Before the development of an oxidising atmosphere through photosynthesis (see *oxygen), iron was present in the more soluble Fe^{2+} state. Its concentration in sea water was probably much higher than now, and its general availability much greater. As atmospheric O_2 increased, several changes took place. Many of the iron-containing proteins were concentrated inside cells, where they can continue to act at relatively low redox potentials, protected from the more oxidising conditions. Fe^{2+} in the external environment oxidised to insoluble Fe^{3+}, and some bacteria developed the ability to catalyse this oxidation reaction and use it as a free energy source (see §4.2). *Chemosynthetic* bacteria of this kind include species that live under remarkable conditions near hydrothermal vents in the sea floor. By oxidising iron sulfides they are able to obtain energy to reduce CO_2 to organic compounds, and thus to live without a normal supply of organic 'food'. But in many situations, such as surface waters in the ocean, iron is in very short supply. Many species of bacteria excrete special molecules which chelate iron, thus capturing it in a form which they can absorb through their cell walls. In more complex organisms, a variety of proteins called *transferrins* perform a rather similar function in transporting iron to cells where it is needed. Other proteins, *ferritins*, store iron in the form of $Fe(OH)_3$ solids. Another solid found in some species is magnetite, Fe_3O_4, which can be used for orientation in the Earth's magnetic field. This was first discovered in *magnetotactic* bacteria, which use it to sense 'up' and 'down' through the vertical component of the field; it may be present in species such as pigeons, and even humans.

Some of the functions of iron proteins are summarised in Table II.27. They are very diverse, although redox reactions figure prominently in the list. The simplest type involves electron transfer, converting Fe^{2+} to Fe^{3+} or vice versa. There are many proteins, haem-containing cytochromes and iron–sulfur types, which participate in

Table II.27 Examples of iron-containing proteins

Type of Fe centre	Function	Examples
Haem	O_2 transport and storage	myoglobin, haemoglobin
	Electron transfer	cytochromes *a*, *b*, *c*
	$O_2 \rightarrow H_2O$	cytochrome oxidase (with Cu)
	NO receptor	GMP cyclase
	$RH + H_2O_2 \rightarrow ROH + H_2O$	peroxidases
	$H_2O_2 \rightarrow H_2O + O_2$	catalase
	Oxidation $RH \rightarrow ROH$	cytochrome P-450
Iron–sulfur	Electron transfer	rubredoxins (1 Fe), ferrodoxins (2, 3 or 4 Fe)
	$N_2 \rightarrow NH_3$	nitrogenase (with Mo)
Fe–O–Fe	O_2 transport	haemerythrin
	Free radical redox reactions	ribonucleotide reductase, methane oxidase
Other non-haem, non-Fe–S	$O_2^- \rightarrow H_2O_2 + O_2$	superoxide dismutase
	Misc. oxidation reactions	prostaglandin oxidase
	Fe uptake and storage	transferrin, ferritin

this process, important both in aerobic metabolism and in photosynthesis (see ⋆oxygen). By small alterations of ligand environment it is possible to 'tune' the potential of the Fe^{3+}/Fe^{2+} redox couple. Thus there is a range of proteins which allow electrons to be transferred along a chain with carefully controlled potential differences, so that energy available from the overall redox process can be extracted in the most efficient way.

Many other types of redox reaction, involving the making and breaking of covalent bonds, are catalysed by enzymes containing iron. Several of these involve dioxygen or species derived from it, such as peroxide and superoxide. Often it appears that an unusually higher oxidation state, Fe^{4+} present as the *ferryl* group, $[Fe{=}O]^{2+}$, is an intermediate. This is the case in cytochrome oxidase, which also contains copper, and which performs the crucial step

$$O_2 + 4e^- + 4H^+ \rightarrow 2H_2O$$

in the oxidation of organic molecules to produce energy. Other redox reactions illustrated in Table II.27 are used either to generate molecules needed in biology, or to detoxify unwanted products of metabolism.

The transport of oxygen by haemoglobin in red blood cells is of course the best-known function of iron in the body. A simpler protein, myoglobin, is used to store oxygen in muscle cells. This has a single haem unit which can reversibly bind a dioxygen molecule. Haemoglobin itself has four haem units. The way in which O_2 coordinates to each one is illustrated in Fig. II.16. Two interesting features are worth noting. In the first place, the upper histidine residue in the figure is not bonded to the iron, but it is located in such a way that oxygen must coordinate in a tilted configuration as shown. This is the preferred mode of attachment for O_2, because of the particular location of its bonding electrons. But poisons such as carbon monoxide prefer an 'upright' mode of attachment. Being forced to tilt as oxygen does, their binding is therefore weaker than it would otherwise be. CO does bind strongly to haemoglobin, and is therefore extremely poisonous as it inhibits the transport of oxygen in the blood. Without the particular structural feature just described, however, it would act much more strongly.

Another aspect illustrated in Fig. II.16 is the small change in position of the iron when oxygen binds to the haem. Attachment of O_2 changes the configuration of electrons in the Fe^{2+} ion from the 'high-spin' to the 'low-spin' state, and makes the ion slightly smaller. Thus it can move more into the plane of the organic ring, and in doing so it changes the position of the attached histidine. The fact that haemoglobin contains four sub-units each with a haem now becomes important, as the protein chain communicates the small structural change from one sub-unit to another. This is known as an *allosteric interaction*, and it results in a cooperative effect, whereby binding of one O_2 molecule increases the affinity of the molecule for another one.

This is important in the function of haemoglobin, as it allows the reversible binding to respond in a sensitive way to small changes in the effective O_2 concentration.

The affinity of iron-containing proteins for O_2 has certain dangers for biological systems. Dioxygen is not an especially good ligand molecule, compared with others such as CO. Although it is normally regarded as a pollutant, CO is formed in natural biological processes in the body. The binding mechanism described for haemoglobin must have evolved to cope with this. Another notorious poison is cyanide (CN^-). This can also bind to haemoglobin, but its major toxic effect (it causes death very rapidly) comes from its strong interaction with cytochrome oxidase. It binds to this and strongly stabilises the Fe^{3+} form, which prevents the enzyme from participating in the catalytic cycle which reduces O_2.

In other cases, the binding ability for other molecules has been utilised. Nitric oxide (NO) is, like CO, very toxic. It is however generated by cells in the body, and is used as a messenger to stimulate the relaxation of smooth muscle (see *nitrogen). It does this by binding to a haem group in a receptor enzyme *GMP cyclase* found on the cell surface. It is likely that an allosteric change takes place when NO binds, as with O2 in haemoglobin; this change of conformation 'switches on' the catalytic function, producing cyclic GMP (see *phosphorus), which in its turn causes the muscle to relax.

Krypton ($_{36}Kr$)

Krypton is a chemically inert noble gas element, occurring at about 1 ppm concentration in the atmosphere. It is extracted and used in small amounts for gas-filled lamps. Radioactive isotopes, including ^{85}Kr with a half-life of 11 years, are formed as fission products in nuclear reactors. As with *xenon, the chemical inertness of krypton makes it very hard to contain, and ^{85}Kr represents a substantial fraction of the radioactive substances emitted from nuclear power stations and during the reprocessing of *uranium reactor fuel. Its rapid dispersion in the atmosphere probably renders it relatively harmless.

Lanthanum and lanthanides ($_{57}La$–$_{71}Lu$)

Lanthanum is a strongly lithophilic element of fairly low abundance in the crust. It is always found in association with 13 of the 14 following *lanthanide* elements, cerium to lutetium, and with yttrium, the element above lanthanum in the periodic table. These elements all have a very similar chemistry. In the geological and the older chemical literature they are known as the *rare earth elements*. The main ores are the phosphate mineral *monazite*, $LnPO_4$, which also contains *thorium, and *bastnaesite*, $LnCO_3F$; the symbol Ln is used here to represent a combination of lanthanides. Lanthanides are used in some alloys, especially in the mixed form known as *mischmetal* which is used in lighter flints. Individual lanthanide ions, Ln^{3+}, have distinctive optical properties, and are used in lasers (especially neodymium) and in colour television screens.

Lanthanide minerals are very insoluble in water, so that the abundance of these elements in all natural waters is very low. They are not essential for life and are generally non-toxic, except for *promethium, which is a radioactive element that does not occur naturally. Radioactive isotopes of other lanthanides are formed as fission products from *uranium, and are therefore present in high-level radioactive waste from nuclear power stations, and in fallout from nuclear weapons. The most important isotopes involved are shown in Table II.28. In general, the relatively short half-lives of most of these, and the involatility and insolubility of compounds of the lanthanide elements, make these isotopes a less significant environmental hazard than those of some other elements.

Natural abundance patterns of the lanthanides

The graph in Fig. II.17 (a) shows the abundance of the lanthanide elements in the Earth's crust. Promethium is missing because of its radioactive instability. Otherwise there is a regular alternation in abundance, the elements of even atomic number being more common than those of odd number. This is a feature of the overall abundance of elements in the Universe (see §2.3), and is a result of the processes by which elements are made by nuclear reactions in stars. Nuclei with even numbers of protons are generally more stable than those with odd numbers, and are therefore made in greater yield.

Although they always occur together, the lanthanides show chemical trends which lead to some degree of chemical fractionation between different rocks. These are normally shown by plotting, not the absolute concentrations found, but the values

Table II.28 Lanthanides formed as fission products

Element	Isotope	Half-life
Lanthanum	^{140}La	1.7 d
Cerium	^{141}Ce	23 d
	^{143}Ce	1.4 d
	^{144}Ce	284 d
Praesodymium	^{143}Pr	13.7 d
Neodymium	^{147}Nd	11 d
Promethium	^{147}Pm	2.6 years
Samarium	^{151}Sm	93 years
Europium	^{154}Eu	16 years

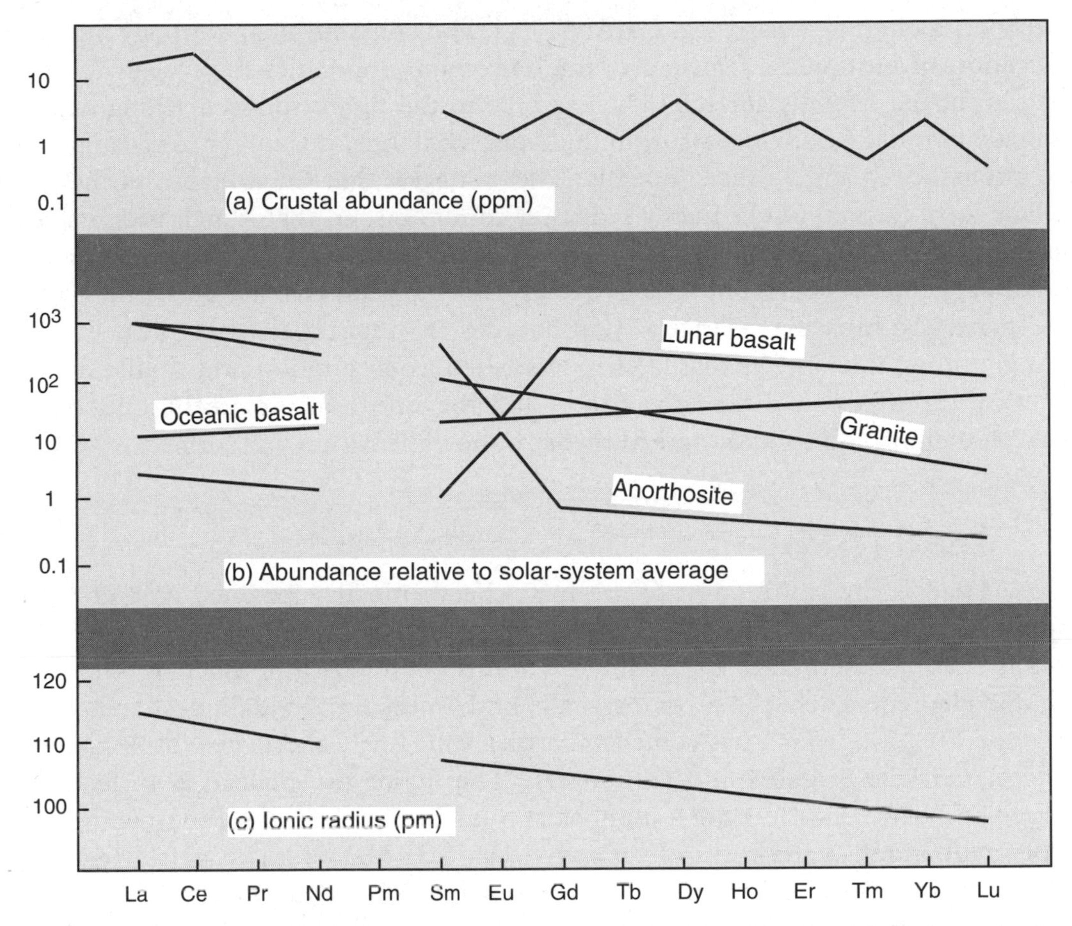

Fig. II.17 Abundance of the lanthanides. (a) The abundance of lanthanides in ppm in the continental crust. (b) Abundance curves showing the concentrations in different rocks, relative to the overall solar-system abundance deduced from meteorites. (c) Ionic radii of the Ln^{3+} ions. (Based on Henderson, 1982.)

'normalised' by dividing by the overall solar-system abundance as found in meteorites. Four plots of these relative abundances are shown in Fig. II.17 (b), for different rocks. The main trend in chemistry arises from a steady contraction in the size of the Ln^{3+} ions with increasing atomic number. The effect of this can be seen by comparing the distributions in oceanic basalt and in granite. The increasing fractionation of the earlier (and larger) lanthanides into granite is a manifestation of a general trend in the chemistry of the continental crust, shown also by the concentration of other large ions (e.g. of barium, thorium, and uranium; see §4.6).

Europium is peculiar among lanthanides in that it is relatively easily converted to the Eu^{2+} oxidation state under strongly reducing conditions. Eu^{2+} is similar in size to Sr^{2+}, and enters minerals as a replacement for Ca^{2+} and Sr^{2+}. Figure II.17 (b) shows two examples of so-called 'europium anomalies'. *Anorthosite* is a rock composed largely of calcium feldspars, e.g. $Ca[Al_2Si_2O_8]$, and contains an abnormally high concentration of europium. Especially large europium anomalies have been found in rocks from the Moon's surface. The example in the figure shows a strong negative anomaly in potassium feldspars from the lunar highlands; it is matched by large positive anomalies in some other minerals. These suggest that the minerals of the lunar surface were formed under highly reducing conditions, so that a high proportion of europium was in the Eu^{2+} form. Although this is understandable given the absence of an atmosphere on the Moon, there are many other aspects that are harder to explain. The generally high abundance of lanthanides in lunar rocks is typical of other strongly lithophilic and involatile elements, such as aluminium and titanium. Such chemical differences between the Moon and the Earth are not really understood, because there is no generally agreed theory of how the Moon was formed.

Lead ($_{82}Pb$)

Lead is the commonest of the toxic 'heavy metal' elements. A strong chalcophile, it is normally found in sulfides and forms the common mineral *galena*, PbS. This is the main source of the element, and also contains a number of other chalcophilic elements such as ⋆silver, ⋆arsenic, and ⋆mercury. Oxidation of sulfide ores gives the Pb^{2+} ion, which has some similarities with Ca^{2+}, and forms insoluble oxide minerals such as sulfates and carbonates. The major use of lead is in lead–acid storage batteries, but there are a number of other applications, as a roofing material, in pigments, and in the form of tetraethyl lead, $(C_2H_5)_4Pb$, as a petrol (gasoline) additive to improve the smoothness of combustion. The natural levels of lead in sea water and rivers is very low, but there is widespread pollution, as is reflected in the range of values shown in Table II.29.

Probably no region on the Earth's surface in free from lead pollution. Atmospheric aerosols deposited with snow in remote polar regions show a lead enrichment of up to 1000 times the normal crustal concentration. Atmospheric concentrations are particularly large in urban and industrial areas, and result partly from smelting and refuse incineration. By far the largest source, however, is from tetraethyl lead additives in petrol. 1,2-Dibromoethane (CH_2BrCH_2Br) is also added to scavenge the lead, and exhaust emissions contain $PbBr_2$ and many other lead compounds. Fortunately, the use of leaded petrol is declining.

Although atmospheric dusts form the major route for transporting lead in the environment, another source of exposure comes from its use in pipes for domestic water systems. This has been abandoned, but many such pipes remain in older dwellings. Soluble Pb^{2+} comes from the oxidation reaction

$$2Pb + O_2 + 4H^+ \rightarrow 2Pb^{2+} + 2H_2O.$$

Table II.29 Lead concentrations in the environment

Location	Concentration
Crust	14 ppm
Ocean:	
surface	0.02 ppb
deep	0.002 ppb
Fresh waters:	
natural	1–10 ppb
polluted	20–100 ppb
mining areas	100–1000 ppb
Drinking water (limit)	50 ppb
Soils:	
normal	2–200 ppb
polluted	1–30 ppm
Human body:	
average	0.5–2 ppm
blood	1–25 ppm
bone	0.2–10 ppm

The absorption of lead is larger with soft water supplies which contain little calcium bicarbonate. This is partly because the greater acidity of these waters facilitates the above reaction, and partly because hard waters contain anions which give insoluble lead salts, such as $PbCO_3$ and $PbSO_4$.

Lead is strongly held in soils, in the form of insoluble compounds such as carbonates, sulfates, and phosphates. High lead levels are still found in soils that were contaminated by lead mining in the fourth century BC. Some of this lead is taken up by plants, and food forms a significant source of the element in the diet.

Table II.30 shows some estimates of lead intake and absorption from various sources by adults and children in industrialised countries. The major source in both cases is from food, but children absorb more lead than adults. Smoking cigarettes can form a significant source for adults, and for children, street and house dust on the hands makes a contribution. These are average values, and particularly large intakes can occur from the air and dust near busy roads, and from water if lead pipes are still used in soft-water areas.

The toxicity of lead has become notorious, and is certainly not a recent problem. The element has been used widely for many centuries, in water supplies and as a component of pewter and glazes used for drinking vessels. The basic carbonate $Pb_3(CO_3)_2(OH)_2$, known as 'white lead', was also a commonly used white pigment, and a constituent of make-up worn in Elizabethan times. Lead poisoning may have been a common source of illness through much of European history, and could have

Table II.30 Sources of lead intake and absorption, with average daily uptake figures for inhabitants of industrial areas (from Fergusson, 1990)

	Source	Intake ($\mu g\ d^{-1}$)	Absorption ($\mu g\ d^{-1}$)
Adults	air	11	4.4
	food	150	15
	water	15	1.5
	smoking 30 cigarettes	15	6
Children	air	3.3	1.3
	food	100	50
	water	10	5
	dust	50	25

contributed to the downfall of the Roman Empire. Symptoms include anaemia, anorexia, abdominal pains, as well as neurological effects such as irritability, mood disturbance, and loss of coordination. There is now strong evidence that chronic lead exposure in children can lead to behaviour disturbance and impairment of learning, and that there can be long-term effects even if exposure to high lead levels is only temporary. The residence time of lead in the body is very long, one reason being that a significant amount is taken up in bones, where Pb^{2+} replaces some Ca^{2+} in calcium phosphates.

As with other chalcophilic heavy metals such as *cadmium and *mercury, lead combines strongly with –SH groups in enzymes, as well as competing with essential metals such as *calcium, *zinc, and *copper. Lead inhibits the biosynthesis of *iron-containing haem. This causes anaemia through a deficiency of haemoglobin in the blood, and may also disturb the respiratory metabolism performed in mitochondria, where haem proteins are also involved (see under *oxygen). A number of specific points of interference with haem synthesis have been identified. One of these is the incorporation of Fe^{2+} into the porphyrin system by the enzyme *ferrochelatase*. Another is the inhibition of the *δ-aminolevulinic acid dehydrase*, and a consequent rise in the concentration of ALA (δ-aminolevulinic acid, $NH_2CH_2COCH_2CH_2CO_2H$), which is the substrate for this enzyme. Although the basis for the neurotoxic effects of lead is not well understood, it is possible that ALA may interfere with the neurotransmitter GABA (γ-aminobutyric acid, $H_2NCH_2CH_2CH_2CO_2H$), which has a similar chemical structure.

Severe lead poisoning can be treated with compounds known as chelating agents, that form strong complexes with Pb^{2+} and assist its excretion from the body. These include EDTA (ethylenediamine tetraacetic acid, $(HO_2C)_2NCH_2CH_2N(CO_2H)_2$), normally supplied as the salt $CaNa_2EDTA$, 2,3-dimercaptopropanol ($HSCH_2C(SH)CH_2OH$), and D-penicillamine ($HSC(CH_3)_2CH(NH_2)CO_2H$). Repeated treatment is sometimes necessary, because of the slow release of lead stored in bone.

Lead isotopes and the age of the Earth

Lead has four stable isotopes, ^{204}Pb, ^{206}Pb, ^{207}Pb, and ^{208}Pb. The proportions of these present in natural lead vary appreciably according to the source, and this gives rise to a variation in the relative atomic mass of the element. Such a variability of atomic mass was known before the discovery of isotopes allowed its origin to be understood. It now known that ^{208}Pb is formed by the radioactive decay of *thorium, and ^{206}Pb and ^{207}Pb from the *uranium isotopes ^{238}U and ^{235}U, respectively. Lead from a region where it has been in contact with uranium minerals therefore contains a higher proportion of ^{206}Pb and ^{207}Pb. In principle, the amounts of these different isotopes present can be used to date the origin of the rocks concerned, although the application of this idea in practice is complicated, because of geochemical processes that may have caused the loss of some lead, or its mixing with lead from another source, since the formation of the rocks.[72] The data plotted in Fig. II.18 illustrate a relatively simple application.

The coordinates in Fig. II.18 show the amounts of ^{207}Pb and ^{206}Pb in various samples, relative to the non-radiogenic isotope ^{204}Pb. One of the points represents the concentrations found in iron meteorites, which contain lead in the form of sulfide, but no uranium. The isotope concentrations should not have changed with time therefore, and these ratios presumably give the original composition of lead in the solar system. Other points in the figure are for lead taken from stony meteorites, which contain oxide minerals and have some uranium present. The increased ^{206}Pb and ^{207}Pb content has come from the decay of ^{238}U and ^{235}U, respectively, since the formation of the meteorite. The larger the uranium content, the more radiogenic lead has been produced. The two uranium isotopes have dif-

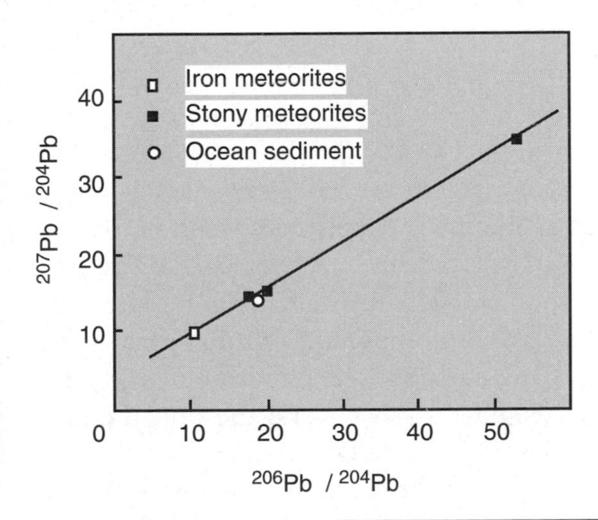

Fig. II.18 Lead isotope ratios for some samples from meteorites and terrestrial sediments. The slope of the line suggests a common age of about 4.55×10^9 years. (See Faure, 1986).

72 Mason and Moore (1982) and Faure (1986) give details of dating methods using lead isotopes.

ferent half-lives, and the relative amounts of ^{206}Pb and ^{207}Pb depend on how long the radioactive decay has been going on. Thus the slope of the line in Fig. II.18 can be used to determine the age of the meteorites, which is found to be 4.55 billion years. Another point on the graph shows the composition of lead from modern sediments on Earth; as this falls on the same line, it indicates the same age for the Earth itself. In reality, not all lead on Earth fits this, because it comes from various sources, some (such as PbS) containing very little uranium, some from oxide minerals which contain much more. But although the general interpretation of lead isotope data is complicated, they are consistent with the same overall age for the Earth. It is generally thought that this represents the age of the whole solar system, including the sun.

Lithium ($_3Li$)

A relatively uncommon element, lithium is chemically similar to both sodium and magnesium, and is widely though thinly distributed in magnesium minerals. It occurs in the rare mineral *spodumene*, $LiAlSi_2O_6$, and is moderately abundant in sea water. Lithium and its compounds are used in batteries, industrial lubricants, glasses, lightweight alloys, and in nuclear weapons. It is not essential for life, but has an effect on the nervous system, and has been found effective in the long-term treatment of manic–depressive illness.

Lutetium ($_{71}Lu$)

The heaviest, and rarest, member of the lanthanide (rare earth) series. See under *lanthanum.

Magnesium ($_{12}Mg$)

Magnesium is the third most abundant element on the Earth overall (above silicon) and the sixth in the crust. The Earth's mantle is largely composed of magnesium silicates such as olivine, $(Mg,Fe)_2SiO_4$, and pyroxene, $(Mg,Fe)SiO_3$. The preponderance of these exerts a control over the distribution of elements between the mantle and crust (see §4.6). Magnesium is a major component of most rock types, and occurs especially as sedimentary and metamorphosed deposits of *dolomite*, $MgCa(CO_3)_2$. Evaporite minerals such as the sulfate *epsomite* and the mixed halide *carnallite* are also found. Some examples are listed in Table II.31.

The fibrous *serpentine* minerals are formed by the weathering of igneous rocks containing olivine, and are an important source of asbestos. Soils containing these minerals, found in areas of California, Scotland, and Zimbabwe, also have high

Table II.31 Some magnesium-containing minerals

Name	Chemical formula
Olivines: forsterite	Mg_2SiO_4
Pyroxenes: enstatite	$MgSiO_3$
Talc	$Mg_6[Si_8O_{20}](OH)_4$
Serpentine (chrysotile)	$Mg_3[Si_2O_5](OH)_4$
Dolomite	$MgCa(CO_3)_2$
Epsomite	$MgSO_4.7H_2O$
Carnallite	$K_2MgCl_4.6H_2O$

Table II.32 Magnesium concentrations in the environment

Location	Concentration
Crust	2.3 %
Sea water	1200 ppm
Rivers (average)	4 ppm
Human body:	
average	300 ppm
blood	38 ppm

concentrations of elements such as *chromium and *nickel which are toxic to many plants. Serpentine soils are known for their impoverished flora, which is confined to a few species that have developed a resistance to these elements.

Magnesium is a very electropositive metal, and must be extracted, like aluminium, by electrolysis of molten salts. It is used in lightweight alloys, especially for aircraft production. The oxide MgO is a very useful refractory material, and finds wide application as a lining for industrial plant which operates at high temperatures, for example blast furnaces for iron production.

Table II.32 shows some typical magnesium concentrations found in the environment. Mg^{2+} ions are readily released in the weathering of rocks, and occur in freshwater at a concentration comparable to that of the alkali metals Na^+ and K^+. The residence time for Mg^{2+} in the ocean is very long (10^7 years), and it is the second most abundant cation (after Na^+) in sea water.

Magnesium is an essential element for all life, the Mg^{2+} ions being (unlike Na^+ and Ca^{2+}) fairly evenly distributed within cells and in extracellular fluids. Many biological phosphates, including ATP (see under *phosphorus) are associated with Mg^{2+} ions. Magnesium is also a component of chlorophyll, essential for photosynthesis by green plants. The structure of this molecule, shown in Fig. II.19, has Mg^{2+} coordinated by four nitrogen atoms in a tetrapyrrole ring similar to that found in *iron-containing

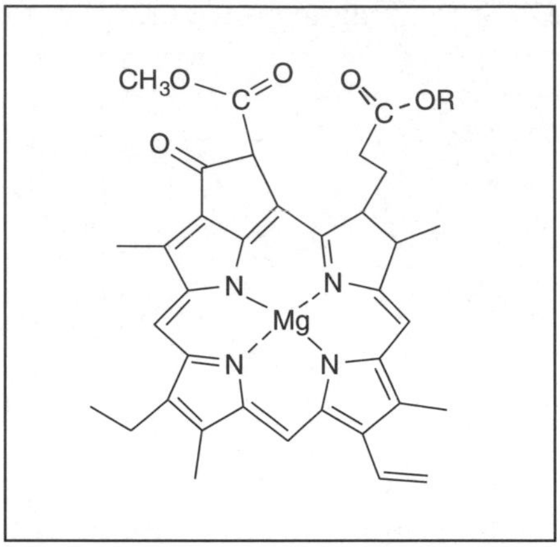

Fig. II.19 Structure of chlorophyll.

haem. This is unusual for a strongly lithophilic element from the early groups in the periodic table, where oxygen is generally the preferred coordinating atom. The function of Mg^{2+} in chlorophyll appears to be to change the wavelength at which light is absorbed by the molecule, so that it is in an optimal part of the visible spectrum. Photosynthesis involves other elements such as *manganese, and is described in more detail under *oxygen.

Manganese ($_{25}Mn$)

Manganese is a relatively common element on Earth. A lithophilic metal, it is widely distributed as a component of many silicate minerals. Weathering may release the soluble Mn^{2+} ion, which is occasionally found as the carbonate $MnCO_3$; normally, Mn^{2+} is oxidised by air to the 3+ or 4+ state, which forms very insoluble compounds such as *manganite*, MnO(OH), and *pyrolusite*, MnO_2. These are important constituents of soils, where like iron oxides and clay minerals, they retain many minor elements by adsorption (see §4.4). The dissolved concentration of manganese in natural waters is generally low, although it can rise when Mn^{2+} is formed under anoxic conditions. Large deposits of 'manganese nodules' occur on the sea bed, containing manganese oxides and other desirable elements. The soluble Mn^{2+} from which these are derived comes partly from hydrothermal vents in the sea floor. Although the oxidation to insoluble oxides is thermodynamically favourable, it is thought to be very slow under inorganic conditions and is probably assisted biologically. Manganese is used in large quantities, mainly as a component of steels.

Manganese is an essential trace element in life. Its best-known role is in photosynthesis (see *oxygen), where the enzyme which oxidises water to O_2 contains a cluster of four manganese atoms. Figure II.20 shows the structural transformation thought to be involved. It is a component of several other metalloenzymes, some of which (as in photosynthesis) utilise its ability to change oxygen states. The functions include the decomposition of superoxides and peroxides.

Mercury ($_{80}$Hg)

A very rare element in the crust, mercury is a strong chalcophile and is found as its sulfide mineral *cinnabar*; as the element is rather volatile, deposits of this mineral are frequently found associated with volcanoes. Mercury is the only metallic element liquid at room temperature, and has been known since ancient times as 'quicksilver'. It readily forms alloys known as *amalgams* with many other metals, and has been used for the extraction of precious metals such as gold. A major industrial use is as an electrode in the *Castner–Kellner process* for the electrolytic manufacture of NaOH and chlorine from brine, but this application has declined owing to its toxicity. Mercury is used in thermometers and barometers, in dental amalgams (mostly with tin) for filling teeth, and for electrical and electronic applications such as street lights. Mercury compounds are also used as germicides and fungicides, and as industrial catalysts in the manufacture of plastics such as PVC.

Natural mercury concentrations in the environment are very low (see Table II.33). A significant amount of cycling may occur through the atmosphere, through the release of the volatile element from volcanoes, and by the production of dimethyl mercury, $(CH_3)_2Hg$, from marine life. Some concentration in the food chain occurs, and even the natural concentrations found in some fish such as tuna are appreciable. However, these have been greatly enhanced in some localised areas by the industrial use of mercury and its compounds. Coastal pollution, for example in Minamata Bay in Japan, results from industrial processes using mercury.

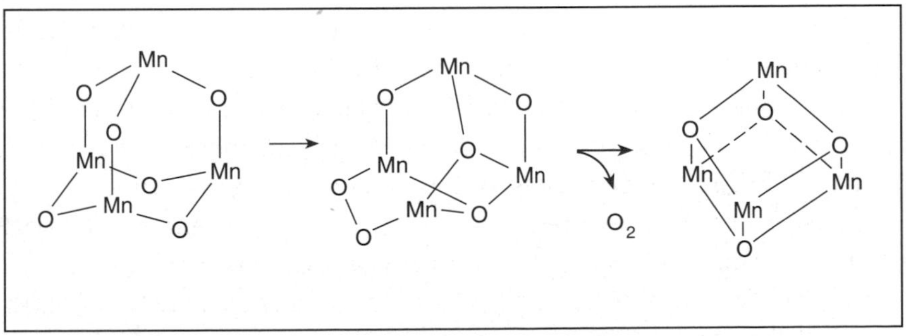

Fig. II.20 Manganese in photosynthesis, showing how a cluster containing four manganese atoms is thought to rearrange as oxygen is evolved.

Table II.33 Mercury concentrations in the environment

Location	Concentration
Crust	50 ppb
Sea water:	
normal	< 0.01 ppb
polluted coasts	0.05–2000 ppb
Fresh waters	0.02–0.1 ppb
Drinking water (permitted limit)	1 ppb
Soils:	
normal	0.01–0.5 ppb
contaminated	0.5–50 ppb
Food:	
vegetables and cereals	1–50 ppb
meat	1–50 ppb
fish	30–1500 ppb
Human body	0.2–2 ppm

Significant soil pollution has also occurred from the use of mercury compounds as pesticides.

Mercury is an extremely toxic element. The methyl mercury compounds $[CH_3Hg]^+$, $(CH_3)_2Hg$, and CH_3HgSCH_3 are especially dangerous, as they are very easily absorbed by the body and can pass into the brain. The vapour of elemental mercury can also be absorbed through the skin. Inorganic compounds such as *calomel*, $HgCl_2$, have been used widely as medicines, including the (probably ineffective) treatment of syphilis in the eighteenth and nineteenth centuries. Although they are toxic, much less mercury is absorbed than from some other sources. The most characteristic and serious symptoms of mercury poisoning are neurological ones resulting from severe brain damage: they include disturbed coordination, tremors, deafness, and failure of peripheral vision. Mercury poisoning was common among goldsmiths in previous centuries, when gold amalgam was widely used for gilding copper and silver objects. Hat-makers also used mercury, and the expression 'mad as a hatter' may have come from their symptoms. In the twentieth century several outbreaks of mercury poisoning have affected hundreds or even thousands of people at a time. Some of these, such as at Minamata in Japan, occurred among populations having a diet rich in fish from heavily polluted coasts. Other serious cases, in Iraq and several other countries, have been the result of eating grain treated with methyl mercury pesticides: this grain was intended for planting, not eating directly, and was treated to prevent fungal attack. There is a worry that dentists, who make and administer dental fillings containing mercury, could receive elevated doses; the direct absorption of mercury from fillings in the mouth is not thought to be significant.

Molybdenum ($_{42}Mo$)

Molybdenum is a relatively uncommon element in the crust. It has chalcophilic tendencies, the most important source being the sulfide *molybdenite*, MoS_2, which occurs in association with copper ores. However, on oxidation the molybdate ion MoO_4^{2-} is formed, which is also found as insoluble compounds such as $CaMoO_4$ and $PbMoO_4$. Molybdenum is an important metal, being used in alloys, as an electrode, and in catalysts for the *dehydrosulfurisation* reaction used to remove *sulfur compounds from oil. The soluble molybdate ion has a fairly long residence time in the ocean.

Molybdenum is an essential trace element, as a constituent of several enzymes where the possibility of cycling between Mo(+4) and Mo(+6) states is important. These catalyse oxidation and reduction reactions of non-metallic elements with two electrons being transferred simultaneously. The best-known example is *nitrogenase*, the enzyme used by bacteria that fix atmospheric N_2 into organic nitrogen; the reduction of sulfate and nitrate are other examples. Molybdenum is thus essential for the biological utilisation of *nitrogen and *sulfur, and in the environmental cycling of these elements.

Neodymium ($_{60}Nd$)

An element of the lanthanide series, discussed under *lanthanum.

Neon ($_{10}Ne$)

Neon is found at a concentration of 18 ppm in the atmosphere, and is extracted in small quantities for use in ornamental lighting ('neon lights'). It is chemically and biologically inert.

Neptunium ($_{93}Np$)

Neptunium is the first of the transuranium elements, which do not occur naturally because they are radioactive with no long-lived isotopes. ^{237}Np and ^{239}Np are produced by the neutron bombardment of *uranium in nuclear reactors. The second of these isotopes has a half-life of only 2.35 days, and decays to ^{239}Pu, the most important isotope of *plutonium. ^{237}Np has a longer half-life (2.14×10^6 years), and is obtained in small quantities from the reprocessing of nuclear fuel. Subjected to further neutron bombardment, it is used to produce ^{238}Pu for use as an energy source. Although neptunium is present in nuclear fallout, and can potentially be released from accidents in nuclear reactors and reprocessing plants, the relatively small amounts present make it a less important hazard than many other radioactive elements.

Nickel ($_{28}$Ni)

Nickel is probably the seventh most abundant element on Earth overall, making up about 10 per cent of the core. A strong siderophile, it is much less abundant in the crust. It occurs as both sulfide and oxide minerals. One major source is *pentlandite*, of approximate composition $(Ni,Fe)_9S_8$, but also containing significant amounts of *cobalt. Weathering of sulfide ores liberates the Ni^{2+} ion, which is similar in size to Mg^{2+} and can be found in *magnesium-containing minerals, especially silicates. Nickel is therefore relatively abundant in highly basic rocks (see §4.6), and in *serpentine* soils derived from them by weathering. The silicate mineral *garnierite*, $(Mg,Ni)_6[Si_4O_{10}](OH)_8$, also forms an important source. Nickel is an important component of many ferrous and non-ferrous alloys, and is also used for electroplating.

Normal environmental concentrations of nickel are low (see Table II.34), but large amounts can be found in some naturally derived serpentine soils, and surrounding mining and smelting areas, such as the major nickel-producing site at Sudbury, Ontario.

Nickel is an essential trace element, although its role in mammals (including humans) is very limited, and may be confined to its presence in a single metalloenzyme, *urease*, which catalyses the decomposition of urea, $(NH_2)_2CO$, to ammonia. It is more important in many anaerobic bacteria which derive their energy supply by metabolising dihydrogen, H_2, and liberating methane, CH_4. The coenzyme F-430 contains nickel coordinated in a tetrapyrrole ring similar to that in haem (see *iron). It is involved in catalysing the reaction

$$4\,H_2 + CO_2 \rightarrow CH_4 + 2\,H_2O.$$

Nickel is a fairly toxic element. Many plant species cannot grow on contaminated soils, although there are especially tolerant species, some of which can accumulate high concentrations. In humans, nickel-containing dust has been recognised as a

Table II.34 Nickel concentrations in the environment

Location	Concentration
Crust	80 ppm
Sea water:	
surface	0.1 ppb
deep	0.5 ppb
Fresh waters	1 ppb
Soils:	
normal average	50 ppm
serpentine	0.1–2 %
mining areas	100–500 ppm
Human body	0.1–1 ppm

cause of occupational cancers, including lung cancer. The metal is sometimes used in jewellery such as ear-rings, and can cause dermatitis; it appears that some people are especially sensitive, or can become sensitised by prolonged contact with the element. The volatile compound nickel tetracarbonyl, $Ni(CO)_4$, which is used in the extraction of the element by the *Mond process*, is especially poisonous.

Niobium ($_{41}$Nb)

A relatively uncommon element in the Earth's crust, niobium is a lithophile and is found in the mineral *columbite*, $(Fe, Mn)Nb_2O_6$. This also contains the chemically similar element *tantalum, and deposits are often found in association with those of *tin. Niobium is used in steels for high-temperature applications, and in superconducting magnets. Its natural minerals are extremely insoluble in water, and the dissolved concentrations are very low. It has no biological role.

Nitrogen ($_7$N)

Nitrogen is not a common element overall on Earth (see Table II.35) although it is important in life and is the major constituent of the atmosphere. It occurs in the crust mainly as the ammonium ion, NH_4^+, which is found in low concentrations in some minerals in place of alkali ions such as K^+. The ion is very soluble in water and is liberated in weathering, but no concentrated deposits of ammonium compounds are found, probably because nitrogen is taken up avidly by living organisms. Decomposition at high temperatures within the crust gives the very stable dinitrogen molecule, N_2, which is abundant in many volcanic gases, and

Table II.35 Nitrogen concentrations in the environment

Location	Concentration
Crust	25 ppm
Sea water:	
surface	0.1 ppb
deep ocean	0.5 ppm
Fresh waters:	
unpolluted	< 0.5 ppm
agricultural land	1–100 ppm
Drinking water (limit as NO_3^-)[a]	11 ppm
Atmosphere	78 %
Human body	2.5 %

[a]The recommended limit is 50 ppm NO_3^-, the value shown in the table being the corresponding N content.

makes up 79 per cent by volume of the Earth's atmosphere. Nitrogen oxides are also made by natural processes. They form the nitrate ion, NO_3^-, in solution, and hence some deposits of the nitrate minerals *nitre* (or *saltpetre*), KNO_3, and *Chile saltpetre*, $NaNO_3$. Nitrate concentrations in unpolluted natural waters are very low, especially in the surface layers of the ocean, because of the demand for this element by life. The common use of nitrogen-containing fertilisers in agriculture can however lead to much higher nitrate concentrations in some areas, and there is concern about its level in drinking water.

Nitrates are the strongest naturally occurring oxidising agents and have been used for many centuries in explosives such as gunpowder. Most nitrogen used today is obtained from the atmosphere by liquefaction and fractional distillation. Liquid nitrogen (boiling point 77 K, or –196°C) is used for applications requiring extreme cold, but the major use of the uncombined element is as an inert gas to prevent oxidation by air, in steel making and in many other metallurgic and chemical industries. Synthetic ammonia, produced by the catalysed reaction of N_2 with H_2 in the *Haber process*, and nitric acid, manufactured by oxidation of ammonia, are among the most important of all industrial chemicals. Their uses include the production of explosives, rocket fuels, plastics, and other petrochemicals, but some 80 per cent of all industrially produced nitrogen compounds are fertilisers such as ammonium nitrate, NH_4NO_3.

The redox chemistry of nitrogen

Nitrogen, like *carbon and *sulfur, is one of the elements that can exist in the environment in a wide range of oxidation states (see also §§3.1 and 4.2). Typical compounds in different oxidation states are listed in Table II.36. States intermediate

Table II.36 Oxidation states of nitrogen

Oxidation state	Name of compound	Formula
–3	ammonia	NH_3
	ammonium ion	NH_4^+
	amino acid, e.g. glycine	$H_2NCH_2CO_2H$
–2[a]	hydrazine	N_2H_4
–1[a]	hydroxylamine	NH_2OH
0	dinitrogen	N_2
+1	nitrous oxide	N_2O
+2	nitric oxide	NO
+3	nitrite ion	NO_2^-
+4	nitrogen dioxide	NO_2 (N_2O_4)
+5	nitrate ion	NO_3^-

[a]Compounds with the –2 and –1 states are not normally found in the natural environment.

between zero (N_2) and –3 (NH_3) are not normally encountered, but all the other forms shown are found in nature, and the transformations between them form an important part of the nitrogen cycle.

Figure II.21 shows some redox reactions of nitrogen, on a scale indicating their oxidising or reducing power. The left-hand scale gives the free energies of oxidation of compounds in dry air, and the right-hand one the electrode potentials in water at pH 7. They are arranged with more oxidising conditions towards the top; reactions shown lower down are more reducing, the molecules concerned being more readily oxidised. The top shaded region, where $\Delta G > 0$, corresponds to conditions even more oxidising than atmospheric O_2. All reactions below this are thermodynamically favoured in the atmosphere. For example the N_2/NH_3 reaction is shown at $\Delta G =$ –210 kJ per mole, or –0.25 V, so that ammonia should oxidise to dinitrogen in air. In dry air no oxides of nitrogen are thermodynamically stable, as the NO_2/N_2 reaction

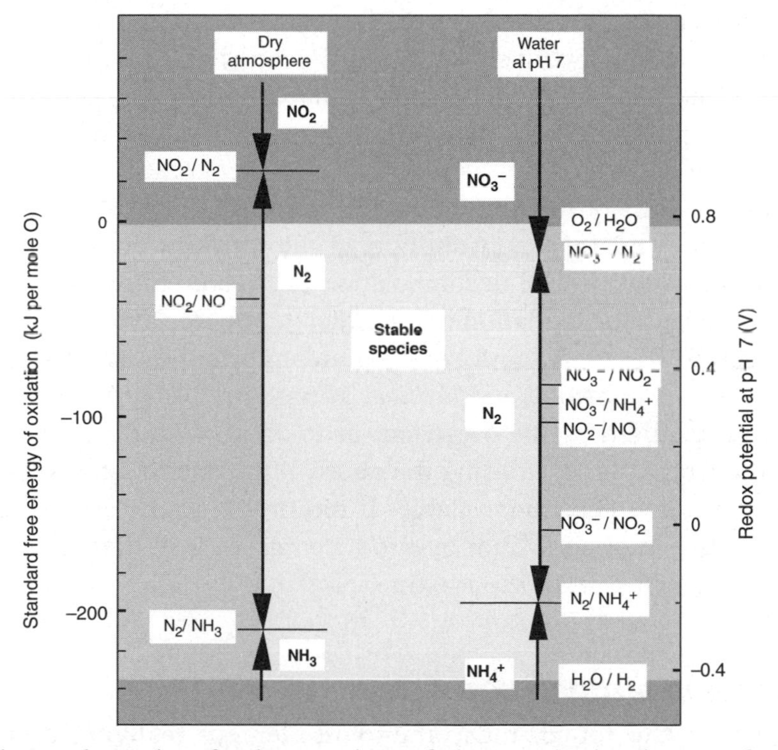

Fig. II.21 Thermodynamics of redox reactions of nitrogen, in the absence of water (left) and in water at pH 7 (right). The left-hand scale shows the Gibbs free energy change, ΔG, per mole of O consumed in the oxidation reactions shown; the right-hand scale shows the corresponding electrode potential. The heavy vertical lines show the regions of stability of the species listed in the central region. Reactions between thermodynamically unstable species are shown at the side. The dark-shaded regions at the top and bottom indicate conditions outside the region of thermodynamic stability in the natural environment.

has positive free energy. In water however, the NO_3^-/N_2 potential is below that for O_2/H_2O, so that nitrogen should be oxidised to nitrate.

These thermodynamic predictions are important, but they give no indication of how *fast* the predicted reactions take place. Many of the redox reactions of nitrogen have large activation barriers, and are extremely slow under normal conditions. One reason for this is the unusual strength of the triple bond in the dinitrogen molecule (955 kJ per mole); another is that oxidation of N_2 to NO_3^- is impossible in a single step, and must take place through intermediate species such NO and NO_2, which are not thermodynamically stable in air. Small amounts of these oxides are produced at high temperatures, such as in lightning and in internal combustion engines, and their chemistry is important in the atmosphere. NO_x compounds are rapidly oxidised to nitrate, which is very soluble in water and so quickly washed out of the atmosphere. Deposits of nitrate salts $NaNO_3$ and KNO_3 are formed by evaporation of nitrate in solution, but relatively little nitrogen occurs in this form, as it is an essential element for life, and generally in rare supply—again a consequence of the stability of dinitrogen, and of the difficulty of converting this to the reduced form necessary in biology. As discussed below, the biological fixation of nitrogen is a very important, and still imperfectly understood reaction. When nitrate is available, it is often used in preference to dinitrogen.

Many other reactions of nitrogen compounds are performed by life, especially by specialised bacteria, and their contribution to the nitrogen cycle greatly outweighs the purely inorganic parts. The decomposition of organic nitrogen compounds initially yields ammonia. The oxidation of ammonia to nitrate (bypassing the thermodynamically favoured $NH_3 \rightarrow N_2$ conversion) is an important energy-giving reaction for some bacteria, and is known as *nitrification*. Nitrate itself may be reduced, either to organic nitrogen required by the organism, or to dinitrogen; this *denitrification* reaction is also an energy source, utilising the ability of nitrate to oxidise organic matter in situations where free O_2 is unavailable. It returns N_2 to the atmosphere and thus completes the nitrogen cycle. Minor by-products can arise in these reactions, including the nitrite ion, NO_2^-, and nitrous oxide, N_2O.

Biological nitrogen chemistry

Nitrogen is the fourth most abundant element in living organisms, after carbon, oxygen, and hydrogen: a person weighing 70 kg contains about 1.8 kg of nitrogen. A small selection of biological molecules containing nitrogen is shown in Fig. II.22. Amino acids, which have the general formula $H_2NCHRCO_2H$, are an essential component of all proteins. The three illustrated in Fig. II.22 have extra nitrogen in their side-chains. Also shown are the bases which form part of the genetic code in DNA.

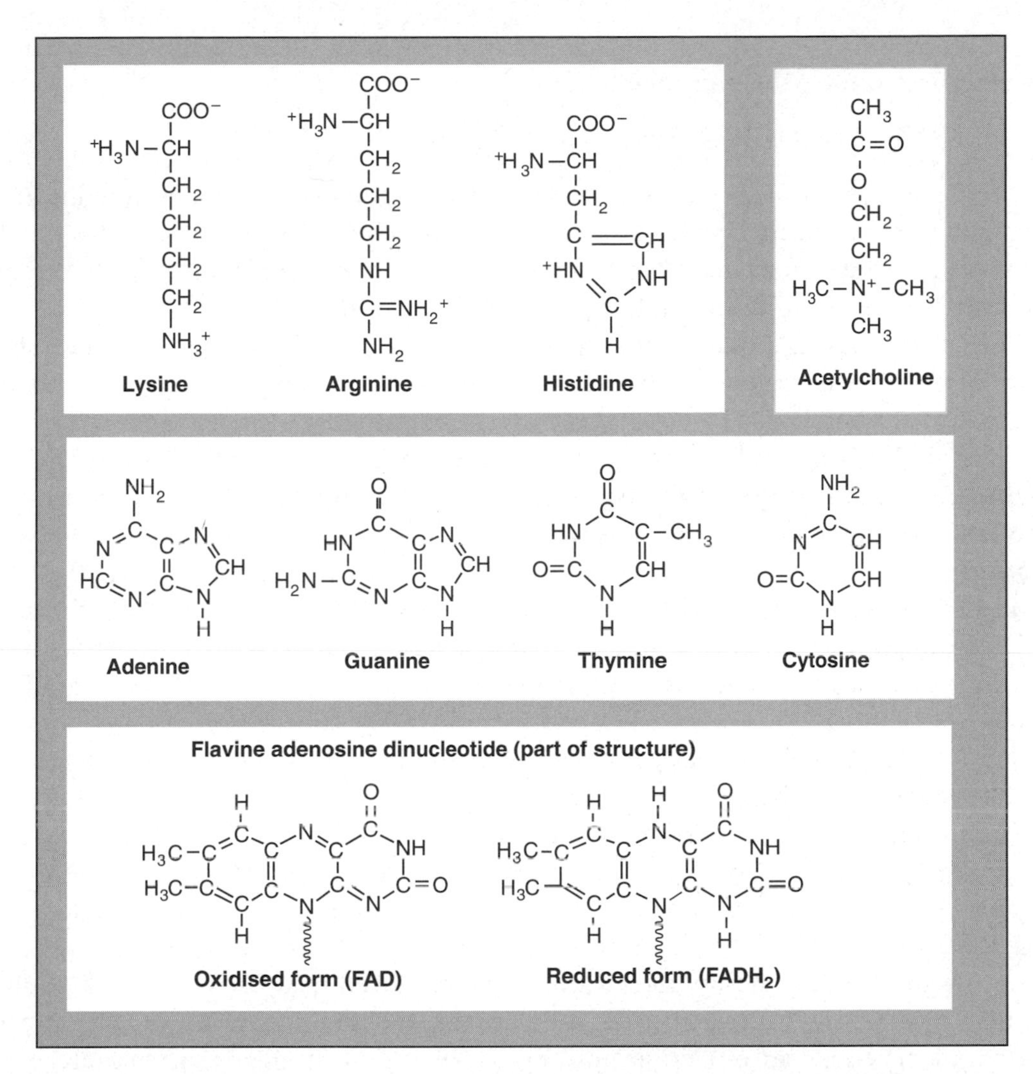

Fig. II.22 Some important biological nitrogen compounds. Showing three amino acids (lysine, arginine, and histidine) with extra nitrogen atoms in side-chains; the four bases (the purines adenine and guanine, the pyrimidines thymine and cytosine) forming the genetic code in DNA; the neurotransmitter acetylcholine; and part of FAD, a molecule performing an important hydrogen transfer function in biological redox reactions.

Most biological nitrogen is found in the reduced (–3) oxidation state, bound directly to carbon and hydrogen. The resulting amine group ($-RNH_2$) is basic and generally protonated to give $-RNH_3^+$ at neutral pH. Thus nitrogen provides one of the few sources of positively charged groups in biological molecules, as most other ionisable groups—e.g. RCO_2H, $ROPO_3H_2$, and $ROSO_3H$—are acidic and generally carry negative charges at pH 7. The basic properties of nitrogen are important in other ways. For example, proteins are formed by the condensation of amino acids, the basic amino group of one molecule combining with the acid carboxyl group of the next one to form the peptide bond:

$$RNH_2 + HOOCR' \rightarrow RNH\text{–}COR' + H_2O$$

(see Fig. I.22 for a protein structure). Nitrogen also acts as a good coordinating atom for many metal ions, such as iron, copper, and zinc. For example, *iron in haemoglobin is bound to four nitrogen atoms in the planar haem molecule, and a fifth nitrogen from a histidine group in the globin (protein) chain (see Fig. II.16).

The N–H bond is polar, with a partial positive charge on the hydrogen. This enables it to form hydrogen bonds to other polar groups, especially carbonyls (>C=O); such bonds have a crucial role in determining the overall shapes of proteins and nucleic acid, and indeed the operation of the genetic code depends on a specific pairing through hydrogen bonding of adenine with thymine, and of guanine with cytosine (see *hydrogen for more details).

Nitrogen compounds are important in a wide variety of metabolic processes. The purine bases are constituents of molecules such as ATP and GMP (see *phosphorus). Although the nitrogen-containing parts of ATP itself do not have reactive function, in other cases nitrogen is crucial. Fig. II.22 shows part of the structure of the flavine molecule FAD, and its reduced form $FADH_2$. These compounds, and the analogous NAD^+–NADH pair, have an essential role in metabolism as 'hydrogen transfer' agents in energy-producing redox reactions.

The molecule acetylcholine performs another type of function: that of a transmitter, or 'signalling' molecule. It is released when a nerve impulse arrives at the synaptic junction with another nerve cell or with a muscle. The transmitter molecule diffuses across the space between the two cells, and is recognised by special receptors on the next cell, which is thereby stimulated into activity. Acetylcholine is only one such neurotransmitter, but among the most important. The proper control of nervous and muscular activity requires that the transmitter molecule is inactivated rapidly when the stimulus has disappeared. In the case of acetylcholine, a particular enzyme—*acetylcholinesterase*—performs this function by removing the acetyl (CH_3CO–) group. A number of poisons and drugs, including some natural alkaloid poisons and organic *phosphorus pesticides and nerve gases, act by combining with the acetylcholinesterase itself and preventing it from acting.

Recently another, and surprising, messenger molecule containing nitrogen has been recognised. This is the reactive molecule nitric oxide, NO, the most important function of which is to cause *smooth muscle* to relax.[73] Such muscle is important in functions outside conscious control, such as the lining of the digestive tract and of

73 Young (1993).

blood vessels. By controlling the relaxation of blood vessels, NO is thus involved in the regulation of blood pressure. Another function is to control the supply of blood to erectile tissue such as the penis. NO must count as the simplest molecule with a sophisticated biological function. It acts by binding to an *iron atom in the enzyme *GMP cyclase*. Unlike many messenger molecules, it is non-polar and can diffuse easily through cell membranes. It is also very reactive, combining with oxygen to make NO_2, which then generates nitrite and nitrate. Thus no special enzyme is needed to get rid of it after it has performed its messenger role.

One of the most important reactions performed by biological systems is that of *nitrogen fixation*. The conversion of dinitrogen to reduced (NH_3 or RNH_2) form is not demanding in a thermodynamic sense: as shown in Fig. II.21 it occurs at a redox potential of about −0.2 V at pH 7, a potential typical of many biological redox systems. There is however an extremely large activation barrier required to break the bond in N_2, and very few organisms have acquired the ability to perform this reaction. Those that have are all micro-organisms—bacteria and blue-green algae—and include ones that live symbiotically in special nodules in the roots of nitrogenous plants. Nitrogen-fixing organisms are globally important in controlling the supply of nitrogen to all life, and it is something of a puzzle that this ability has not evolved more widely. It may be that limitations in the supply of other crucial elements, such as phosphorus, have prevented this.

Nitrogen fixation is performed by the enzyme *nitrogenase*. This contains *molybdenum, as well as iron and sulfur, but many details of the reaction are still uncertain. Nitrogenase works under strongly reducing conditions and liberates dihydrogen. Possibly one reason for this is that the nitrogen-fixing reactions are strongly inhibited by dioxygen, which binds to the active site. Leguminous plants maintain a very low O_2 concentration in their root nodules by synthesising a special haemoglobin molecule, *leghaemoglobin*, which has a very high affinity for oxygen. Strongly reducing conditions may act to remove any remaining free O_2. But the major difficulty with nitrogen fixation is the kinetic barrier resulting primarily from the strong bond in dinitrogen, and this may be the main reason for the presence of a very low redox potential (around −0.4 V, much less than the thermodynamically required −0.2 V).

Nitrate is assimilated more easily than N_2, and is used in preference if it is available, either from natural sources (nitrification) or as a fertiliser. The reduction of nitrate to organic compounds can give the toxic nitrite ion as an intermediate step. High nitrate levels in drinking water are thought to be harmful to young infants, through the production of nitrite in their stomachs; this ion is capable of oxidising the Fe^{2+} ion in haemoglobin to the Fe^{3+} form (methaemoglobin), which is incapable of transporting oxygen in the blood, resulting in the so-called 'blue baby syndrome'.

Atmospheric nitrogen chemistry

Dinitrogen is the dominant constituent of the Earth's atmosphere, making up 78.1 per cent by volume of dry air. Table II.37 shows the concentrations of other nitrogen compounds in the atmosphere, together with rough estimates of the annual inputs and the residence times. The annual flux of nitrogen and its compounds is similar to that for sulfur, and orders of magnitude less than for oxygen and carbon. Nitrogen dominates the atmosphere partly because the very unreactive N_2 molecule has an abnormally long residence time, of the order of 100 million years. Other nitrogen compounds enter the atmosphere in quantities comparable to N_2, but survive only briefly.

Nitrogen in the atmosphere came originally from volcanic outgassing, through the decomposition of ammonium compounds under heat in the crust. N_2 is an important constituent of the atmospheres of Venus and Mars, and must derive from a similar source. Volcanic emissions still contain significant amounts of nitrogen, but this source is now much less important than biological denitrification, which occurs both on land and in the deep sea.

Only UV radiation of exceptionally short wavelength is able to break the strong triple bond in N_2 (bond energy 945 kJ per mole), and this process is insignificant even in the stratosphere (see §4.3). Only at very high altitudes (above 100 km) can dissociation or ionisation occur, leading to the formation of some NO. Some NO is produced in mixtures of N_2 and O_2 at high temperatures, which occur naturally in lightning flashes. The other main natural route by which dinitrogen is removed from the atmosphere is in biological fixation.

The next compound listed in Table II.37 is nitrous oxide, N_2O, which is produced mostly as a minor side-product by nitrifying and denitrifying bacteria in soils and oceans. The concentration is currently increasing, possibly because of the increased global use of nitrogenous fertilisers. The molecule is also rather inert, and undergoes no chemical reactions in the troposphere. It is however a potential 'greenhouse gas'

Table II.37 The principal atmospheric nitrogen species

Molecule	Annual input (10^9 kg N year^{-1})[a]	Average concentration[b]	Residence time[b]	Main sources
N_2	200	78 %	10^8 years	denitrification
N_2O	10	0.3 ppm	100 years	nitrification, denitrification
$NO + NO_2$ (NO_x)	100	0.05 ppb	1 d	industrial, lightning
NH_3	50	< 1 ppb	10 d	biological excretion

[a]Very approximate values, quoted by Schlesinger (1991). [b]From Ramanathan *et al.* (1987).

like CO_2, although it makes only a minor contribution. Nitrous oxide passes eventually into the stratosphere, where it reacts with excited *oxygen atoms produced by the photolysis of ozone:

$$(O_3 + h\nu \rightarrow O_2 + O^\star);$$

$$O^\star + N_2O \rightarrow 2NO.$$

N_2O is thus an important source of reactive NO_x species in the stratosphere. One should be concerned about an increase in its concentration, because of its contribution to global warming, and its influence on the ozone layer.

Nitric oxide (NO) and nitrogen dioxide (NO_2) are much more reactive. Because of the facile interconversion of these two molecules (see below) they are generally grouped together, and called NO_x (note that this designation does *not* include the more inert N_2O). The main natural source of NO_x in the troposphere is from lightning, but these oxides are also produced during high-temperature combustion, especially in power stations and internal combustion engines.

NO_2 is readily dissociated by sunlight (see §4.3). Oxygen atoms are produced which react rapidly to form ozone:

$$NO_2 + h\nu \rightarrow NO + O;$$

$$(O + O_2 \rightarrow O_3).$$

This is the main source of ozone in the troposphere, and hence also of OH radicals (see *oxygen). The conversion of NO back to NO_2 is also rapid, but more complicated. The direction reaction with oxygen,

$$2NO + O_2 \rightarrow 2NO_2,$$

is familiar in the laboratory, but its rate is proportional to the square of the NO concentration, and is insignificant in the normal atmosphere. The major route is through reaction with peroxo radicals, such as HO_2 and CH_3O_2, which are themselves formed from other radicals and dioxygen:

$$(R + O_2 \rightarrow RO_2);$$

$$RO_2 + NO \rightarrow RO + NO.$$

Through steps such as this, NO_x participates, often having a catalytic effect, in many reactions involving the production of hydroxyl radicals and the oxidation of hydrocarbons and CO (see *carbon) and of sulfur compounds.

Another important molecule that can be grouped with NO_x is the nitrate radical, NO_3, formed by the reaction with ozone:

$$NO_2 + O_3 \rightarrow NO_3 + O_2.$$

During the day, NO_3 is rapidly dissociated by light, thus re-forming NO or NO_2, but at night-time it may be an important oxidising agent in the troposphere. It readily extracts hydrogen from organic compounds,

$$NO_3 + RH \rightarrow HNO_3 + R,$$

thus acting as a source of nitric acid, and of radicals that undergo further reaction.

Low concentrations of NO_x are normal, indeed necessary, in the lower atmosphere. But in elevated levels arising from traffic fumes in urban areas, these species are serious pollutants. They make substantial contributions to acid rain, through their rapid oxidation to nitric acid, either via NO_3, or by the reaction with hydroxyl radicals:

$$NO_2 + OH \rightarrow HNO_3.$$

Nitric acid is extremely soluble in water, and is a strong acid, present as H^+ and NO_3^-. It is rapidly washed out of the atmosphere, and this forms the main route by which NO_x is removed from the atmosphere. The global contribution to rainwater acidity from NO_x is probably less than from oxidation of *sulfur compounds, but is still very significant.

NO_x compounds are also involved in photochemical reaction with hydrocarbons, giving rise to several compounds which are unpleasant or toxic: notable among these is the peroxyacetyl nitrate (PAN):

$$CH_3CO.O_2 + NO_2 \rightleftharpoons \underset{\text{PAN}}{CH_3CO.O_2.NO_2}.$$

This equilibrium shifts according to the temperature, so that PAN is not only a serious pollutant, but can act as a reservoir for NO_x in the atmosphere.

Together with unburned hydrocarbons and CO, NO_x emissions from vehicle exhausts contribute to the formation of photochemical smog in areas such as Los Angeles. Attempts to control these emissions, through 'lean-burn' engines and catalytic converters, have reduced somewhat the concentrations of reactive carbon compounds, but not of NO_x; the 'lean-burn' cycle, which uses higher combustion temperatures in the engine, significantly increases NO_x production. Ultimately, the only solution may be to decrease car traffic in the affected areas. Power stations are another source of NO_x, globally more important than vehicle exhausts although not generally concentrated in urban areas. Their contribution is less to localised air pollution than to acid rain. 'Scrubbing' the smoke emissions to remove acid products (including SO_2) may help; in coal-fired stations improvements can also be made by using 'fluidised-bed' technology. This involves the combustion of finely powdered fuel; lower temperatures are possible, and less NO_x is produced.

NO_x species are removed from the lower atmosphere so rapidly that none can reach the stratosphere directly. These molecules are however important in upper-atmosphere chemistry, and are produced naturally by the reaction of longer-lived N_2O with oxygen atoms as mentioned above. NO is one of the species (along with H, OH, and Cl) that can have a catalytic effect in destroying stratospheric ozone (see *oxygen). The simplest reaction sequence is

$$NO + O_3 \rightarrow NO_2 + O_2,$$

$$NO_2 + O \rightarrow NO + O_2,$$

with NO being regenerated, and the net conversion

$$O + O_3 \rightarrow 2O_2.$$

NO_x is produced by high-flying aircraft, and the possibility of destroying the ozone layer was one of the environmental objections put forward to the development of supersonic passenger transport in the 1970s. (In fact, it has not so far proved to be an economic proposition, and has been confined to a small number of Concorde aircraft.) The influence of nitrogen species on stratospheric chemistry is now recognised to be much more complicated, and although there is still some cause for concern—both from possible future supersonic transport developments and from the currently increasing N_2O levels—it is even possible that NO_x may reduce the rate of ozone removal, rather than accelerating it. The reason is that different pollutants do not act independently, but can also react together with unexpected effects. Some important reactions of NO_x include

$$HO_2 + NO \rightarrow OH + NO_2;$$

$$ClO + NO \rightarrow Cl + NO_2;$$

$$ClO + NO_2 \rightleftharpoons ClONO_2.$$

These perturb the influence of the other compounds which can remove ozone. The formation of chlorine nitrate ($ClONO_2$) illustrates a reaction which acts as a temporary sink for pollutant species. Thus under some conditions NO_x may act to counteract the presence of *chlorine radicals produced from CFC pollution, and now thought to be a more serious threat to the ozone layer. The reactive species are not permanently removed, and can be regenerated as the temperature rises. Although the detailed reactions are very complicated, and involve the surfaces of ice crystals as well as purely gas-phase processes, the formation of chlorine nitrate is thought to be involved in the formation of the 'Antarctic ozone hole'.

The last compound mentioned in Table II.37 is ammonia, which is a variable and very minor constituent of the atmosphere. Reduced nitrogen liberated by decaying organic matter is converted to NH_3 by some bacteria, but under normal conditions it is immediately taken up by other organisms, as nitrogen is often in short supply.

Most ammonia volatilisation probably comes from the decomposition of urea in animal urine, under rather dry conditions such as semi-arid grazing areas. NH_3, like CH_4, is susceptible to oxidation by hydroxyl radicals, but unlike CH_4 it is soluble in water, forming an alkaline solution:

$$NH_3 + H_2O \rightleftharpoons NH_4^+ + OH^-.$$

Ammonia also reacts rapidly with sulfuric and nitric acids, forming aerosols of $(NH_4)_2SO_4$ and NH_4NO_3. It is therefore quickly removed from the atmosphere, and acts partly to neutralise some of the acidity generated by oxidised nitrogen and sulfur compounds. It is interesting as the only naturally produced volatile molecule with such alkaline properties. The natural emission of some 50×10^9 kg NH_3 per year is capable of neutralising 3×10^9 kg hydrogen ions: this should be compared with an estimated 8×10^9 kg of H^+ entering the atmosphere from natural sources, and about the same again from anthropogenic N and S emissions (see Table II.23 under *hydrogen).

The nitrogen cycle

The cycling of nitrogen through the environment is complex, and involves the redox, biological, and atmospheric reactions discussed in previous sections. A summary of these reactions is shown in Fig. II.23. Figure II.24 gives estimates of the fluxes involved. Many of the magnitudes are somewhat uncertain, but it is clear that biological processes play a dominant role. Fixation transfers nitrogen from the atmosphere to the biosphere, and denitrification returns it to the atmosphere. Atmospheric oxidation through natural processes (e.g. lightning) makes some contribution to the cycle, but this is now outweighed by anthropogenic sources, both deliberate (use of synthetic nitrogenous fertilisers) and accidental (NO_x produced by transport and industry).

One very important aspect of the nitrogen cycle is the internal cycling by land and marine biota. Nitrogen released by the decay of organisms is avidly taken up by other life forms. The natural concentrations of inorganic nitrogen compounds in water are thereby maintained at a very low level, especially in the surface ocean layers where photosynthesis occurs and life predominates (see data in Table II.35). Because it is lost by denitrification, the supply of nitrogen to the ocean may be crucial in controlling its biological productivity. Marine biota have about one N atom per six of carbon, and given the rate of primary photosynthetic production in the sea, it can be estimated that only a small fraction of the nitrogen required could be supplied by extraneous sources, e.g. fixation, rivers, and NO_x washout. Upwelling ocean currents also supply only a small fraction, and as much as 90 per cent of the needs of surface organisms may come from rapid recycling.

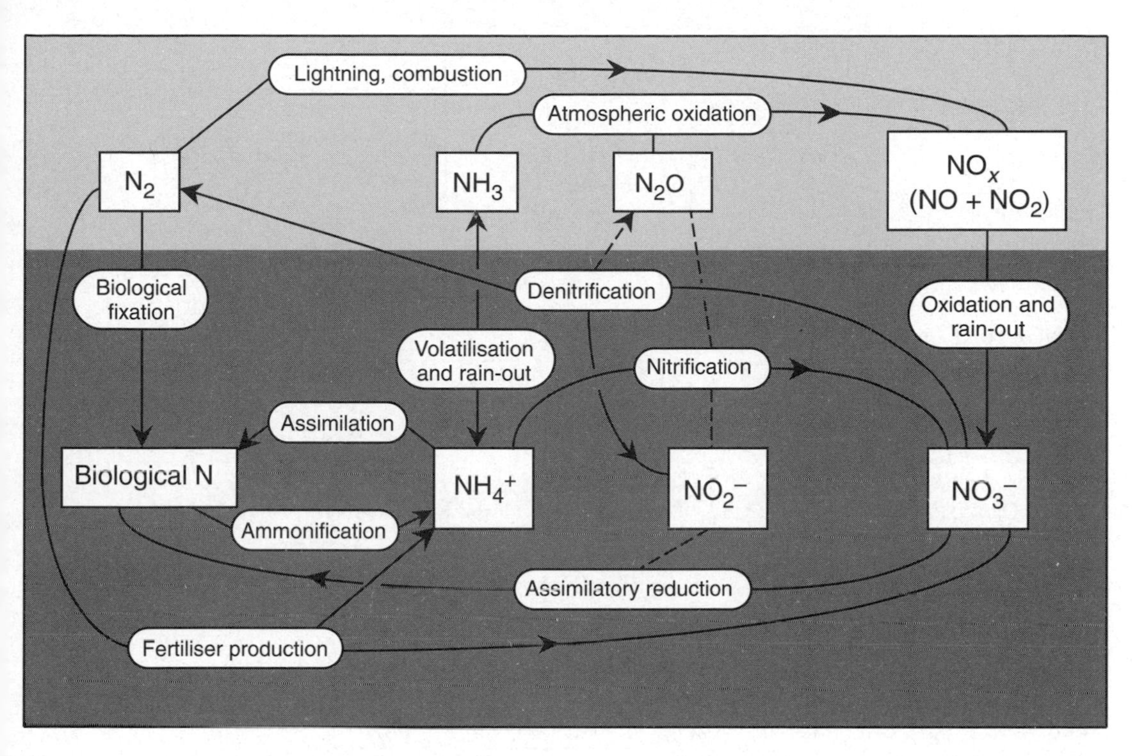

Fig. II.23 Chemical transformations in the nitrogen cycle.

Nitrogen is also important for life on land, and nitrogenous fertilisers now supply a significant fraction of the needs of agriculture. They represent an anthropogenic perturbation to the nitrogen cycle which is important for various reasons. It leads to elevated nitrate levels in rivers and lakes which can disturb their natural ecology, although *phosphorus is thought to be even more important in this respect. Fertiliser use also may be contributing to an increase in the atmospheric concentration of nitrous oxide (N_2O).

The other important human input to the nitrogen cycle comes from the production of NO_x by transport and industry. As discussed above, this not only contributes to local air pollution caused by traffic fumes, but also to acid rain (see under *hydrogen).

Osmium ($_{76}Os$)

Osmium is very rare, and occurs in association with other *platinum group elements. It is separated and used in small quantities in catalysts and in very hard alloys. It is not essential for life. Some of its compounds, especially the volatile tetroxide, OsO_4, are very toxic.

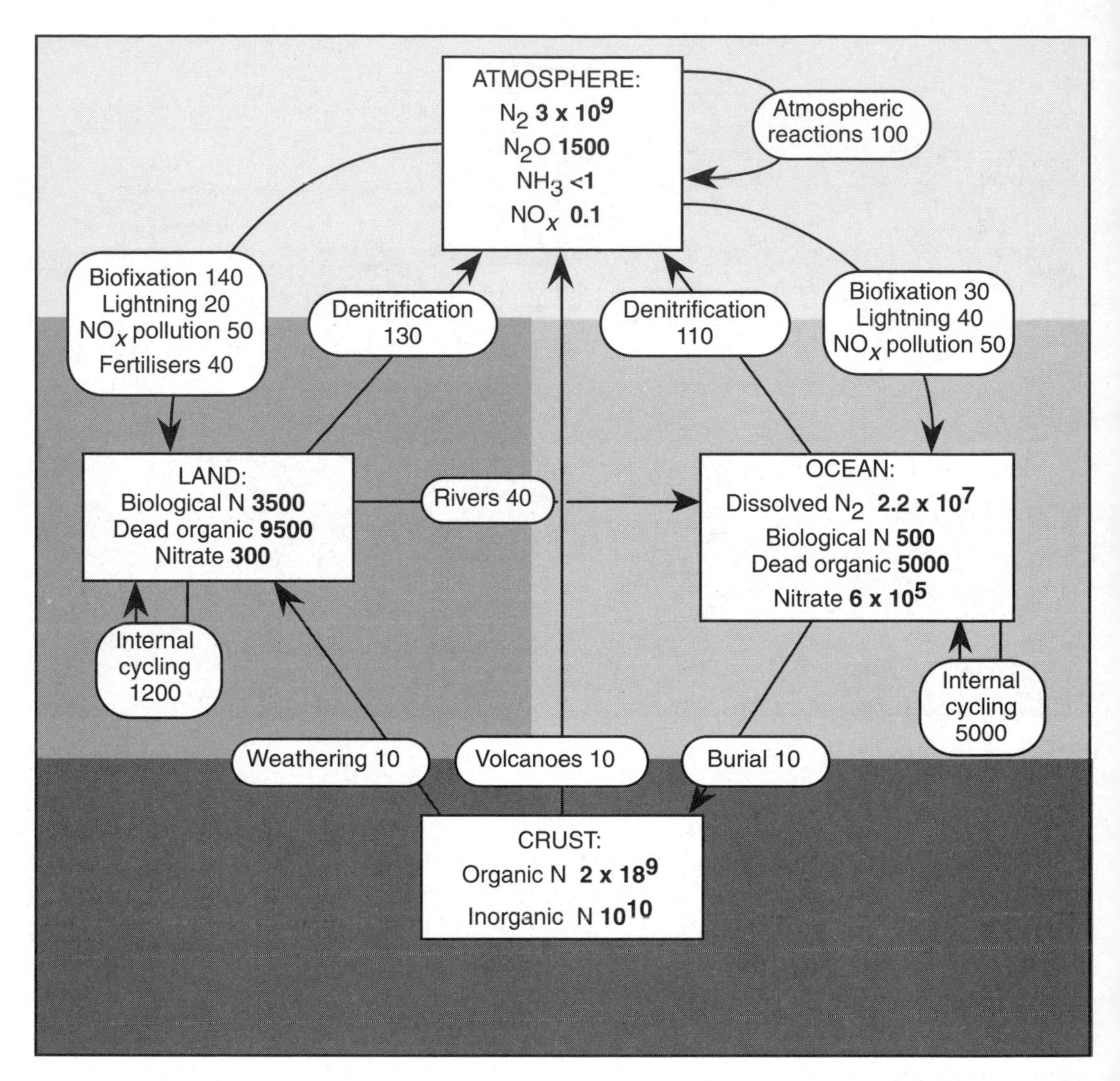

Fig. II.24 The nitrogen cycle. Reservoirs (square-cornered boxes) and annual fluxes (round-cornered boxes) are in given units of 10^{12} kg N. (Based on data in Schlesinger,

Oxygen ($_8$O)

Oxygen is the most important element at the Earth's surface. It is the third most abundant element in the Universe (after hydrogen and helium), and with its highly electronegative character, forms stable compounds with a majority of other elements. As described in §3.1, reactions of the elements with oxygen were important in the formation of the Earth, and especially in the differentiation of solids into the dense metallic core and the surrounding mantle and crust. Oxygen makes up 47 per cent of the crust by mass, and over 60 per cent of the atoms present. Nearly all elements can be found as oxides, generally of complex formulae containing several elements. For a major class of elements, known as *lithophiles*, oxides are the predominant form in the crust (see §§2.2 and 3.1). Oxygen also forms small covalent molecules with some non-metallic elements, water (H_2O) and carbon dioxide (CO_2) being the commonest natural examples. Over 50 per cent by mass of living things is also composed of oxygen, in the form of water and as constituents of nearly all the compounds of life.

The Earth is unique among the planets of the solar system in having a high concentration (20 per cent by volume) of the uncombined dioxygen molecule, O_2, in its atmosphere. Dioxygen is almost entirely derived from photosynthesis, and was not present in the early history of the Earth. The presence of this strongly oxidising molecule dominates many important chemical processes, both biochemical and inorganic. Reactions involving oxidation and reduction are important in the environmental cycling of many elements, such as *carbon, *nitrogen, and *sulfur (see also §4.2).

Many high-volume industrial chemicals (acids, alkalis, etc.) contain oxygen in combined form. The element itself is obtained by liquefying air, and separating the other constituents by fractional distillation. Oxygen is used in many industrial processes, especially metallurgy (steel-making, etc.), and as an oxidiser for rocket fuels. The allotropic form ozone (O_3), which is a minor constituent of the atmosphere, is produced from O_2 by the action of a high-voltage electric discharge, or by UV irradiation. It is strongly oxidising, and is used in the manufacture of some organic chemicals, and sometimes in the treatment of drinking water, and in the preservation of foods.

Oxygen plays an important part in the extraction and industrial uses of many other elements. Several important metals are obtained from oxide ores, and the strength of combination with oxygen largely determines the method of extraction which must be used (see Table I.13, §5.2). Exposure to atmospheric oxygen tends to reverse this extraction process, and some metals such as iron are susceptible to corrosion (rusting). In some cases, such as aluminium or titanium, the metals are potentially so reactive that they can only be used because of the natural formation of an oxide film, which is sufficiently robust and impervious that further oxidation is prevented.

Oxygen has three stable isotopes, ^{16}O, ^{17}O, and ^{18}O, present in natural oxygen in proportions of 99.76, 0.04, and 0.20 per cent, respectively. This fact caused some confusion in the past, as before 1961 there were two scales of relative atomic mass, each based on oxygen having RAM = 16 exactly; chemists used natural oxygen for this, and physicists pure ^{16}O. Because of the mixture of isotopes present in the chemists' standard, the two scales differed by 0.0275 per cent. In 1961 the scales were unified by adopting the pure isotope ^{12}C as a universal standard with RAM = 12. In the present scale the RAM of natural oxygen is 15.9994, and that of ^{16}O is 15.9950. In fact the exact proportions of oxygen isotopes in water and oxide minerals vary slightly, and interesting information can be deduced from these variations, especially about past climates on the Earth.

Water is discussed in detail under *hydrogen; the other issues touched on above are treated in the sections below.

Oxide minerals

The predominance of oxygen in the crust gives rise to an enormous variety of minerals, a selection of which is shown in Table II.38. Figure II.25 shows the occurrence and importance of these minerals in the periodic table. Nearly half of all the naturally known elements are normally obtained from oxides; many others are found in oxides. Elements 'also found in oxides' in Fig. II.25 vary from ones such as Na and K, where oxide (silicate) forms predominate although chlorides are a more convenient source, to ones such as arsenic and selenium where oxides are uncommon.

Binary oxides containing a single element combined with oxygen are much less common than more complex compounds containing several elements. Even these 'simple' compounds are rarely found pure. The more complex compounds may contain a great variety of elements in small concentrations, and the formulae given in Table II.38 are idealised. For example, the silicate clay mineral kaolinite, shown as

Table II.38 Selected oxide minerals. Some of the formulae are highly idealised, as many other elements may be present in small concentrations. Silicates are described in more detail under *silicon (e.g. Table II.46)

Type	Example	Formula
Binary oxides	Quartz	SiO_2
	Rutile	TiO_2
	Haematite	Fe_2O_3
	Uraninite	U_3O_8
Mixed oxides	Chromite	$CrFe_2O_4$
	Ilmenite	$TiFeO_3$
Hydroxides	Gibbsite	$Al(OH)_3$
	Goethite	$FeO(OH)$
Complex oxides:		
Carbonates	Calcite	$CaCO_3$
	Dolomite	$MgCa(CO_3)_2$
Sulfates	Barite	$BaSO_4$
Phosphates	Fluorapatite	$Ca_5(PO_4)_3F$
Silicates	Olivine	$(Mg,Fe)SiO_4$
	Zircon	$ZrSiO_4$
	Beryl	$Be_3Al_2[Si_6O_{12}]$
	Plagioclase feldspar	$Ca[Al_2Si_2O_8]$
	Kaolinite	$Al_4[Si_4O_{10}](OH)_8$

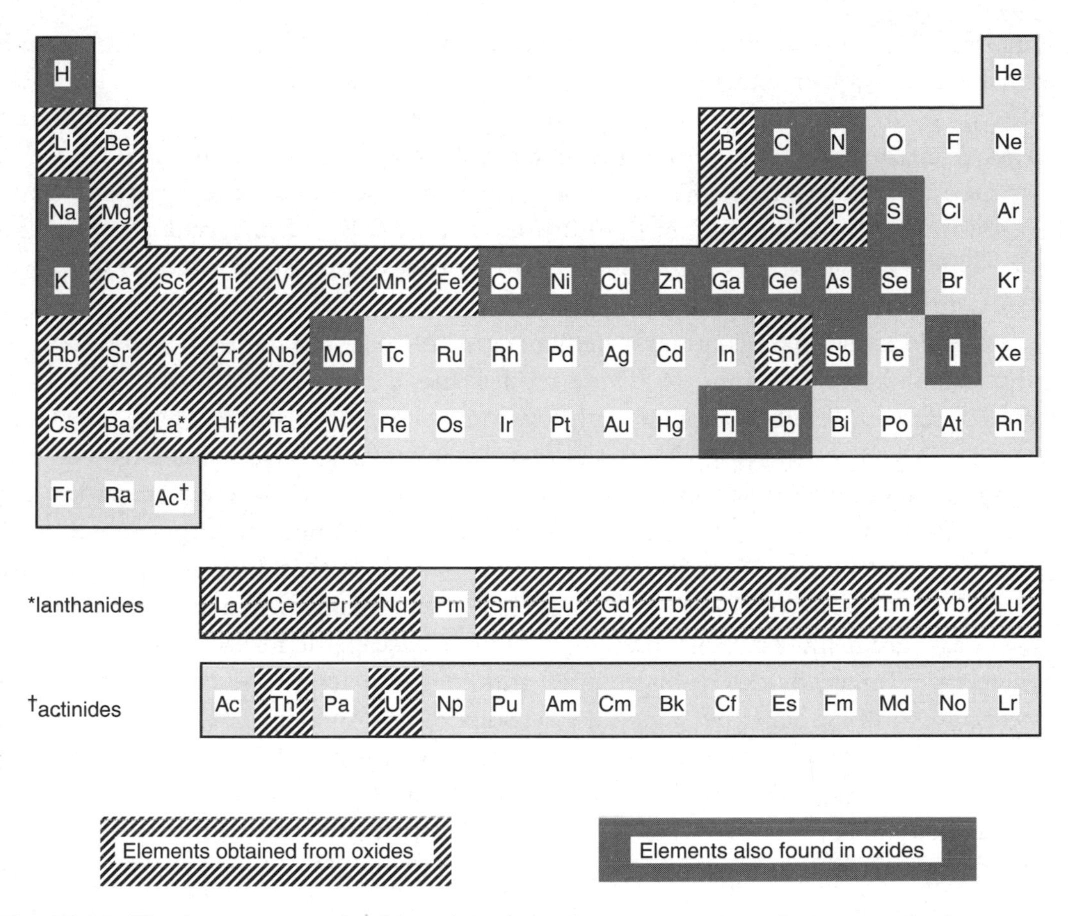

Fig. II.25 The importance of oxide minerals in the occurrence and sources of other elements. Note that most oxides in the crust are not simple binary compounds, but complex oxides such as silicates and carbonates. (See §3.2, and especially Fig. I.17.)

an aluminium silicate, may contain sodium, potassium, magnesium, calcium, iron, and other elements in proportions of a fraction of a per cent. The crystal structures of a few oxide minerals were shown in Fig. I.14; see also *silicon for a discussion of silicate structures.

Most common elements have a fixed oxidation state in oxide minerals, although some variability occurs for elements in the transition series; for example in iron, which is found both as Fe^{2+} and Fe^{3+}, and in manganese, found in the Mn^{2+} and Mn^{4+} states (see §3.1). It is possible to think of the composition of a mineral, not so much in terms of the individual elements it contains, but rather as a combination of simple binary oxides of each of the elements. Thus ilmenite ($TiFeO_3$) could be written $TiO_2.FeO$, zircon ($ZrSiO_4$) as $ZrO_2.SiO_2$, and kaolinite ($Al_4[SiO_{10}](OH)_8$) as $2Al_2O_3.4SiO_2.4H_2O$. As a way of describing the actual structure of minerals, this is highly misleading, as all these compounds are very different from simple combinations of the binary oxides. But a notional breakdown into different simple oxides is frequently used by mineralogists and geochemists as a way of expressing the detailed composition of a mineral or rock. Table II.39 shows the overall composition of the continental crust of the Earth expressed in the same way. Only the commonest elements are shown, and volatile constituents such as H_2O (present in hydroxides) and CO_2 (in carbonates) have been removed before making up the numbers. It must be emphasised that expressing the composition in this form does *not* imply that oxides such as Na_2O are actually present.

The simple binary oxides of most elements are not found in nature because they are highly reactive. It is common to divide oxides into those which are *basic* and those which are *acidic*. Metallic elements with low ionic charge, +1 and +2, have very basic oxides, e.g. Na_2O or CaO, which react or dissolve in water to form metal ions, Na^+ and Ca^{2+}. Elements in high oxidation states (especially non-metals) have acidic oxides, e.g. CO_2 and P_2O_5, which in combination with water give carbonate and phosphate ions, respectively. The oxides of many elements with +3 and +4 oxidation states, except for those with very small 'ions' such as B^{3+} and C^{4+}, have oxides which are very insoluble in water, and are neither strongly basic nor acidic. These elements are sometimes found in the form of binary oxides or hydroxides, but the more acidic and basic oxides naturally tend to combine together, giving the carbonates, phosphates, etc. of elements such as calcium. The very common element silicon has an oxide, SiO_2, which is weakly acidic, and is most often found as the silicates of other elements. Many silicates are however easily decomposed by water, giving soluble metal ions and silica. The consequences of these different patterns of chemistry and solubility of oxides are important

Table II.39 Average composition of the continental crust, showing the proportion of each simple oxide, *present not in general on its own but in combination with other oxides*, making up the minerals of the crust, after removal of volatile components such as water and carbon dioxide. (From Mason and Moore, 1982.) See also Fig. I.29 (§4.5) for the variations in major elements in different igneous rock types

Component	Per cent by mass
SiO_2	61.9
TiO_2	0.8
Al_2O_3	15.6
Fe_2O_3	2.6
FeO	3.9
MnO	0.1
MgO	3.1
CaO	5.7
Na_2O	3.1
K_2O	2.9
P_2O_5	0.3

in controlling the distribution and mobility of many elements, and the processes of weathering and sedimentation (see especially §§3.4 and 4.4).

Oxygen and life: photosynthesis and respiration

Oxygen is a major constituent of all living matter, making up 60 per cent of the mass of the human body. A high proportion of this is in the form of water, but oxygen is also an essential component of nearly all the molecules involved in life. The solids which make bones, shells, or teeth are all oxide minerals (see §3.5). One function of oxygen in biological molecules is to provide a variety of functional groups, such as carboxylates ($-CO_2^-$), carbonyls (>CO), and hydroxyls (–OH), which are important in determining both the structure and reactivity of these molecules.

The most interesting aspects of oxygen in biology are those which involve the dioxygen molecule, O_2.[74] The oxidation of foods—organic molecules such as carbohydrates—provides the essential energy supply for all air-breathing organisms such as ourselves. The typical overall reaction can be written

$$O_2 + [CH_2O] \rightarrow CO_2 + H_2O,$$

where $[CH_2O]$ represents part of a carbohydrate molecule. It generates a supply of free energy amounting to 480 kJ per mole of carbon oxidised. Dioxygen itself is produced by photosynthesis, essentially the reverse reaction, performed by green plants using free energy derived from sunlight. Figure II.26 gives a very schematic view of the transformations involved.

The oxidation of carbohydrates to CO_2 is extremely complicated and needs many reactions in which *phosphorus-containing molecules play a part. The normal overall result is the conversion of NAD^+ to the reduced *hydrogen carrier, NADH:

$$2NAD^+ + H_2O + [CH_2O] \rightarrow 2NADH + 2H^+ + CO_2$$

NADH is then reoxidised by oxygen in a series of steps, represented by the arrows in Fig. II.26. The scale of electrode potentials has been reversed here, so as to give a natural-looking 'downhill' direction to the spontaneous flow of electrons from a more reducing couple (lower E value) to a more oxidising one (high E value). The difference between the electrode potentials for O_2/H_2O and $NAD^+/NADH$ (+0.82 V and –0.32 V, respectively) can give about 110 kJ per mole of electrons transferred. This is used to convert several moles of ADP into ATP, the main energy currency required by living cells. Several steps are required to convert this energy efficiently, rather than 'short-circuiting' it in a single step. Many redox-active molecules are involved, especially the *iron–sulfur proteins, and cytochromes containing haem iron.

74 Stryer (1988) gives a good account of biochemical processes, including those of oxygen. Williams and da Silva (1978) contains articles on specific aspects of dioxygen biochemistry.

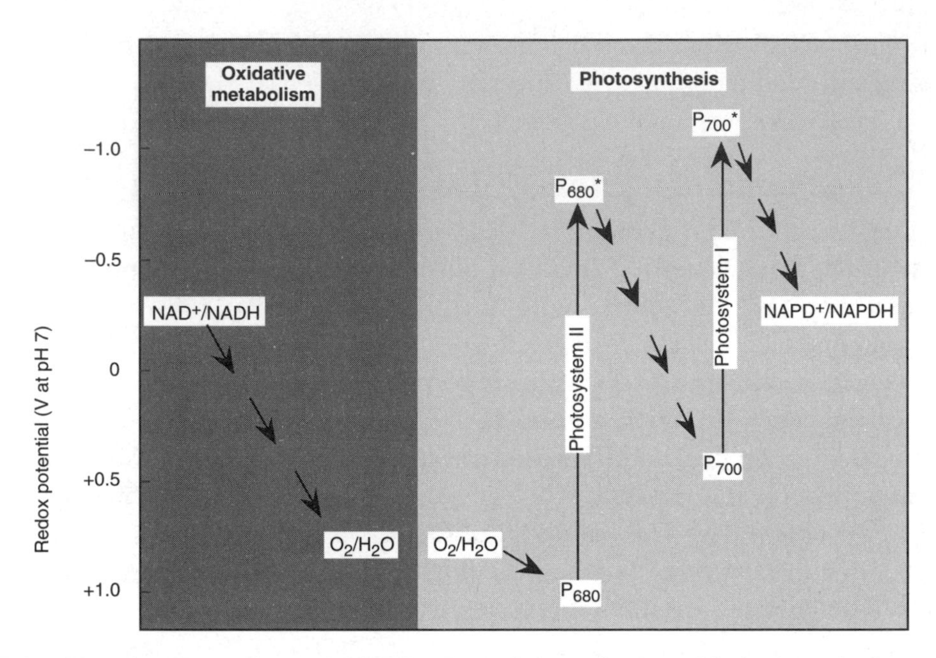

Fig. II.26 Showing very schematically the flow of electrons in oxidative metabolism and in photosynthesis. The vertical scale is the electrode potential; negative values towards the top correspond to 'high-energy' electrons which can act as reducing agents. The normal thermodynamic tendency is for a 'downhill' transfer. As suggested by the successive arrows, many intermediate steps occur. P_{680} and P_{700} are the two chlorophyll-containing photosystems, where the absorption of light raises the energy of an electron to the excited $P_{680}^\star$ and $P_{700}^\star$ states, which are strongly reducing.

The final stage in this electron-transfer chain is a potentially awkward one, as the reduction of an O_2 molecule requires four electrons:

$$O_2 + 4H^+ + 4e^- \rightarrow 2H_2O.$$

Intermediate reduction products of dioxygen—superoxide (O_2^-) and hydrogen peroxide (H_2O_2)—are undesirable: they are of high energy, and therefore would decrease the energy obtainable from the reaction. They are also highly reactive and hence damaging to cells. In respiratory metabolism, the oxygen-reducing step is performed by the enzyme *cytochrome oxidase* (so called because it in turn oxidises the next stage in the chain, *cytochrome c*). This contains an iron and a copper atom in close proximity, so that the O_2 molecule can be coordinated between them. Transition metals are important constituents in this type of process for two reasons: they have variable oxidation states and also they are able to coordinate small molecules. An intermediate step in the reaction of cytochrome oxidase is the formation of a *ferryl* complex containing $[Fe{=}O]^{2+}$, with iron in the unusual oxidation state Fe^{4+}.

In spite of the careful evolutionary design of this reduction step, some unwanted superoxides and peroxides are generated. Special enzymes are present to render these harmless. The copper- and zinc-containing *superoxide dismutase* catalyses the disproportionation of superoxide,

$$2O_2^- + 2H^+ \rightarrow O_2 + H_2O_2,$$

and various peroxidases, including the *selenium-containing *glutathionine peroxidase*, decompose hydrogen peroxide.

The energy input for photosynthesis comes from sunlight. Figure II.26 shows how this is done in two steps. Photosystems I and II absorb light at slightly different wavelengths, having absorption maxima at 700 nm and 680 nm, respectively. Light energy is absorbed by chlorophyll molecules which contain *magnesium, and transmitted to the reactive centres known as P_{700} and P_{680} as shown. In each case an electron is raised to a high-energy state where it can act as a strong reducing agent. When this electron is lost the centre can pick up another one at low energy, thus acting as a good oxidising agent. Thus by means of the light energy absorbed, an 'uphill' redox reaction can be performed. The P_{680} centre in photosystem II takes electrons from water and generates O_2. A reactive site containing four *manganese atoms is involved, which as in cytochrome oxidase is designed to perform the four-electron step without generating intermediates. The other centre, P_{700} in photosystem I, transmits electrons at high energy (i.e. negative redox potential) to generate the active hydrogen carrier NADPH from $NADP^+$; these have similar structures to the NADH and NAD^+ molecules appearing in oxidative metabolism. NADPH is then able, through a variety of complex steps, to reduce CO_2 to carbohydrates. The 'downhill' electron transfers shown in Fig. II.26 also take place in several steps, allowing some ATP to be made.

The two types of reaction, utilising oxygen in one case and making it in the other, have been described in a way intended to bring out some similarities. There are other analogies between them. Both take place in special cell compartments: mitochondria for the oxidation of food, and chloroplasts containing special structures known as thylakoids for photosynthesis. It has been suggested that both these structures first developed as independent single-cell organisms, which later 'invaded' other cells, so as to live with them in a symbiotic manner, beneficial to both the 'parasites' (chloroplasts and mitochondria), and the host organisms. The similarities do not end there. In both processes the electron-transfer steps take place in proteins bound to a membrane. Each electron transfer is also accompanied by the passage of a *hydrogen ion (H^+), which generates a concentration gradient across the membrane. This gradient then provides the energy required for *ATP-synthase* to convert ADP into ATP.

Since the advent of atmospheric dioxygen, many other reactions have evolved to benefit from it. Several *oxygenase* enzymes containing *copper act outside cells: some catalyse the oxidation of small molecules, either to generate desirable metabolites or to detoxify unwanted products; others use oxygen to cross-link connective tissues such as collagen. Another use of oxygen is in the destruction of foreign cells by the body's immune system. Some of the cells involved in this process of *phagocytosis* convert O_2 into the toxic superoxide ion, as a form of 'chemical warfare'. There are serious dangers from such reactive species, which can also be generated as unwanted minor products in all reactions which utilise oxygen. Attack on the body's biochemistry may be responsible for some cancers, and for the general process of ageing. Thus dismutases and catalases which remove these harmful compounds are important protective devices. Other compounds such as Vitamin E also act as antioxidants.

Oxygen used in biochemical reactions needs to be transported from the atmosphere to the appropriate places. Sufficiently small organisms can rely on diffusion to do this, but along with the evolution of larger animals, specific transport molecules developed. The most familiar are the proteins containing *iron bound in haem groups. Myoglobin contains a single such centre, and is used to store oxygen in muscle cells. Haemoglobin, which transports O_2 in the blood, contains four haem groups which act in a cooperative fashion, known as allostery, so as to enhance the transport capabilities. Other proteins are known which perform a similar function. For example, in arthropods oxygen is transported by *haemocyanin* and *haemerythrin*; both compounds are misleadingly named, as the former contains two copper atoms and no iron, whereas the latter has iron but not in a haem group.

Atmospheric oxygen chemistry and the ozone layer

Chemical processes occurring in the Earth's atmosphere are dominated by the presence of dioxygen, yet somewhat paradoxically this molecule itself is rather unreactive, as otherwise it would not be present in such a high concentration. Most oxidation reactions are mediated by less stable species, produced directly or indirectly from O_2 by photochemistry. They include oxygen atoms, ozone (O_3), and hydroxyl radicals (OH). Fig. II.27 summarises the production and reactions of these in the stratosphere (upper atmosphere) and the troposphere (lower atmosphere).

The double bond in O_2 is strong (500 kJ per mole), and can only be dissociated by UV radiation with wavelength less than 240 nm. This radiation does not penetrate into the lower atmosphere, and the reaction

$$O_2 + h\nu \rightarrow 2O$$

occurs only in the upper atmosphere. The oxygen atoms combine with O_2 to give ozone, although a third body (e.g. N_2, but normally represented M) is required to remove excess energy:

$$O + O_2 + M \rightarrow O_3 + M.$$

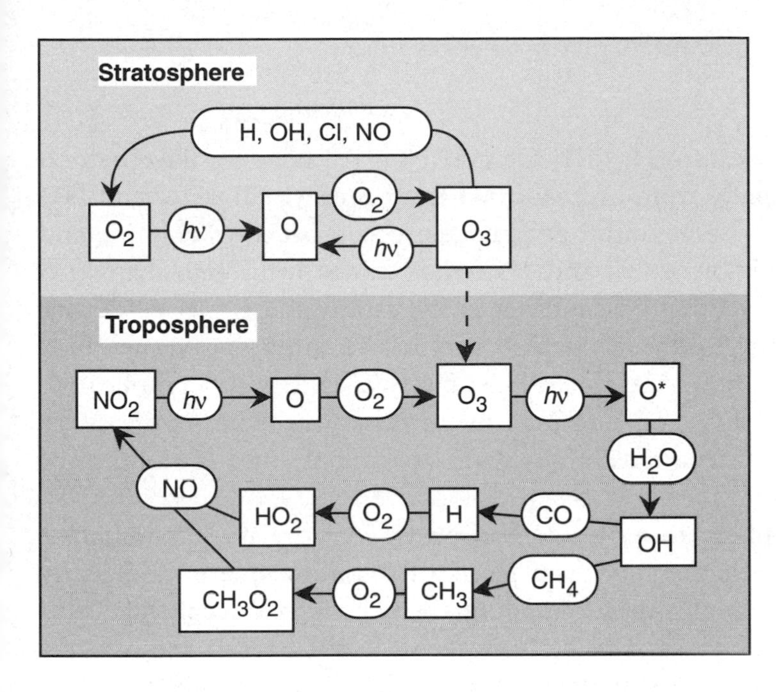

Fig. II.27 Atmospheric reactions involving oxygen. The square boxes show some of the main reactive species, and round boxes the routes to reaction. $h\nu$ indicates photochemical energy from sunlight. O* represents oxygen atoms in an excited (high-energy) state, which are more reactive than normal, ground-state, atoms.

Ozone is itself dissociated easily by solar radiation,

$$O_3 + h\nu \rightarrow O + O_2,$$

and it is present at a concentration such that the rates of production and loss are equal. A further reaction of importance is

$$O + O_3 \rightarrow 2O_2.$$

These three reactions make up the *Chapman cycle*, constituting the most important 'oxygen only' reactions in the stratosphere. The ozone produced in this way is concentrated between 20 and 40 km above the Earth's surface, with a maximum concentration of about 8 ppm, or in absolute terms approaching 10^{13} molecules per cm^3. Its presence is important to life, as it strongly absorbs radiation in the UV wavelength range 230–290 nm, which would otherwise reach the surface and be damaging to biological molecules.

It is now known that the reactions of the Chapman cycle would lead on their own to a much larger ozone concentration than that actually present. Other reactive species in the stratosphere, although present in very small concentrations, can have a catalytic effect in destroying ozone. The most important type of reaction sequence is

$$X + O_3 \rightarrow XO + O_2,$$

$$XO + O \rightarrow X + O_2,$$

where X is a radical that is regenerated, giving a net reaction

$$O + O_3 \rightarrow 2O_2.$$

The most important X species are H, OH, Cl, and NO. All of these have natural sources: H and OH principally from methane, Cl from methyl chloride, and NO from N_2O. But there are also some anthropogenic sources, especially *chlorine and *bromine from CFC and Halon gases, and *nitrogen oxides from high-flying aircraft. Recent evidence suggests that ozone levels in the stratosphere may be declining, and particularly worrying has been the discovery of the 'Antarctic ozone hole'. Recent research has shown that stratospheric chemistry is extremely complicated. Currently it is believed that CFC emissions pose the most serious threat to the ozone layer, and there have been international agreements to limit the use of these compounds (see under *chlorine).

Chemistry in the troposphere is very different, partly because no UV radiation capable of dissociating O_2 reaches the lower levels of the atmosphere, and partly because of the presence of many other compounds which do not reach high altitudes. Some ozone may be transported down from the stratosphere, but another very important source of tropospheric O_3 is from oxides of nitrogen. As shown in Fig. II.27, NO_2 may be dissociated to NO and O atoms by lower-energy UV radiation at ground level.

Ozone in the lower atmosphere is the source of hydroxyl radicals (OH), the species chiefly responsible for the oxidation of many other compounds, especially those of *carbon and *sulfur. There is a photochemical oddity here. The reaction of normal oxygen atoms with water vapour,

$$O + H_2O \rightarrow 2OH,$$

does not proceed at a significant rate, because it is endothermic, the new O–H bond being weaker than the HO–H one broken. But the photolysis of ozone produces O atoms in an excited state, with more energy than normal. These excited atoms are more reactive, and do react with water to give hydroxyl radicals.

As shown in Fig. II.27, OH reacts primarily with the reduced *carbon compounds CO and CH_4. The ultimate oxidation product is mostly CO_2 (not shown), but the radicals (H and CH_3) produced in the first step of reaction can ultimately lead to the regeneration of NO_2. Thus tropospheric ozone chemistry is intimately bound up with that of nitrogen oxides.

Whereas ozone in the stratosphere is essential to our existence, at ground level it can be a serious pollutant. In small concentrations, ozone is a natural constituent of the troposphere, and through its role in the production of OH radicals, plays a part in the oxidation of many molecules. In higher concentrations, it is toxic, and harmful to animal and plant life. It also participates in oxidation reactions of hydrocarbons, some of which give noxious products. Elevated ozone concentrations come especially from the oxides of nitrogen produced in automobile exhausts, and it is an important intermediate in the formation of the photochemical smog which afflicts some polluted cities such as Los Angeles.

The oxygen cycle and its evolution

Because of their abundance on Earth, oxygen compounds take part in all the physical cycling of the crust, of water, and of the atmosphere. Thus a complete description of the oxygen cycle would involve the chemical and physical transformations of nearly everything on Earth. But most oxygen occurs in fixed proportions in the oxides of strongly lithophilic elements, such as silicon, magnesium, aluminium, and calcium. The interesting parts of the oxygen cycle are those where elements are changing their oxidation states, so that oxygen is effectively being transferred from one chemical form to another (see §4.2). Although there are many elements which can cycle between oxidised and reduced forms, most of them are fairly uncommon ones, and make very little contribution to the cycle. The most abundant elements having variable oxidation states in the natural environment are ⋆carbon, ⋆sulfur, and ⋆iron, and it is with these that the chemical cycling of oxygen is closely associated.

Figure II.28 shows some of the reservoirs and fluxes involved in the chemical transfer of oxygen. H_2O and CO_2 are grouped together, as they represent the original sources of oxygen available to photosynthesis, and in this respect are the major reservoirs of chemically available oxygen in the environment.

By far the most rapid transfer of oxygen is that associated with life. Essentially all O_2 in the present atmosphere comes from photosynthesis, and is cycled back into CO_2 by respiration and by the aerobic decomposition of organic matter by bacteria. The turnover gives atmospheric O_2 a residence time of about 3000 years. That is much longer than for CO_2 (3 years), largely because of the much higher concentration of O_2 in the atmosphere. For this reason, short-term perturbations, caused at present by the burning of fossil fuels, have much less influence on atmospheric oxygen than they do on CO_2 (see ⋆carbon).

Photochemical processes in the atmosphere, especially ozone production, also give a rapid turnover, although this is mostly reversible, only a small proportion of the oxygen involved being removed from the atmosphere as oxidation products such as CO_2 and sulfate.

Other parts of the cycle operate more slowly. At the present time there is a net conversion of O_2 to CO_2 through the burning of fossil fuels and land vegetation, but throughout hundreds of millions of years the net flow has been in the reverse direction: of the organic carbon produced by photosynthesis, some 10^{11} kg has not

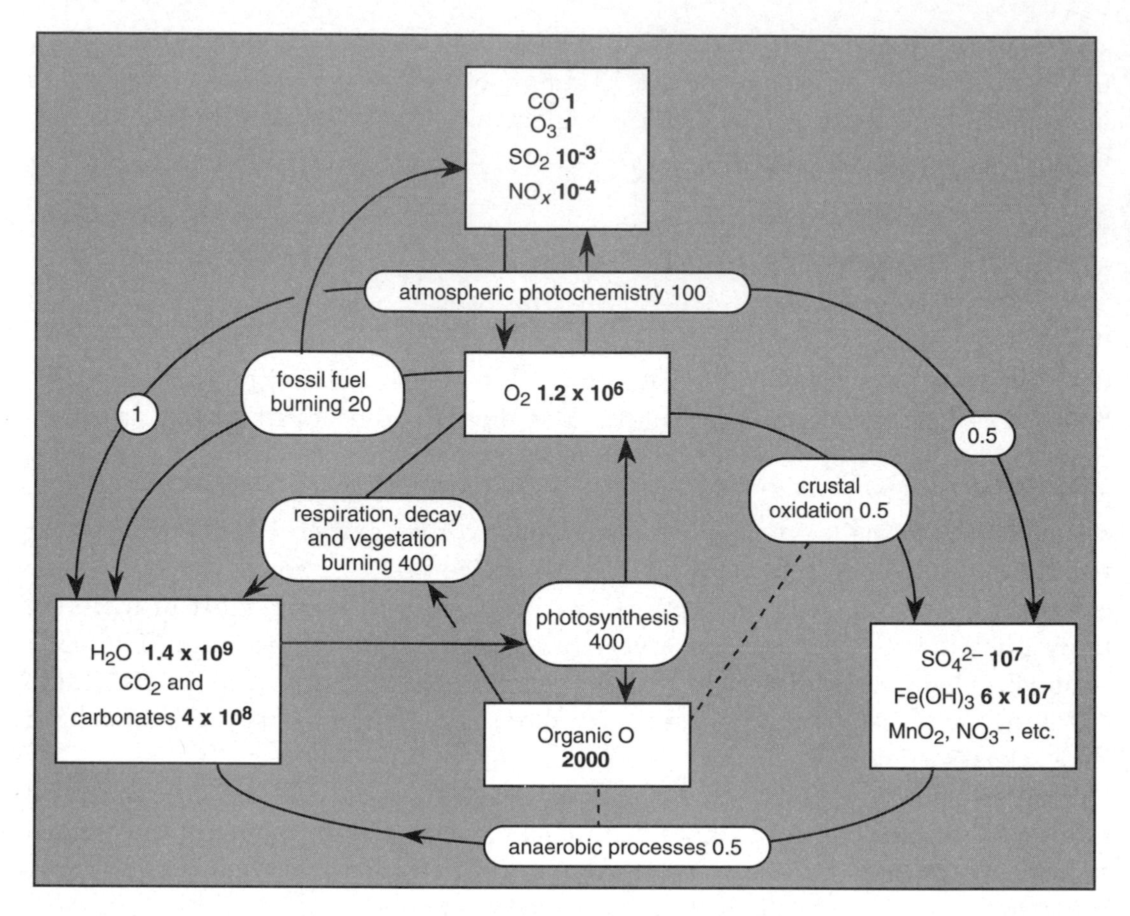

Fig. II.28 Reservoirs and fluxes in the oxygen cycle, showing different chemical forms and not their physical location. Only chemical exchangeable O is shown, out of the total of 2.6×10^{24} kg on Earth. Reservoirs in units 10^{12} kg O, with fluxes in 10^{12} kg per year.

been reoxidised, but has been buried in reduced form, as coal, oil, or natural gas. Over geological history around 10^{19} kg of carbon has been buried before reoxidation, and a corresponding amount of oxygen has been freed from its original combination in CO_2 and carbonate. About 4 per cent of this total forms the current O_2 content of the atmosphere, the remainder having been used up in the oxidation of minerals in the crust. Most iron deep in the crust is present as Fe^{2+}, and most sulfur is in the reduced form as sulfide. In the presence of oxygen, these elements are oxidised, to Fe^{3+} and sulfate, respectively. As much of the reduced iron and sulfur are combined together as pyrites, FeS_2, the net oxidation reaction can be written

$$4FeS_2 + 8H_2O + 15O_2 \rightarrow 2Fe_2O_3 + 8SO_4^{2-} + 16H^+.$$

Sulfate has mostly been precipitated as insoluble compounds in sedimentary rocks, chiefly with *calcium as gypsum, $CaSO_4.2H_2O$. Thus a rough approximation to overall transfer of oxygen is

$$\text{calcium carbonate} + \text{pyrites} \rightarrow \text{reduced carbon} + \text{calcium sulfate} + Fe_2O_3.$$

This overall reaction is *thermodynamically unfavourable*, and could not take place without the free-energy input derived from the sun by photosynthesis.

The above reaction can also operate in reverse. In anaerobic conditions, such as in sediments, and in natural waters where the lack of mixing with surface layers prevents the transfer of dissolved O_2, many bacteria obtain their metabolic energy by reducing sulfate and Fe^{3+}, and depositing FeS_2 or elemental sulfur as a consequence. (Other reactions, involving *nitrogen compounds, are involved here also, but their total contribution to the oxygen cycle is comparatively small.) Thus the oxidation and reduction reactions of minerals form a reversible system, which can in principle buffer the atmospheric concentration of oxygen, although over an extremely long period of around 10 million years, compared with the 3000 year time-scale associated with photosynthesis and respiration.

Almost certainly, the primitive atmosphere on Earth contained very little O_2. The evidence comes not only from other planets, which have atmospheres almost devoid of free oxygen, but also from studies of minerals laid down on Earth billions of years ago. So-called *banded ironstones* contain iron which was probably present as Fe^{2+} in the ocean waters at the time, and not oxidised to Fe^{3+} as it is today. A little oxygen would have been produced from water by the action of hard UV radiation, before the formation of the ozone layer which prevents such radiation from reaching the Earth's surface today. The *photodissociation* of water, where the energy of radiation ($h\nu$) splits a chemical bond, can be written as

$$H_2O + h\nu \rightarrow H + OH.$$

Hydrogen atoms escape into space, and OH radicals undergo reactions leading to O_2, which then reacts with surface rocks, producing some Fe^{3+} and sulfate. Similar processes almost certainly happened on our neighbouring planets Venus and Mars, which have oxidised surfaces: the red colour of Mars is due to Fe^{3+} oxides, and the clouds of Venus are made largely of sulfuric acid. Significant oxygen in the atmosphere only arose, however, when photosynthesis started, between 3 and 4 billion years ago. The oxygen produced initially was used up in oxidising rocks, and enormous deposits of iron ore were produced during the period up to around 2 billion years ago. At this point, the oxygen concentration started to rise, although it may not have reached its present level until about 500 million years ago, the point at which land plants and animals started to appear.

It is interesting to ask what controls the present level of O_2 in the atmosphere. Undoubtedly, as life produces O_2, it exerts a major control. With less oxygen, aerobic respiration and decay would be harder. Photosynthesis, together with anaerobic processes including the burial of fossil fuels, would dominate, and increase the atmospheric O_2 content. As the oxygen concentration rises, the amount dissolved in water and sediments also increases, and the extent of anaerobic processes declines. Among other effects, these reactions produce methane, the oxidation of which in the atmosphere makes a contribution (about 4 per cent) to the consumption of O_2. Another contribution comes from forest fires, which are significant even in the natural environment. If the O_2 content were to increase significantly, fires would burn out of control, and so increase the rate of conversion to CO_2. As with the carbon cycle, however, the direct influence of living matter can only operate in the short term, and the biosphere provides a limited buffer against change. Crustal processes, such as the burial of reduced carbon and the reactions of sulfur and iron compounds, operate much more slowly but are important in the longer term.[75] The oxygen cycle is currently out of balance through human activities, but in the short term we have less to worry about here than with carbon, because of the very different residence times of atmospheric O_2 and CO_2. Even burning all the land plants would only consume a small fraction of the O_2 in the atmosphere; the major short-term effects would come from the enormous increase in CO_2, and the fact that we would have irreparably damaged the biosphere and be without food.

Oxygen isotope variations

Oxygen from different sources on Earth contains slightly different proportions of the rarer oxygen isotopes ^{17}O and ^{18}O. Studies of these variations have been used to give information about the origins of rocks, and of climatic changes on the Earth.[76] The δ scale defined in §2.4 expresses the deviations in the abundance of isotopes, in parts per thousand relative to a convenient standard. For oxygen,

$$\delta^{18}O = 1000\,\{[^{18}O/^{16}O]_{sample} - [^{18}O/^{16}O]_{SMOW}\}/[^{18}O/^{16}O]_{SMOW},$$

where $[^{18}O/^{16}O]$ represents the ratio of isotopic abundance, and SMOW is 'Standard Mean Ocean Water'. Positive or negative δ values indicate an enrichment in the heavier or lighter isotope, respectively. Thus atmospheric O_2 has δ around +23.5, showing a higher proportion of ^{18}O than in sea water. It appears that oxygen produced in photosynthesis has a δ value similar to that of water, but that the reactions

---

75 See discussions in Schneider and Boston (1991).

76 See Faure (1986) for a detailed account.

of aerobic respiration consume the lighter ^{16}O slightly faster, so leaving the atmosphere enriched in the heavier isotope.

When water evaporates, the lighter isotope is more volatile and so is more abundant in the vapour phase. The degree of fractionation is larger at lower temperatures, and is reflected in the precipitation (rain and snow) falling at high latitudes, which has δ as low as –50. This is the same effect, although smaller in magnitude, that is found with *hydrogen isotopes, and a good correlation exists between $\delta^{18}O$ and $\delta^{2}H$ values for many waters. When this relation fails, it is a sign that oxygen in the water has not been derived entirely from precipitation, but has had the opportunity to exchange with oxygen in rocks, where the isotopic ratio may be different. The equilibration of isotopes between different oxide minerals also depends on temperature, and so $\delta^{18}O$ studies can give information about the conditions under which minerals were formed. Magmatic processes, operating at higher temperatures, should produce rather little fractionation, but one surprising observation is that some igneous intrusions, for example in the island of Skye in Scotland, have $\delta^{18}O$ values way out of the range expected. This seems to indicate that these rocks must subsequently have interacted with large amounts of water.

Variations in the composition of sedimentary rocks such as carbonates arise from a combination of factors. Temperature changes alter the equilibrium between ^{18}O in water and carbonate. But during colder periods, there will also be more ice cover, which will be enriched in ^{16}O as it is mostly derived from precipitation. Under these conditions, therefore, the remaining ocean water will have a slightly higher ^{18}O content. Although it is hard to make a quantitative correlation between $\delta^{18}O$ in carbonate sediments and the prevailing temperature, these variations provide a clear qualitative record of changing climates. For example, Fig. II.29 shows some data from marine carbonate sediments over a period of half a million years. The alternation between warmer times (more negative $\delta^{18}O$ in this case) and colder periods of glaciation, can be clearly seen. Other records have been studied, showing climatic variations over periods of millions of years. Such studies can help to understand the causes and effects of climatic change, and are especially relevant to the current concern over the 'greenhouse effect' caused by increasing CO_2 levels (see *carbon). One major driving force for change in the past has been the small periodic variations in the Earth's orbit around the sun, which alter the amount of solar radiation received. However, these changes seem to have been amplified by various feedback effects, which are not yet fully understood.[77]

Variations in $\delta^{17}O$ should parallel those of $\delta^{18}O$, although because the difference in mass, relative to ^{16}O, is only half as much with ^{17}O, the degree of fractionation is

---

77 Henderson-Sellers and Robinson (1986).

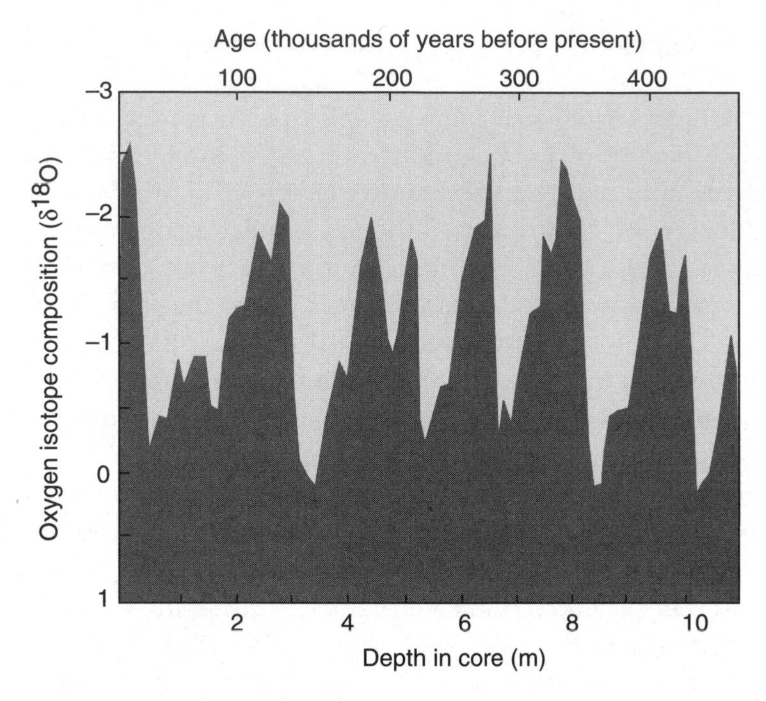

Fig. II.29 Oxygen isotope data from marine carbonate sediments, showing cyclic changes corresponding to periods of global warming and cooling. The $\delta^{18}O$ scale shows how much the ^{18}O content varies from a mean sea water standard, and is defined in the text. (Based on Smith, 1981.)

expected to be correspondingly smaller. All terrestrial samples of oxygen seem to fit this prediction, but anomalies have been found in some meteorites. Oxygen from different minerals, even in the same meteorite, sometimes shows combinations of ^{17}O and ^{18}O composition that could not have arisen from any normal chemical process. The only plausible explanation seems to be that oxygen in the solar system is derived from at least two sources, with different initial isotopic ratios. One of these sources contained almost pure ^{16}O, and may have been a supernova occurring fairly soon before the origin of the solar system. It may even be that shock waves from this supernova triggered the contraction in a cloud of gas in space that gave rise to the solar system (see §1.1).

Palladium ($_{46}Pd$)

Palladium is one of the *platinum group of elements, of very low abundance in the crust and in the environment generally. It is not essential for life, and is generally non-toxic. The major use of palladium is as a catalyst for hydrogenation and dehydrogenation reactions in the petrochemicals industry. The metallic element has the unusual property of absorbing large volumes of hydrogen, and acquired widespread fame in 1989 with the announcement of 'cold fusion'. It was claimed that deuterium (the heavy isotope of *hydrogen) could undergo nuclear fusion when inserted into palladium in an electrolytic process. This gave the promise of a virtually unlimited energy supply, without the enormous temperatures and other technical difficulties normally associated with fusion reactions. Many research groups investigated the cold fusion claims, and generally failed to reproduce them.[78]

Phosphorus ($_{15}P$)

Phosphorus is a relatively abundant element in the Earth's crust, and a crucial element in the environment. At the Earth's surface it occurs exclusively in the oxidised phosphate (PO_4^{3-}) form, although iron phosphides such as Fe_3P_2 are found in meteorites, and some phosphorus may be present in reduced form in the core. Numerous phosphate minerals are known, the most abundant being those of calcium: *apatite* has the general formula $Ca_5(PO_4)_3X$, where X can be OH^- (*hydroxyapatite*), F^- (*fluorapatite*), or less commonly Cl^- (*chlorapatite*). Another phosphate mineral is *monazite*, $LnPO_4$ (where Ln represents one of the lanthanide elements), an important source of lanthanides and of thorium. Enormous quantities of phosphate minerals are extracted, the major use being in fertilisers. Elemental phosphorus is manufactured by the reduction of apatite with carbon. It is used to produce compounds, such as organic phosphates for detergents, phosphorus sulfides used in matches, and toxic organophosphorus compounds used as pesticides and chemical warfare agents ('nerve gases').

The predominant species in water at neutral pH are $H_2PO_4^-$ and HPO_4^{2-}. But as shown in Table II.40 the concentrations are very low, because the phosphates of several common elements, especially calcium, aluminium, and iron, are extremely insoluble. Figure II.30 shows how the solubility of calcium and aluminium phosphates varies with pH. The least soluble calcium phosphate is fluorapatite, and this determines the soluble phosphate concentration under normal conditions. Availability falls with increasing pH, because the overall equilibrium is

$$Ca_5(PO_4)_3F + 6H^+ \rightleftharpoons 5Ca^{2+} + F^- + 3H_2PO_4^-,$$

Table II.40 Phosphorus concentrations in the environment

Location	Concentration
Crust	1000 ppm
Sea water:	
surface	1.5 ppb
deep	60 ppb
Fresh waters	*low*
Human body:	
average	1.1 %
bones and teeth	7 %

78 See Close (1990) for a popular, somewhat journalistic account of this strange affair.

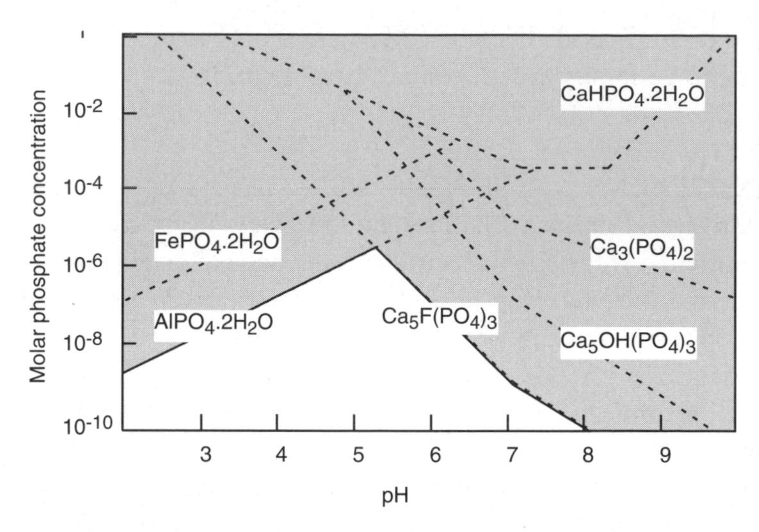

Fig. II.30 Variation in solubility of phosphate compounds with pH. Showing the concentration of soluble phosphate in equilibrium with various calcium and aluminium phosphates as a function of pH. The heavy line shows the maximum equilibrium concentration determined by the least soluble compounds, $Ca_5(PO_4)_3F$ and $AlPO_4$.

which leads to a decrease in the concentration of dissolved phosphate as the H^+ concentration falls. Aluminium phosphate is actually much less soluble than apatite, but is only formed under more acid conditions, because of the insolubility of *aluminium hydroxides in neutral solution (see Fig. II.1). The equilibrium

$$AlPO_4.2H_2O + H_2O \rightleftharpoons Al(OH)_3 + H_2PO_4^- + H^+$$

is forced to the left by increasing H^+ concentration below pH 5, and the phosphate availability falls as the acidity rises. Acid rain may reduce the availability of soluble phosphate in this way.

Phosphorus is a major element of life, being a constituent of DNA, ATP, and numerous other biological molecules which play an essential role in metabolism. Solid calcium phosphate minerals are used to make bones and teeth. Because of its low solubility in water, and also because no volatile phosphorus compounds occur in the atmosphere, phosphorus is often the least available of the essential elements, and in global terms the availability of phosphorus may be the single most important factor in limiting the extent of life on Earth. The natural cycling of phosphorus has been considerably modified by human activities, through its extensive use in fertilisers, and in industrially developed countries as soluble organic compounds for detergents. Excess phosphate inputs have been identified as the major cause of *eutrophication* of lakes, a phenomenon that entails the abnormal growth of algae and other organisms, and which upsets the natural balance of life and leads to oxygen depletion. The effects are exacerbated by untreated sewage, which also increases the oxygen demand. A number of lakes in North America, such as Erie in the Great Lakes and Lake Washington near Seattle, have been affected by eutrophication. Since the effects and their causes were recognised in the 1960s, the inputs of phosphates and of organic sewage have been reduced and some recovery has occurred, especially in Lake Washington.

Elemental phosphorus and many synthetic phosphorus compounds are highly toxic. A selection of toxic organophosphorus compounds is shown in the Fig. II.31. Many of these inhibit the enzyme *acetylcholinesterase*, which performs an essential function in operation of the nervous system of animals. Particularly deadly compounds are the industrial compounds: TOCP; TEPP, which was briefly used as an insecticide before being banned; and the nerve gases *Sarin* and *VX*. *Parathion* and *malathion* are used as insecticides; the latter of these is much less toxic to mammals (including humans), apparently because they have enzymes lacking in insects, which hydrolyse the carboxyester linkages (C_2H_5–O–CO–) giving less toxic products.

The biological chemistry of phosphorus

A human body of 70 kg weight contains 800 g of phosphorus, making it sixth in order of abundance, after nitrogen and calcium, but above potassium and sulfur. Unlike its neighbours nitrogen and sulfur in the periodic table, phosphorus is found in biology exclusively in the oxidised phosphate form. It fulfils some essential roles in all living organisms. All cells contain a concentration of free phosphate, and the equilibrium

$$H_2PO_4^- \rightleftharpoons HPO_4^{2-} + H^+,$$

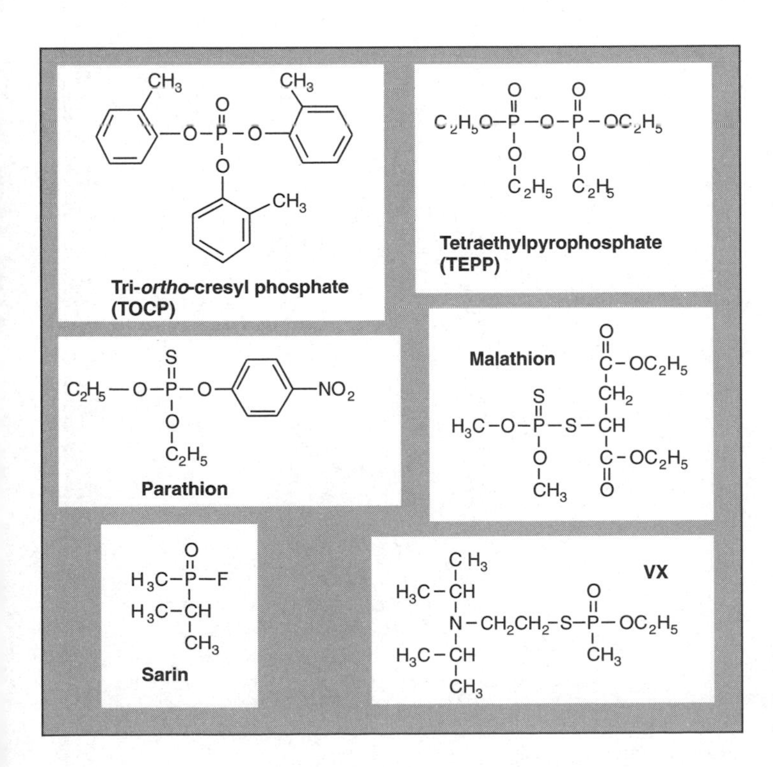

Fig. II.31 Some highly toxic organophosphorus compounds.

together with the similar equilibrium involving bicarbonate (see *carbon), provides a buffering effect which helps maintain the pH at close to a neutral value. A significant proportion of phosphate is combined with biological molecules as phosphate esters. A small selection is shown in Fig. II.32.

Phosphate is often bound to a sugar molecule: in the cases shown, ribose, which together with a nitrogen-containing base constitutes a *nucleoside*. One function of phosphate is to act as a link between such units, as in the portion of the DNA chain shown. In a similar way to *silicon but not nearly to the same degree, the phosphate group may also polymerise to form polyphosphates, or 'pyrophosphates': examples in Fig. II.32 are ATP and ADP, probably the most important phosphorus-containing molecules after nucleic acids. The conversion

$$ATP + H_2O \rightarrow ADP + H_2PO_4^-$$

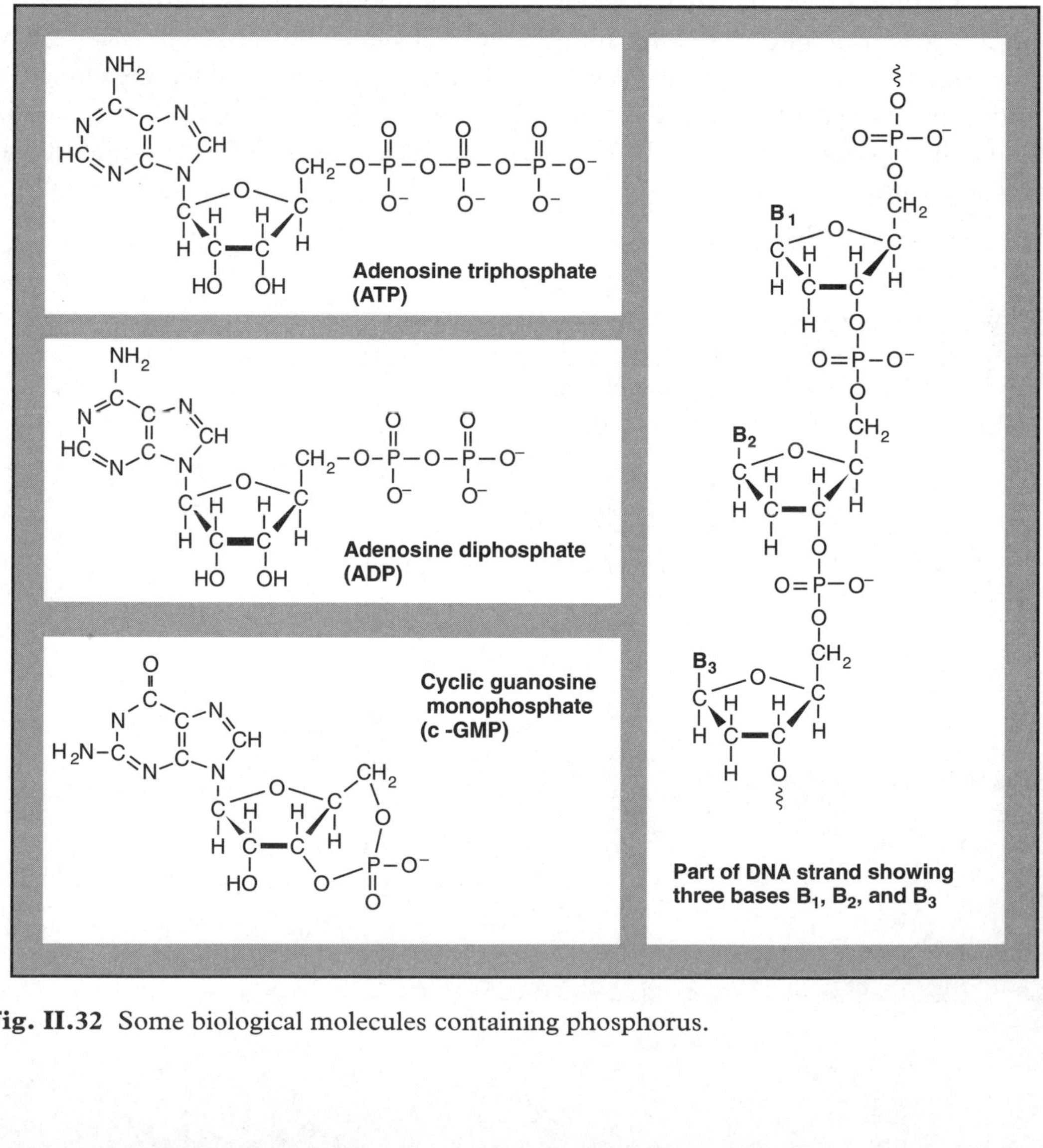

Adenosine triphosphate (ATP)

Adenosine diphosphate (ADP)

Cyclic guanosine monophosphate (c -GMP)

Part of DNA strand showing three bases B_1, B_2, and B_3

Fig. II.32 Some biological molecules containing phosphorus.

is thermodynamically favourable, having a free energy change ΔG of about –40 kJ per mole under biological conditions. The reaction is slow in the absence of a catalyst, however, and is controlled in living cells by a variety of enzymes. ATP is synthesised from ADP by energy-yielding reactions such as the oxidation of glucose; phosphate plays a major role in all of these metabolic reactions, being transferred between different molecules as they are broken down. The hydrolysis of ATP (the forward reaction above) is then used in a wide variety of situations where energy is required; these include the maintenance of ion concentration gradients across membranes, such as those produced by the *sodium pump. A resting human turns over about 40 kg of ATP per day, of which some 40 per cent is used in the sodium pump. During strenuous exercise where ATP is required to work muscles, the turnover may be as much as 0.5 kg per minute.

Another molecule shown in Fig. II.32 is cyclic guanosine monophosphate (c-GMP), which illustrates a different mode of bonding of phosphate—bridging between two parts of the same ribose unit—and a different function for phosphate. Cyclic phosphates such as c-GMP and the analogous c-AMP are formed from the 'open' acyclic forms in reversible reactions catalysed by enzymes. They fulfil important control functions in biological systems, in metabolic reactions, and in processes taking place in membranes. They thus play a role in diverse phenomena such as vision and muscle contraction. One of the enzymes which produces c-GMP from the acyclic molecule is activated by nitric oxide (NO), now known to be an important messenger in the body (see *nitrogen).

Phosphate molecules in biological systems are negatively charged and must have a corresponding concentration of cations present. The Mg^{2+} ion is usually associated with phosphate in cells, and exerts a control over its reactions. Thus DNA and ATP may be in part regarded as magnesium complexes. Calcium is very dangerous to phosphate metabolism because of the insolubility of apatite; this is one reason why the Ca^{2+} concentration is kept at a very low level in most cells. Insoluble hydroxyapatite is used in mammals as the inorganic component of bones and teeth.

The phosphorus cycle

The cycling of phosphorus in the environment differs significantly from that of other important non-metallic elements such as *carbon, *nitrogen, and *sulfur, because phosphates—the only stable form at the Earth's surface—are involatile and of low solubility. Transfers through the atmosphere are limited to small quantities of dusts, and even the transport by rivers is dominated by insoluble sediments. Fig. II.33 shows an estimate of the quantities involved.

Soluble phosphate is avidly taken up by living organisms, and recycled on death. Oceanic life contains about one phosphorus atom per 120 of carbon. Based on this ratio, it may be estimated that about 10^{12} kg P is required annually for the net photosynthetic production of the oceans. As shown in Fig. II.33, only 2 per cent of this can be supplied by rivers, from the weathering of phosphate minerals on land. Somewhat more comes from upwelling ocean currents, and this accounts especially for the high productivity of certain regions, such as off the west coast of South America. Even so, however, no more than 10 per cent of the phosphorus needed can be supplied in this way, the remainder being recycled from dead cells. Phosphorus is thus an element, like nitrogen, with a much smaller free concentration at the surface (where life predominates) than in the deep ocean.

Platinum ($_{78}Pt$)

Platinum is the most abundant of a group of rare elements (*ruthenium, *rhodium, *palladium, *osmium, and *iridium being the others) that generally occur in association, as metallic alloys and occasionally combined with *tellurium. They are mostly obtained as by-products from the production of copper and nickel. Uses of platinum are widespread, although the quantities involved are small. The metallic element is very resistant to chemical attack. It is used in jewellery, as a lining for chemical reaction vessels where very corrosive substances such as HF are involved, in

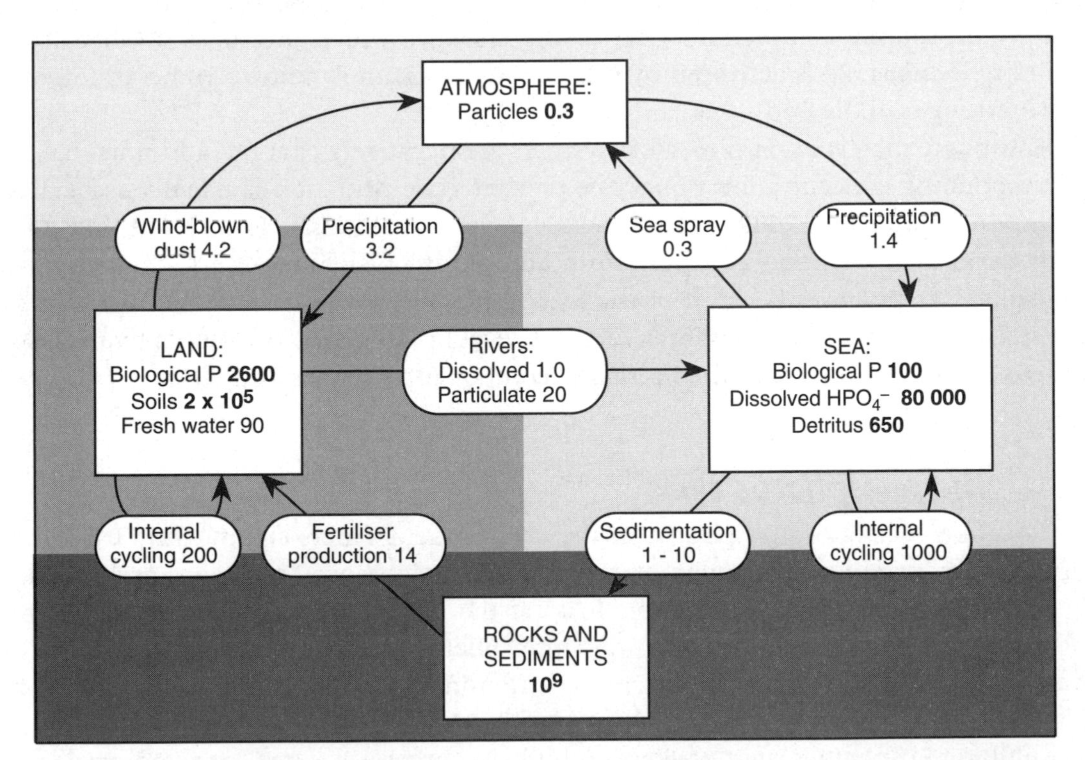

Fig. II.33 The phosphorus cycle. Reservoirs (square-cornered boxes) and annual fluxes (round-cornered boxes) are shown in units of 10^9 kg P. (Based on Schlesinger, 1991.)

electrolysis, and for some electrical wires and contacts. These applications are especially important in scientific research. Platinum is also an important catalyst for many reactions in the petrochemical industry, and for 'catalytic converters' fitted to automobile exhaust systems to reduce pollution by *carbon compounds.

Platinum concentrations in the natural environment are extremely low. The element has no natural role in life. However, the compound *cis-platin* is used in cancer chemotherapy as an anti-tumour drug. It combines with DNA and prevents its duplication during cell division. It is thus (like all potent drugs) very toxic, but affects rapidly dividing cancer cells more than normal ones.

Plutonium ($_{94}Pu$)

Plutonium is the most important of the radioactive transuranium elements. It does not occur naturally, but is produced in large amounts by the neutron bombardment of *uranium in nuclear reactors. Several isotopes are produced (see Table II.41), the most important one ^{239}Pu with a half-life of 24 000 years coming from the series of reactions

$$^{238}U + n \rightarrow {}^{239}U \rightarrow {}^{239}Np \rightarrow {}^{239}Pu.$$

The conversion of ^{239}U into ^{239}Np and then ^{239}Pu occurs by β decay (see §2.4), the first very rapidly, and the second with a half-life of 2.3 days. The plutonium can be separated by chemical processing of spent nuclear fuel, although this is difficult and expensive because of the radioactivity involved.

Like ^{235}U, ^{239}Pu also undergoes neutron-induced fission, and so is capable of sustaining a chain reaction. Only 5.6 kg of pure ^{238}Pu is needed to form a critical mass, and in concentrated aqueous solution as little as half a kilogram is sufficient. Some care is needed in the handling of plutonium, therefore, to avoid an unwanted chain reaction. The element was used in the nuclear bomb dropped on Nagasaki in 1945, although the design of an efficient weapon is more difficult than for ^{235}U. This is partly because some plutonium undergoes spontaneous fission, producing neutrons

Table II.41 Important plutonium isotopes

Isotope	Decay mode	Half-life (years)
^{238}Pu	α	90
^{239}Pu	α, γ	24 000
^{240}Pu	α, γ	6600
^{241}Pu	α, β	13
^{242}Pu	α	3.8×10^5
^{244}Pu	α	7.5×10^7

before the chain reaction starts. It is therefore hard to assemble a critical mass before some of the plutonium reacts, and vaporises the rest. Heavier isotopes such as ^{240}Pu are particularly troublesome in this way, and the production of plutonium for weapons is designed to minimise their concentration. As heavier isotopes are produced by neutron capture from ^{239}Pu itself, they are mostly formed at a later stage in the reactor operation, and can be avoided by removing the fuel elements earlier than would be normal for power production.

^{239}Pu could also be used as a nuclear fuel for power stations, and it has been proposed that uranium stocks could be more efficiently used by converting the otherwise useless ^{238}U into ^{239}Pu. 'Fast breeder' reactors for doing this have been in operation on an experimental basis in the UK, France, and Japan, and large stocks of plutonium are available, although so far unused for a variety of economic, technical, and political reasons. These stocks have increased greatly with the dismantling of large numbers of nuclear weapons in the early 1990s.

Another isotope of plutonium, ^{238}Pu, is obtained by neutron bombardment of ^{237}Np (also obtained from fuel reprocessing; see *neptunium) and is used as a power source in a different way. It is an α emitter with no accompanying γ radiation. As the α particles are easily absorbed, ^{238}Pu is entirely safe so long as it is securely enclosed. The heat generated from its radioactive decay is used on a small scale for some purposes, either directly in deep-sea diving suits, or in combination with a thermoelectric converter, normally a semiconductor compound which generates electricity directly from heat without moving parts. Power units based on ^{238}Pu are very long-lasting, and have been used in spacecraft and in cardiac pacemakers.

Although plutonium has been called 'the most toxic substance known to man', this is rather misleading. It is doubtful whether it has any serious chemical toxicity, as its most stable compound, PuO_2, is extremely insoluble, and chemically analogous elements such as *cerium and *zirconium are fairly harmless. The danger from plutonium comes from its radioactivity. The insolubility of its compounds gives it a very long residence time in the human body, and in soils that have been contaminated from nuclear weapons testing. Airborne particles may lodge in the lungs, but if chemically absorbed by the body, plutonium concentrates in bones. Its α radiation thus has a high potential for damaging the sensitive cells in bone marrow, and so for causing cancers such as leukaemia.

The other danger from plutonium is the possibility that it might be used by terrorists to make nuclear weapons. Even an inefficient device with little explosive power could have serious consequences, by spreading plutonium and other radioactive products over a wide area. Stocks of plutonium therefore need to be carefully guarded, and preferably some way of disposing of them found. One possibility is to mix PuO_2 with UO_2 to make a 'mixed oxide' fuel for nuclear power stations.

Polonium ($_{84}Po$)

A short-lived radioactive element, polonium has an exceedingly low natural abundance. The isotopes ^{210}Po and ^{218}Po, with half-lives of 138 days and 3 min respectively, are formed in the *uranium decay series, and so the element occurs in small amounts (0.1 mg per tonne) in uranium ores. The direct precursor of ^{218}Po is the isotope ^{222}Rn of the volatile gas *radon. Polonium may be therefore formed in the body from inhaled radon; it is relatively immobile and so remains to undergo further radioactive decay. ^{210}Po is also produced artificially by neutron bombardment of bismuth, and used as a small-scale heat source (in some spacecraft) and as a source of α particles.

Potassium ($_{19}K$)

Potassium is an abundant and widely distributed element in the Earth's crust, in life, and in natural waters. Most of it occurs in silicate minerals such as *alkali feldspars*, $(Na,K)AlSi_3O_8$, and many clay minerals which are common constituents of the continental crust. The soluble K^+ ion is dissolved in weathering, and is abundant in all freshwater and in the ocean. Although the crustal abundance of potassium is nearly the same as that of the other common alkali metal sodium, however, the concentration of K^+ is less than that of Na^+, by a factor of around 2 in rivers, and 40 in sea water. K^+ is slightly less readily removed from rocks in weathering, and is also more easily taken up in reactions on the sea floor between ocean water and sediments; thus its residence time in the ocean (5×10^6 years) is less than that of sodium (10^8 years). Potassium, like sodium, forms evaporite minerals such as *sylvite* (KCl), *sylvinite* (mixed KCl and NaCl), *carnallite* ($KMgCl_3.6H_2O$), and *saltpetre* (*nitre*; KNO_3). Large quantities of potassium salts are extracted for use in fertilisers. The element itself is obtained by reduction of KCl with sodium metal. It is used only in small amounts, chiefly for the manufacture of potassium superoxide, KO_2. This compound reacts with CO_2 to liberate dioxygen,

$$2KO_2 + CO_2 \rightarrow K_2CO_3 + 2O_2,$$

and is used to regenerate O_2 in breathing masks, in mines, submarines, and manned space vehicles.

Potassium is an essential element for all life. It occurs exclusively in solution as the K^+ ion; unlike Na^+ and Ca^{2+}, this is concentrated inside cells rather than in extracellular fluids. As shown in Table II.42, the potassium concentration in red cells is 10 times higher than in blood plasma. K^+ plays a part complementary to that of Na^+ in maintaining the ionic and osmotic balance in cells, and in the transmission of nerve impulses in animals. The mechanisms which maintain and utilise the gradients of these ions are described under *sodium.

Table II.42 Potassium concentrations in the environment

Location	Concentration
Crust	2.1 %
Sea water	380 ppm
Fresh waters	2.3 ppm
Human body:	
average	2000 ppm
blood plasma	400 ppm
red cells	4000 ppm
muscle	16 000 ppm

Potassium is a major nutrient for plants. In natural ecosystems, potassium taken up by plants is later returned to the soil, where K^+ is held in clay minerals. Intensive agriculture upsets this cycle, as the potassium removed from the soil ends up in rivers and in the sea. Potassium salts are therefore required as a fertiliser. Wood ash contains a significant concentration of potassium carbonate (K_2CO_3, 'potash'—hence the name of the element) and has traditionally been used as a way of returning potassium to the soil. Industrially extracted salts such as KCl are now more widely used.

The isotope ^{40}K, which makes up a small fraction (0.012 per cent) of all natural potassium, is radioactive with a half-life of 1.25×10^9 years. It can decay either into ^{40}Ca or ^{40}Ar, the latter possibility being important as the source of most *argon present in the Earth's atmosphere. The radioactive decay of ^{40}K contributes to the natural background radiation to which all living things are exposed, and also to the heat source within the Earth which causes the convection of the mantle and the tectonic cycling of the crust. Although potassium is normally a strongly lithophilic element, it has been speculated that under very high pressures it might behave as a siderophile, and so occur in small concentrations in the Earth's metallic core. If so, its radioactive decay could form the heat source responsible for the convection currents in the core, which are believed to generate the Earth's magnetic field. A more likely possibility, however, is that this heat comes from the slow solidification of iron in the core, liberating latent heat and gravitational energy.

Praseodymium ($_{59}Pr$)

A lanthanide element, discussed under *lanthanum.

Promethium ($_{61}Pm$)

Promethium is one of two elements with atomic number less than 84 (*technetium being the other) which do not occur naturally because all their isotopes are radioactive. Along with other lanthanides (see under *lanthanum), it is formed as a fission product in nuclear reactors and explosions. The most important isotope is ^{147}Pm, with a half-life of 2.6 years.

Protactinium ($_{91}$Pa)

Protactinium is a short-lived radioactive element, formed in the *uranium decay series. The longest-lived isotope is ^{231}Pa, with a half-life of 33 000 years, but as this comes from the decay of the relatively rare ^{235}U, its concentration in uranium ores is very small. Traces of protactinium have been detected in sea water, but at a concentration so low (a few parts per 10^{17}) as to be unimportant.

Radium ($_{88}$Ra)

Radium is the most abundant of the short-lived elements formed in the decay series of *uranium. The isotope ^{226}Ra is an α emitter with a half-life of 1600 years, and is found at a concentration of 1 mg per 3 kg of uranium in natural ores. It is separated from uranium in very small quantities. It was once used in luminous dials for clocks and watches, but this was abandoned when the dangers owing to its radioactivity were realised. More recently, it has been used in radiotherapy (as 'radium needles') but has now largely been superseded by artificially produced isotopes.

Radium occurs in sea water at a concentration of one part per 10^{17}, and in the human body at a few parts per 10^{15}. It is strongly held in the body, concentrating in bones. It must therefore make some contribution to the natural background radiation, but one which is normally insignificant compared with other sources.

Radon ($_{86}$Rn)

Radon is a short-lived radioactive element coming from the decay of *uranium. The principal isotope is ^{222}Rn, which has a half-life of 3.8 days. Although far from the most abundant of the uranium decay products, it probably contributes the largest source of natural radiation to human populations. This is because radon is an inert noble gas, which does not remain chemically combined in the uranium minerals where it is formed. Some radon diffuses out, and is present in the atmosphere close to areas where uranium-containing rocks are present. Radon atoms are fairly soluble in fat and can thus dissolve in body tissues. If they undergo radioactive decay while in the body, they are converted to other radioactive elements such as *polonium which are retained, and thus are likely to form a continued source of radiation. Radon exposure is highest for workers involved in uranium mining, but areas where the main rocks are granite are also particularly at risk, because of the relatively large concentration of uranium in these. Poorly ventilated houses may accumulate radon seeping upwards from the rock. The best prevention is a combination of good ventilation beneath the house with an impermeable plastic membrane.

Rhenium ($_{75}Re$)

Rhenium is a very rare element, generally found in association with molybdenum in its sulfide ore MoS_2. Small amounts are extracted during the processing of molybdenum ores, and the metallic element is used as a catalyst (often as an alloy with platinum) and in a few other specialised scientific applications.

Rhodium ($_{45}Rh$)

One of the *platinum group of elements, rhodium is very rare. It is obtained along with other elements of the group, chiefly as a by-product of the production of nickel. Metallic rhodium and its compounds are important catalysts, and it is used in catalytic converters for automobile exhaust systems.

Rubidium ($_{37}Rb$)

A member of the alkali metal group, rubidium is chemically akin to *potassium but considerably less abundant on Earth. It is widely distributed along with potassium in silicate minerals such as alkali feldspars, but its principal source is from lithium-containing minerals such as *lepidolite*, $K_2Li_3Al_4Si_7O_{21}(OH,F)_3$. It has very little use. As with other alkali metals, rubidium compounds are generally soluble in water, and it is not a rare element in the ocean. Although it is present in the body at a concentration of one part in 10^5 (higher than many essential elements) it has no known biological role.

The isotope ^{87}Rb, which makes up 27.8 per cent of all natural rubidium, is radioactive and decays to ^{87}Sr with a half-life of 4.9×10^{10} years. Because of the lower abundance of rubidium, it does not make nearly such an important contribution to natural radioactivity as some other elements such as *potassium, *thorium, and *uranium. Rocks contain an excess of ^{87}Sr derived from ^{87}Rb, to an extent which depends both on their rubidium content and on their age. This forms the basis of the important *rubidium/strontium* technique for dating rocks, and obtaining other information about their formation.[79]

Ruthenium ($_{44}Ru$)

A member of the *platinum metal group, ruthenium is extracted along with the other elements in this group during the refining of nickel. Its main use is as a cat-

79 See Faure (1986).

alyst, and in electrodes for the manufacture of chlorine by electrolysis. It is not essential for life, and with the exception of the volatile compound RuO_4, is non-toxic. However, the radioactive isotope ^{106}Ru, which has a half-life of 372 days, occurs in the fission products formed from *uranium in nuclear reactors and explosions. It makes a large contribution to the radioactivity in nuclear fuel elements, and has been released into the environment—possibly in the form of RuO_4—during reprocessing.[80] It has been found that ^{106}Ru appearing in the Irish Sea from the reprocessing plant at Sellafield in the UK is concentrated by the edible seaweed *Porphyra*. This seaweed is made into laverbread, a delicacy eaten in South Wales. Collection of *Porphyra* from the coast near Sellafield was discontinued in the mid-1970s, but prior to this, ^{106}Ru may have been the most significant isotope derived from nuclear reprocessing to find its way into the human food chain.

Samarium ($_{62}Sm$)

A lanthanide element, discussed under *lanthanum.

Scandium ($_{21}Sc$)

Scandium is a strongly lithophilic element which is rather evenly, though thinly, distributed in many types of rock. Although by no means rare, it forms very few minerals. It can be obtained from the rare silicate *thortveitite* ($Sc_2Si_2O_7$), and as a by-product of the processing of uranium ores, but has very little use. The Sc^{3+} ion forms very insoluble hydroxides at neutral pH, and so its concentration in natural waters is extremely low. Scandium has no known biological role.

Selenium ($_{34}Se$)

Selenium is chemically similar to sulfur although much rarer. Small concentrations occur in many sulfide minerals, and the element is obtained as a by-product of the refining of copper and other chalcophilic metals. It is used in the manufacture of glass, in photocopiers, and (in the form of the mixed cadmium compound Cd(S,Se)) as a pigment.

Like *sulfur, selenium has a complex redox chemistry, summarised in Table II.43. The normal form in rocks is selenide (Se^{2-}), which may replace the sulfide ion; organic selenides analogous to sulfides can also be formed. In the presence of atmospheric O_2, oxidation to the selenite (+4) and selenate (+6) ions can take place, although these require more strongly oxidising conditions than the formation of sulfate, SO_4^{2-}.

80 Clark (1989).

Table II.43 Oxidation states of selenium in the environment. R represents an organic group, e.g. methyl or part of a biological molecule

Oxidation state	Examples
–2 (selenide)	HSe^-, metal selenides, R_2Se, RSeH
–1 (diselenide)	RSeSeR
0	Se
4 (selenite)	SeO_3^{2-}, $Fe_2(SeO_3)_3$
6 (selenate)	SeO_4^{2-}, $CaSeO_4$

Table II.44 Selenium concentrations in the environment

Location	Concentration
Crust	50 ppb
Sea water:	
surface	0.00003 ppb
deep	0.00018 ppb
Fresh waters	0.02–1 ppb
Soils	10–1000 ppb
Plants	1–1000 ppb
Atmosphere	traces of $(CH_3)_2Se$
Human body	200 ppb

Normal concentrations in natural waters are very low (see Table II.44), because of the insolubility of many compounds, especially the selenides of elements such as copper. High concentrations are occasionally found in soils, especially in association with zinc. Oxidation gives the generally more soluble selenite and selenate ions. However, the mobility of selenium is limited by formation of insoluble compounds such as $Fe_2(SeO_3)_3$; as Fe^{3+} is itself very insoluble at neutral and alkaline pH, however, it is generally in more alkaline soils that the availability of selenium is greatest. These factors are important because selenium is both an essential element, and toxic at higher concentrations. Some plants have a tolerance to selenium-rich soils, and a few act as *accumulators*, acquiring selenium concentrations of up to 1 per cent.

Although the thermodynamically stable form of selenium present in oxygenated water is the selenate ion, studies on sea water have shown that selenite is present in almost equal concentrations with this. In surface waters the element is taken up by marine life, and converted to organic forms such as dimethyl selenide, $(CH_3)_2Se$. Emissions of this gas occur both from the ocean and from soils and plants; the species that accumulate selenium have an unpleasant smell as a result. The cycling of selenium through the environment is summarised in Fig. II.34. It can be seen that

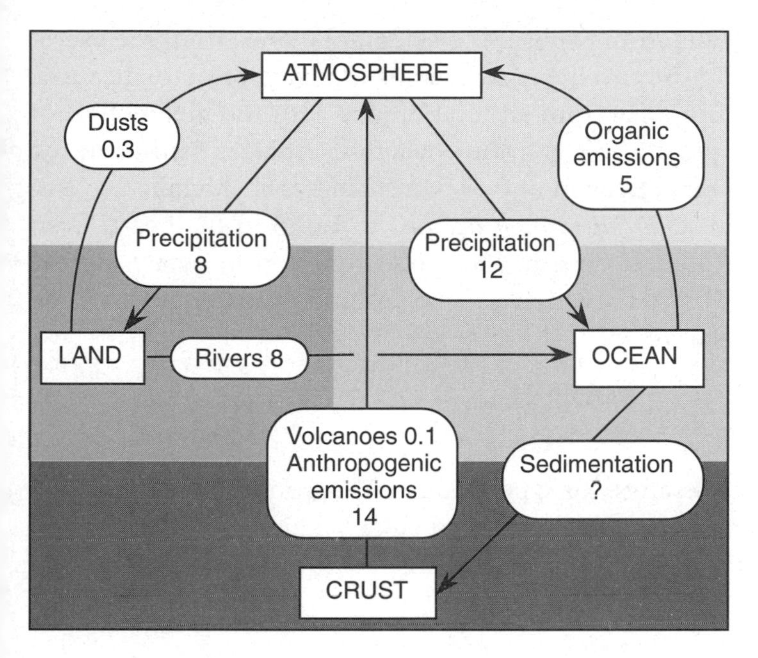

Fig. II.34 Cycling of selenium in the environment. The fluxes are given in units of 10^6 kg Se per year.

anthropogenic emissions, in the form of dusts arising from the processing of chalcophilic metals such as zinc, now dominate the cycling through the atmosphere.

Selenium is the rarest of all the elements known to be essential for human life. A normal diet provides around 100–200 μg per day, mostly from wheat and cereals. In some areas of the world selenium is deficient in the soil, and diets may provide substantially less than this. The clearest evidence of a deficiency disease is the so-called *Keshan disease*, a congestive heart failure in children once common in some areas of China.[81] This has been virtually eliminated by regular dietary supplements of sodium selenite, Na_2SeO_3. Grazing animals in some countries are also provided with supplements, either as long-lasting pellets placed in the rumen, or as crop sprays or additions to fertilisers. Apart from the acute effects of deficiency, there is some evidence that selenium can provide some protection against heart disease, cancer, and against the toxic effects of some heavy metals such as *cadmium and *mercury. The fashion of taking supplementary selenium in tablet form is controversial, however, and there is no definite consensus on the maximum safe intake. Excess selenium is definitely toxic, and animals grazing on the accumulator plant *Astralagus* develop a severe disorientation called 'blind staggers'. This plant is found in Wyoming, and was known to cowboys in the nineteenth century as 'locoweed'. Chronic industrial exposure causes a variety of symptoms, ranging from sore throats and 'garlic breath', to respiratory, liver, and kidney malfunction, and dental caries.

81 See Lenihan (1988).

The 'garlic odour' which results from exposure to selenium comes from the excretion of dimethyl selenide, $(CH_3)_2Se$, probably formed as a means of detoxification. The toxic effects are presumed to arise from an interference with the metabolism of sulfur. Selenium is a constituent of a few enzymes, where it replaces *sulfur in the amino acids *cysteine* and *methionine*. Only in one or two cases has the detailed role of these enzymes been established. *Glutathionine peroxidase* is found in the heart, liver, and some other organs, where it performs an important function in catalysing the decomposition of peroxides (ROOH), which can be formed 'accidentally' in the metabolism of *oxygen. The reaction is

$$\underset{\text{glutathionine}}{2\ \text{GSH}} + \text{ROOH} \rightarrow \text{GSSG} + \text{ROH} + H_2O,$$

replacing the very reactive and therefore toxic peroxide –OO– group by the stable and harmless –SS– group common in the biochemistry of sulfur. The antioxidant role of selenium may be responsible for giving protection against heart disease and cancer.

Silicon ($_{14}$Si)

Silicon is one of the dominant elements on Earth, forming some 27 per cent by mass of the crust. It is invariably combined with oxygen, forming the binary oxide SiO_2 and an enormous variety of silicates of lithophilic metals. Silicon and oxygen together make up four out of every five atoms near the Earth's surface. The chemistry of silicates dominates the crust, and some of the consequences of this are discussed below (see also §4.5). Silica is used for the manufacture of glass and cement. The element itself is obtained by reduction of SiO_2 with carbon, and most is used for the production of resistant *ferrosilicon* steels. Much smaller quantities (a few thousand tons per year) of highly purified silicon are used for the manufacture of semiconductor 'silicon chips' for electronic applications.

Silicates dissolve in water to give 'silicic acid', $Si(OH)_4$, and some 2×10^{11} kg of silicon is cycled annually through the hydrosphere in this form. Its solubility is lower than for other common elements such as calcium, and a much larger amount of silica, around 4×10^{12} kg annually, is carried by rivers as suspended sediments. Crystalline SiO_2, *quartz*, is particularly resistant to weathering, and is a major constituent of sand, formed by the mechanical sorting of weathered sediments by rivers. In the ocean the dissolved concentration of silica is particularly low in surface waters (see Table II.45), as it is taken up by marine organisms such as diatoms to make skeletons. These sink to the bottom and form sediments.

Silicon is believed to be essential for many living species, including possibly humans, although its role is not understood except in species which make skeletons

Table II.45 Silicon concentrations in the environment

Location	Concentration
Crust	28 %
Sea water:	
surface	0.03 ppm
deep	2 ppm
Fresh waters	10 ppm
Human body:	
blood	4 ppm
muscle	100–200 ppm

from silica. It has been suggested that dissolved silica can help protect against the toxicity of *aluminium, by sequestering the latter element through the formation of aluminosilicate polymers. Silicates are normally non-toxic, although some forms, as fibres (*asbestos*) and fine dusts, can cause lung diseases and cancer. This type of toxicity is related to the particle shape and size, and is thought to be due to mechanical irritation of tissues, rather than to a specifically chemical effect.

Silicate minerals

Silicon always has the oxidation state +4 in the environment, as in SiO_2; furthermore under normal conditions of temperature and pressure each silicon atom is nearly always found bonded to four oxygen atoms. Thus the basic unit found in silicates can be written $[SiO_4]^{4-}$. However, this tetrahedral unit has a strong tendency to polymerise. Two units can link by eliminating one oxide ion and sharing another one:

$$2\ [SiO_4]^{4-} \rightarrow [Si_2O_7]^{6-} + O^{2-},$$

but more commonly this process proceeds further, giving rings, chains, 2D sheets, or 3D networks, as illustrated in Fig. II.35.

Progressive polymerisation increases the ratio of Si to O, from 1:4 for orthosilicates containing single $[SiO_4]^{4-}$ units, to 1:2 for the fully polymerised 3D structure of quartz; the latter structure is not shown in Fig. II.36, but is essentially a network in which every oxygen atom is shared between two silicon atoms. Another important feature is that as polymerisation increases, it becomes possible to replace a fraction of the silicon atoms by aluminium. The negative charge on the network is thereby increased; for example the layer $[Si_2O_5]^{2-}$ becomes $[Si_3AlO_{10}]^{5-}$ when one in four silicon atoms is replaced by aluminium. A similar substitution in SiO_2 itself gives $[Si_3AlO_8]^-$. Metal cations which compensate this charge occupy spaces between the Si–Al–O groups, giving rise to an enormous variety of possible formulae and

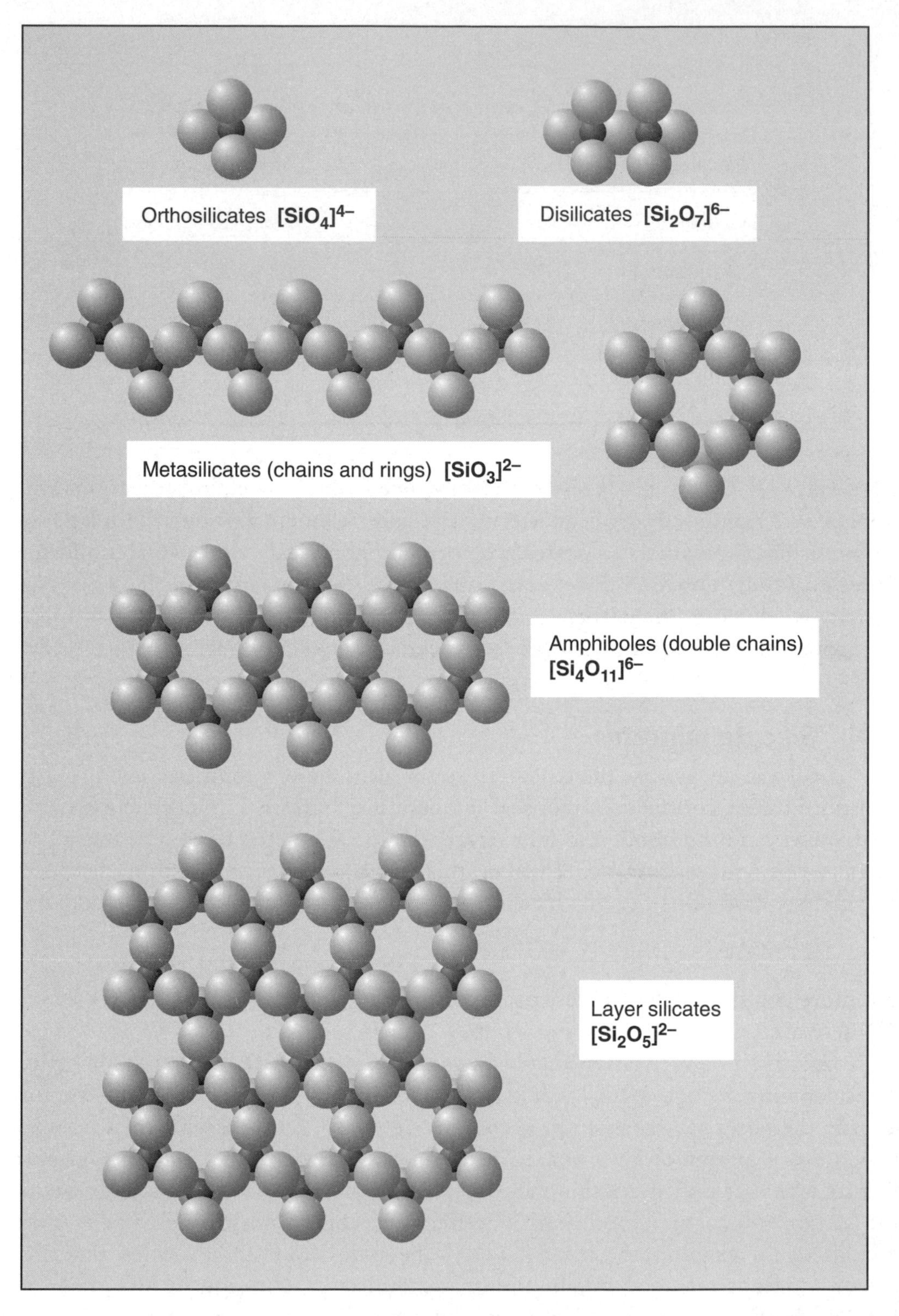

Fig. II.35 Silicate structures. Illustrating the structures formed by sharing oxygen atoms between silicon atoms, to give rings, chains, and sheets. See Table II.45 for typical formulae of these different types.

structures. Examples of this range are given in Table II.46. Most of the formulae given are highly idealised, and many silicates contain large numbers of other elements in low concentration. A major fraction of the Earth's content of most lithophilic elements is contained in silicates. Relatively few of these are important ores, however, as many elements are available in more concentrated sources from other compounds, formed by sedimentary and hydrothermal processes (see §4.6).

Igneous rocks are formed by successive melting and solidification driven by tectonic cycling discussed in Part I (see §§1.3 and 4.5). The ways in which different types of silicates melt and solidify are therefore important in determining the composition of these rocks. Fig. II.36 shows the so-called *reaction sequences* of solidification of a silicate melt. Minerals at the top solidify first, and are succeeded, as the melt cools, by ones lower down. In the discontinuous sequence on the left, the least polymerised silicates are the first to solidify. Conversely, when rocks are heated the most polymerised forms such as silica are the first to melt. Thus olivine and pyroxene, the relatively silica-poor constituents of very basic rocks, are retained in the mantle, whereas feldspars and quartz—highly polymerised silica-rich minerals making up acid rocks—are common in the crust.

There is also a continuous sequence of reactions involving feldspar. Plagioclase has a continuous range of composition from the calcium-containing *anorthite* ($Ca[Si_2Al_2O_8]$) to the sodium-containing *albite* ($Na[Si_3AlO_8]$); this is an example of the variable replacement of silicon by aluminium, accompanied by the replacement of other elements, Na^+ and Ca^{2+}, which have similar sizes (see §3.2). The consequences of these sequences for the typical composition of different types of rock are

Table II.46 Silicate minerals: types and examples. See Fig. II.35 for an illustration of the structures formed by silicon–oxygen linkages. The formulae given are often idealised, as many other metallic elements are present in small proportions

Class	Si–O formula	Examples	Typical formulae
Orthosilicates	$[SiO_4]^{4-}$	olivine	$(Mg,Fe)_2SiO_4$
		zircon	$ZrSiO_4$
		garnet	$Ca_3Al_2Si_3O_{12}$
Disilicates	$[Si_2O_7]^{6-}$	thortvietite	$Sc_2Si_2O_7$
Metasilicates	$[SiO_3]^{2-}$	beryl	$Be_3Al_2[Si_6O_{12}]$
		pyroxene	$(Mg,Fe)[SiO_3]$
Amphiboles	$[Si_4O_{11}]^{6-}$	glaucophane	$Na_2Mg_3Al_2[Si_8O_{22}](OH)_2$
	$[Si_3AlO_{11}]^{7-}$	hornblende	$(Na,Mg,Al,Ca,Fe)_4[Si_3AlO_{11}](OH)$
Sheet silicates	$[Si_2O_5]^{2-}$	serpentine	$Mg_3[Si_2O_5](OH)_4$
		kaolinite	$Al_2[Si_2O_5](OH)_5$
	$[Si_3AlO_{10}]^{5-}$	phlogopite	$K(Mg,Fe)_3[Si_3AlO_{10}](OH,F)_2$
Framework silicates	$[SiO_2]$	quartz	SiO_2
	$[Si_3AlO_8]^-$	alkali feldspar	$K[Si_3AlO_8]$
	$[SiAlO_4]^-$	plagioclase feldpsar	$Ca[Si_2Al_2O_8]$

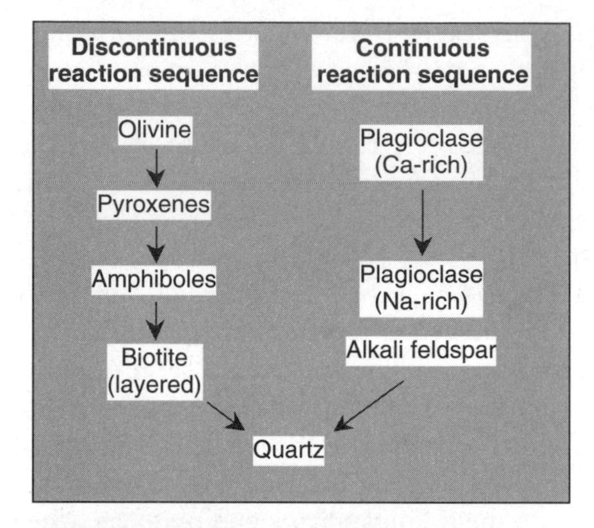

Fig. II.36 Reaction sequences showing the minerals formed in the cooling and solidification of a silicate melt. Minerals at the top solidify earlier than ones lower down.

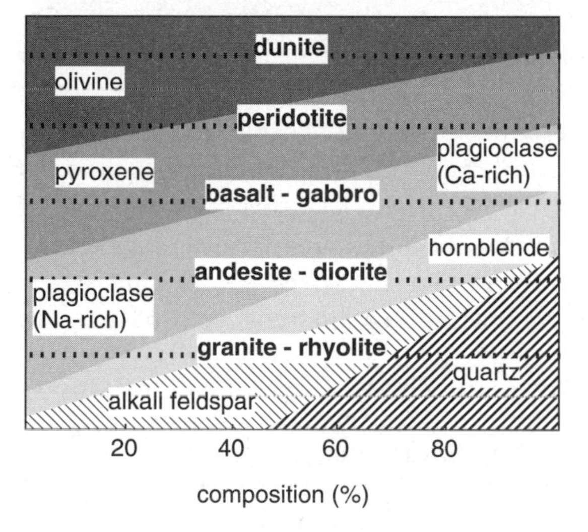

Fig. II.37 The distribution of silicate minerals in igneous rocks. Different rock types are listed, from the top downwards, in order of increasing fractionation from the mantle. The shading shows their approximate make-up in terms of silicate minerals with different degrees of polymerisation.

shown in Fig. II.37. Dunite and peridotite, shown at the top, are ultrabasic rocks characteristic of the mantle, and composed largely of olivine and pyroxene. *Magnesium predominates in these minerals, and is more abundant in the mantle than in the crust. Basalt is typical of oceanic crust and granite of the more evolved continental crust; andesite is intermediate in character.

The distribution of other elements in silicate rocks exerts a major control on the composition of the crust. The occurrence of sodium and potassium in feldspars forming late in the reaction sequence accounts for their fractionation into the crust, discussed in §4.5. As also mentioned there, many *incompatible* elements are retained in the molten silicate phase, and come out late in the solidification process, in rocks known as *pegmatites*.

Silver ($_{47}$Ag)

Silver is a relatively rare element, although known to the ancients because it can occur in 'native' (metallic) form. Such deposits are less common than those of *gold, however, and most silver is found as sulfides, in the mineral *argentite*, Ag_2S, or more commonly as a minor constituent of the ores of copper, lead, or zinc. It is separated during the processing of these elements, and used in photography (as halides such as AgBr), in jewellery, electrical work, and in mirrors. Concentrations in rivers and in the ocean (where it probably exists as the chloride complex $[AgCl_2]^-$) are low. Silver is not essential for life. It is toxic to micro-organisms.

Sodium ($_{11}$Na)

The most abundant of the alkali elements, sodium is common in the crust, and is the most abundant metal ion in sea water. It is a constituent of many silicate minerals, such as *plagioclase feldspar*, $(Na,Ca)(Al,Si)_4O_8$, but as most compounds of sodium are very soluable in water, Na^+ is easily washed out in weathering. The ion is relatively abundant in fresh water, and dominates in the oceans because it has one of the longest residence times (10^8 years) of any element. It is ultimately removed by the formation of evaporite deposits, chiefly of *halite* ('rock-salt' or 'common salt'; NaCl), but also of carbonates, sulfates, borates, and other compounds. NaCl is used in larger quantities than any other industrial inorganic chemical. It is the main source of *chlorine, the sodium being converted to *caustic soda*, NaOH, of major industrial importance. Elemental sodium is manufactured by the electrolysis of molten NaCl. It is a powerful reducing agent, used for the manufac ture of some other metals such as *titanium and *zirconium, and of tetraethyl *lead.

Sodium is essential for life. It is present as the Na^+ ion, which as illustrated in Table II.47, has a considerably lower concentration within cells than in extracellular

Table II.47 Sodium concentrations in the environment

Location	Concentration
Crust	2.3 %
Sea water	1.05 %
Fresh waters	6 ppm
Human body:	
average	1400 ppm
blood plasma	3500 ppm
red cells	250 ppm

fluids. Control of the sodium concentration may have been a feature of the earliest living things. It is necessary to exclude some ions from the cell, relative to the surrounding sea water, as otherwise the osmotic pressure of the cell constituents would cause water to enter and burst the cell membrane. As the most abundant constituent of sea water, Na^+ was the natural ion to discriminate in this way. Even in non-marine animals such as ourselves, overall sodium concentrations are still controlled carefully by the kidneys, and are important in maintaining the osmotic balance. Ions are expelled from cells by special 'sodium pumps' in membranes. An important function of sodium in animals is in the conduction of messages by nerves. In this process, the permeability of the nerve membrane changes, and Na^+ passes into the cell, with potassium ions moving in the opposite direction. Na^+ is quickly expelled again by the sodium pump, and the maintenance of the correct concentration gradients consumes some 40 per cent of the entire metabolic energy supply in a resting human.

Deficiency of sodium in animals and humans causes muscular spasm ('cramps'). As the element is not required in such large amounts by plants, land animals may find it hard to obtain enough. Rocksalt has traditionally been a prized commodity, important in commerce. The English word *salary* is derived from the Latin *salarium*, a payment to Roman soldiers intended specifically for the purchase of salt (*sal*). It has been speculated that the availability of sodium is important in limiting the number of land animals. For example, there was significant increase in the rodent population of the Eastern USA following the 'dust bowl' phenomena of the 1930s, and it may be that this was made possible by the increased availability of sodium, present in minerals in the blown dust. Animals have been seen to lick salt deposits, and a craving for salt may be deeply rooted in our genetics. On the other hand, excess sodium is certainly harmful. The agonising and lethal effects of drinking sea water when no fresh water is available illustrate this most dramatically. Excess salt in the diet may contribute to high blood pressure, and there is a trend in Western societies to consume less salt as a dietary supplement, or to replace it by 'low sodium' alternatives which contain potassium chloride.

Strontium ($_{38}Sr$)

Chemically similar to calcium but less abundant, strontium is widespread in silicates, and is found concentrated in two minerals, *celestite* ($SrSO_4$) and *strontianite* ($SrCO_3$). It is used to give a red colour to fireworks and flares, and in some glass and TV screens. Strontium is relatively common in sea water, and is present in all living things. Some microscopic marine organisms use strontium to make a skeleton composed of $SrSO_4$, but the element is not otherwise essential; together with bromine and rubidium, it is the most abundant element in the human body with no known function there.

Sulfur ($_{16}S$)

Sulfur is very abundant on the Earth overall, but a major part is concentrated in the core, where it occurs at a level of around 5 per cent, combined with metals such as iron and nickel. Although it is less common in the crust (see Table II.48), it has a degree of chemical versatility that enables it to combine with many other elements, and makes it very important in the natural environment.

Sulfur can occur in reduced form as sulfide (S_2^-), as the uncombined element, and in various positive states of oxidation, especially the fully oxidised sulfate (SO_4^{2-}). Table II.49 shows a selection of sulfide minerals, and Fig. II.38 shows the elements in the periodic table found naturally as sulfides. These elements are geochemically classified as

Table II.48 Sulfur concentrations in the environment

Location	Concentration
Crust	260 ppm
Sea water	870 ppm
Fresh waters	5 ppm
Human body	2000 ppm

Table II.49 Some sulfide minerals

Name	Formula
Pyrites	FeS_2
Arsenopyrite	FeAsS
Chalcopyrite	$FeCuS_2$
Pentlandite	$(Fe,Ni,Co)_9S_8$
Sphalerite	ZnS
Molybdenite	MoS_2
Realgar	As_4S_4
Argentite	Ag_2S
Greenockite	CdS
Stibnite	Sb_2S_3
Cinnabar	HgS
Galena	PbS

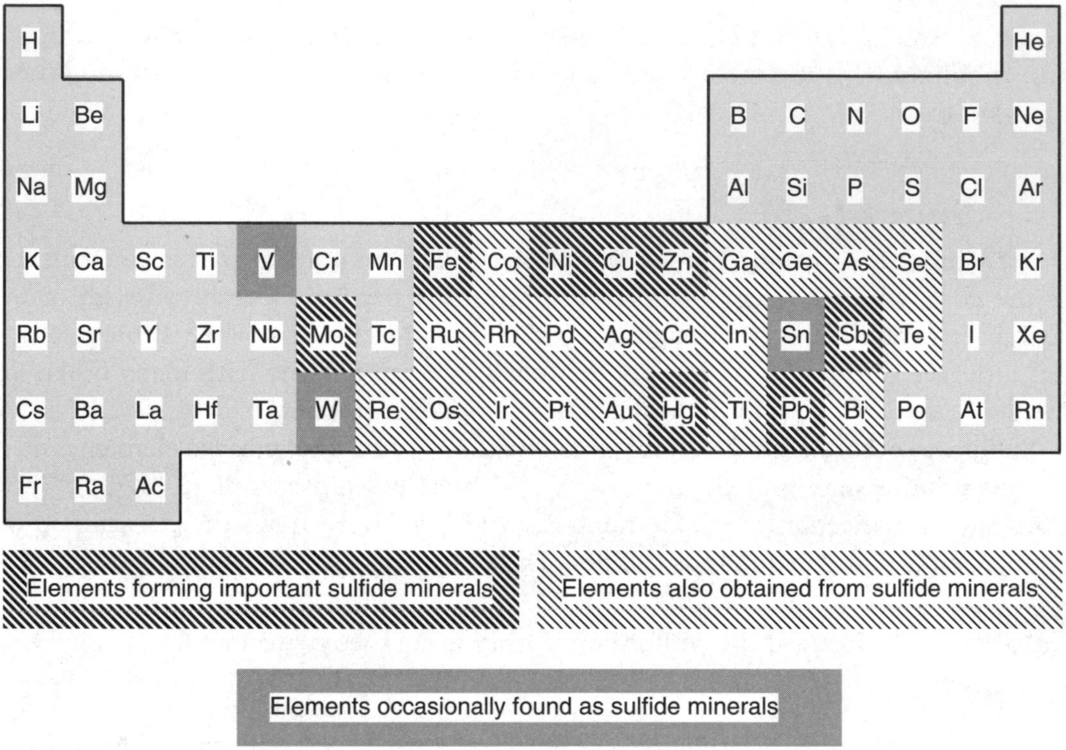

Fig. II.38 Elements occurring in sulfide minerals.

chalcophiles. For many such elements, sulfides form the principal source. These minerals have diverse geological origins (see §4.6): *layered deposits* of nickel sulfide, which also form an important source of the *platinum-group metals, were mostly formed early in the solidification of the crust; copper and molybdenum sulfides are derived largely from *hydrothermal processes*; and other sulfides may be formed under reducing conditions caused by decaying organic matter in lakes and deep ocean trenches.

The sulfate ion is abundant in natural waters, and deposits of sulfate minerals, especially *gypsum*, $CaSO_4.2H_2O$, are formed by evaporation. The element itself is produced naturally by volcanic activity and by biological redox reactions. It is extracted in large quantities, and used mostly to make sulfuric acid (H_2SO_4), one of the most important industrial chemicals.

Sulfur is an essential element for all life, but unlike *nitrogen and *phosphorus it never seems to be in short supply. Some sulfide compounds, for example hydrogen sulfide (H_2S) and organic sulfides, are very toxic. These volatile compounds also have exceptionally strong odours; the 'smell of the sea', popularly attributed to ozone, is partly due to dimethyl sulfide, $(CH_3)_2S$, produced by marine life. This compound is only one of a number containing sulfur that are present in low concentrations in the atmosphere. Others include sulfur dioxide, SO_2, liberated by volcanoes, and from industrial processes and fossil fuel burning. Sulfur compounds in the atmosphere are rapidly oxidised to sulfuric acid, which is a major source of acid rain.

The redox chemistry of sulfur

Table II.50 demonstrates the unusually large range of oxidation states exhibited by sulfur in nature. In addition to the simple sulfide ion S^{2-}, the disulfide ion S_2^{2-} is also stable: it is a constituent of the most abundant of all sulfide minerals, *pyrites* (FeS_2), and as an organic form it plays an important part in biology. The most important oxidised forms are SO_2 and sulfate.

Figure II.39 shows the thermodynamics involved in the redox transformations of sulfur, in the absence of water (left) and in water at pH 7 (right). In the presence of atmospheric oxygen, the highest oxidation state +6 is stable. Under dry conditions, H_2S may be progressively oxidised to the element, to SO_2, and to sulfur trioxide, SO_3. In the presence of water, the range of stability of uncombined sulfur is less. The main differences however are that the +6 oxidation state becomes much more stable when SO_3 reacts with water to form sulfate, and that the +4 oxidation state is no longer thermodynamically stable. Sulfite, SO_3^{2-}, is formed when SO_2 dissolves in water, but it is a strong reducing agent.

The oxidation of sulfide to sulfate on exposure to atmospheric oxygen has important consequences for the elements which occur as sulfide minerals. The obvious oxidation product, a metal sulfate, is in fact rarely formed. In situations where a sulfate might be produced under dry conditions, e.g. $ZnS \rightarrow ZnSO_4$, the product is usually soluble in water. As shown in Table II.51, serveral sulfides oxidise to soluble products, giving either cations such as Zn^{2+} or oxyanions such as molybdate, MoO_4^{2-}.

Table II.50 Oxidation states of sulfur

Oxidation state	Name of compound	Formula
–2	hydrogen sulfide	H_2S
	sulfide:	S^{2-}
	e.g. galena	PbS
	organic sulfide (thiol)	RHS or R_2S, e.g. $(CH_3)_2S$
–1	disulfide:	$(S_2)^{2-}$
	e.g. pyrites	FeS_2
	organic disulfide (dithiol)	RSSR, e.g. CH_3SSCH_3
0	native sulfur	S
4	sulfur dioxide	SO_2
	sulfite, bisulfite	SO_3^{2-}, HSO^{3-}
6	sulfate:	SO_4^{2-}
	e.g. gypsum	$CaSO_4.2H_2O$

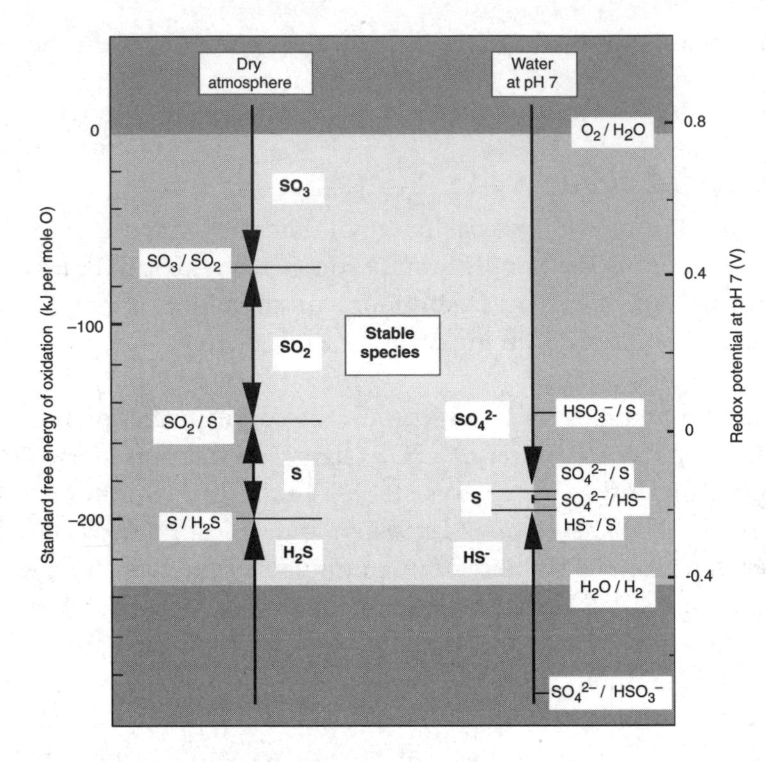

Fig. II.39 Thermodynamics of redox reactions of sulfur, in the absence of water (left) and in water at pH 7 (right). The left-hand scale shows the Gibbs free energy change, ΔG, per mole of O consumed in the oxidation reactions shown; the right-hand scale shows the corresponding electrode potential. The heavy vertical lines show the regions of stability of the species listed in the central region. Reactions between thermodynamically unstable species are shown at the side. The dark-shaded regions at the top and bottom indicate conditions outside the region of thermodynamic stability in the natural environment.

Table II.51 Oxidation of sulfide minerals, showing the type of product expected when oxidation occurs in air

Oxidation product	Elements
Soluble cation or oxyanion	Co, Ni, (Cu), Zn, As, Mo, Cd, Tl
Insoluble oxide or hydroxide	Fe, Sn
Metallic element	Cu, Ru, Rh, Pd, Ag, Os, Ir, Pt, Au, Hg

These may later produce insoluble salts with other ions, such as carbonate, silicate, or a cation as in $CaMoO_4$.

There are other possible outcomes. In the case of pyrites, oxidation of sulfide is accompanied by oxidation of Fe^{2+} to Fe^{3+}, which is normally left as the very insoluble hydroxide $Fe(OH)_3$:

$$4FeS_2 + 15O_2 + 14H_2O \rightarrow 4Fe(OH)_3 + 16H^+ + 8SO_4^{2-}.$$

Sulfides of less reactive metals behave differently, as the redox potential at which sulfide is oxidised is often less than that required to convert the metallic element to an oxide or aqueous cation. Thus we have a reaction such as

$$2CuS + 3O_2 + 2H_2O \rightarrow 2Cu + 4H^+ + 2SO_4^{2-},$$

which leaves the metallic element in native, uncombined, form. Sometimes (as in the case of copper) further oxidation is possible if oxygen is abundant. However, the native deposits of several elements may have originated in this way.

The redox chemistry of sulfur is somewhat analogous to that of *nitrogen. One difference is the much greater range of thermodynamic stability of dinitrogen compared with uncombined S; another is that the redox reactions of sulfur generally happen more rapidly, with smaller activation barriers that those of nitrogen. This is important both in the atmosphere (where most sulfur compounds are rapidly oxidised) and in biology.

One interesting resemblance to nitrogen chemistry is that redox reactions of sulfur compounds are used to obtain energy by some micro-organisms. The *dissimilatory reduction* of sulfate to sulfide (as opposed to the *assimilatory reduction* in the actual uptake of sulfur from the environment) is a source of oxidising power under anaerobic conditions when O_2 is unavailable.[82] It is used by bacteria in the digestive tracts of animals, and in sediments in lakes and some deep ocean trenches. When oxygen is available, together with a sulfide mineral such as FeS_2, the oxidation of sulfide to sulfur or sulfate can be used by *chemosynthetic* bacteria as an energy supply that requires neither a source of organic carbon nor photosynthesis. Species that make use of this include ones that live under extraordinary conditions near hydrothermal vents in the sea floor, where neither ordinary 'food' nor light is available. Yet another possibility—and probably the first one utilised by biology—is sulfur-based photosynthesis. The reaction

$$CO_2 + 2H_2S + h\nu \rightarrow C_{org} + 2H_2O + 2S$$

is much less demanding energetically than the photosynthetic splitting of water carried out by green plants. It is carried out by some photosynthetic bacteria, and has a mechanism simpler than that of conventional photosynthesis, in that only one photosystem is required (see discussion under *oxygen). It seems very likely that sulfur-based photosynthesis evolved before the more complicated oxygen-based reactions.

The biological chemistry of sulfur

Sulfur, like nitrogen, is normally regarded as an 'organic' element in biology, as it forms a constituent of carbon-based organic molecules. A few important ones are illustrated in Fig. II.40.

Most sulfur assimilated by organisms is reduced to the –2 oxidation state, and incorporated into the two amino acids cysteine or methionine, which are constituents of peptides. Figure II.40 shows how two cysteine (–SH) groups in different parts of a protein can link to form a disulfide (–S–S–) bridge. These bridges are important in

82 Levett (1990) gives a general account of anaerobic bacteria.

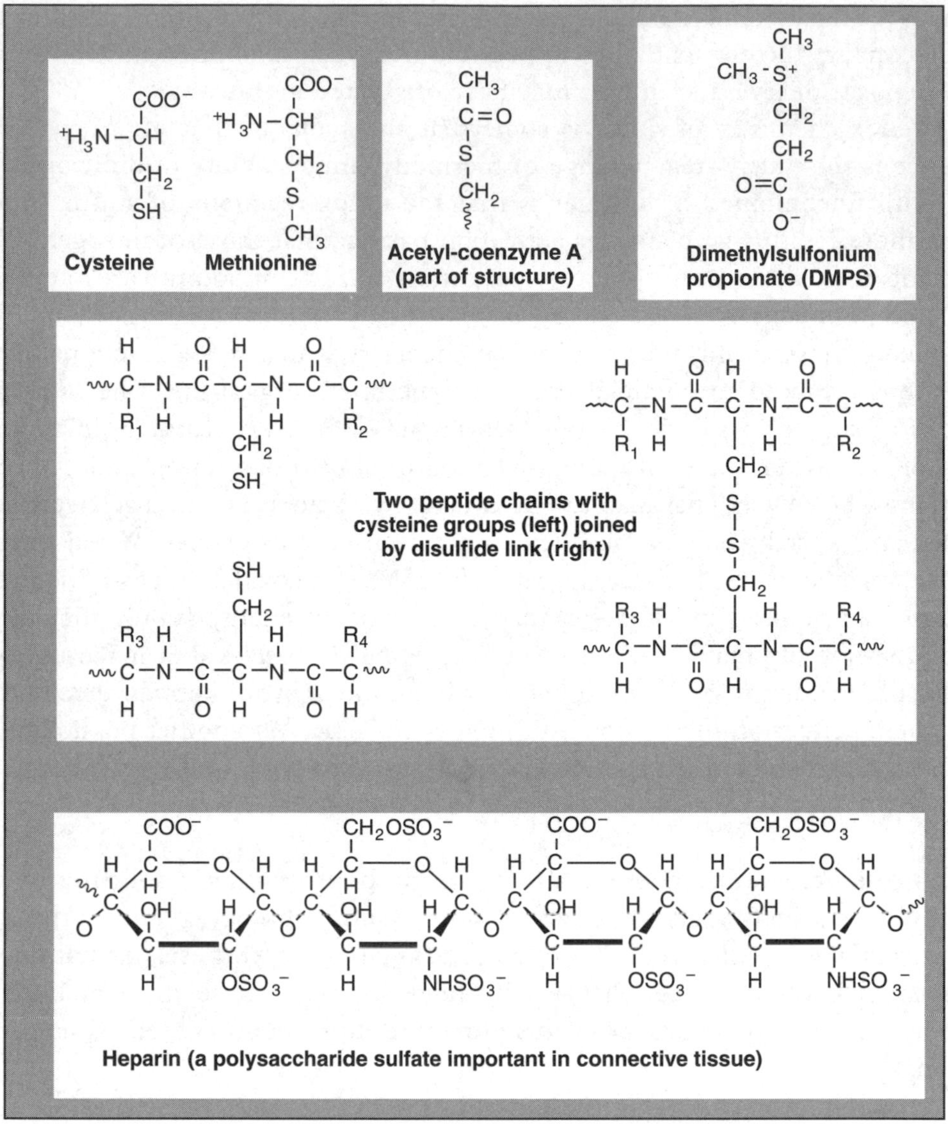

Cysteine Methionine

Acetyl-coenzyme A (part of structure)

Dimethylsulfonium propionate (DMPS)

Two peptide chains with cysteine groups (left) joined by disulfide link (right)

Heparin (a polysaccharide sulfate important in connective tissue)

Fig. II.40 Some biological compounds containing sulfur.

many types of structural protein, and are very abundant in *keratin*, the main constituent of nails and hair. The formation of disulfide requires an oxidation reaction,

$$2R\text{–}SH \rightarrow R\text{–}S\text{–}S\text{–}R + 2H^+ + 2e^-,$$

and illustrates the rather facile redox chemistry of sulfur. A cysteine-containing tripeptide, *glutathionine*, is an important constituent of cells. It cycles between the reduced and oxidised forms, and maintains a 'buffer' of –SH groups, which play a role in various redox processes. One interesting example is the removal of the toxic peroxides, produced as a by-product in the reactions of *oxygen. The reaction

$$2\ GSH + ROOH \rightarrow GSSG + ROH + H_2O$$

glutathionine

is catalysed by the enzyme *glutathionine peroxidase*, which contains *selenium.
Many chalcophilic metals have a strong affinity for thiolate (RS^-) groups, which are therefore common binding sites in metalloproteins. Elements bound in this way include *iron, *copper, *zinc, and *molybdenum. Figure II.16 shows part of the structure of some iron–sulfur proteins, including examples with metal–sulfide clusters; vanadium and molybdenum can also form similar structures. Some elements bind even more strongly to sulfur, and so can displace these essential elements from their active sites. The toxicity of the heavy metals *cadmium, *mercury, and *lead comes largely from their affinity for sulfur, either at sites where no metal is normally combined or in competition with other metals. The special protein *metallothionein* contains an unusually large number of cysteine groups and can bind several heavy metal atoms per molecule. It is manufactured in the liver in order to sequester toxic elements, including excesses of some essential ones such as copper.

Figure II.40 also shows part of the structure of acetyl-coenzyme A (CoA), which illustrates a different role for sulfur. The thioester group (–CO.S–) is less stable in aqueous solution than the conventional ester (–CO.O–). Hydrolysis of acetyl CoA,

$$RCO\text{–}S\text{–}CoA + H_2O \rightarrow RCOOH + HS\text{–}CoA,$$

has a negative free-energy of reaction comparable to that of the hydrolysis of ATP (see *phosphorus). CoA is important in metabolism as a carrier of 'activated' acetyl groups, in the same way that ATP carries activated phosphate.

In most organic compounds, a sulfur atom forms two covalent bonds, and carries no charge. Other molecules in Fig. II.40 illustrate different types of bonding. The *sulfonium ion* occurs in dimethylsulfonium propionate (DMSP), found in marine algae. One of its functions may be simply as an ionic constituent which helps to regulate the osmotic pressure; it also forms the source of dimethyl sulfide (DMS, $(CH_3)_2S$), a marine emission with a role in the global cycling of sulfur.

The other charged sulfur-containing group illustrated in Fig. II.40 is the oxidised sulfate ($ROSO_3^-$) form. Sulfated polysaccharides occur in connective tissue such as collagen. Repulsion between the negatively charged groups keeps the polymers in an elongated configuration, so that they form gel-like structures which are fairly rigid.

Atmospheric sulfur chemistry

Table II.52 shows some data on the input of sulfur compounds to the atmosphere. The atmospheric chemistry of sulfur is dominated by redox processes. Reduced sulfur compounds, especially organic ones such as DMS, are much less soluble in water than the fully oxidised SO_3. Although a substantial amount of sulfate enters the atmosphere in particulate form, from wind-blown dust and evaporated sea-spray, the volatile emissions all contain sulfur in lower oxidation states. Oxidation to sulfate is fairly rapid, so that the mean residence time of sulfur compounds entering the atmosphere is only a few days. A comparison with *nitrogen is again instructive. Whereas N_2 is the major constituent of the atmosphere, the average concentration of all sulfur compounds together totals about one part per billion. Yet the annual fluxes of nitrogen and sulfur are similar (about 10^{11} kg per year) and exceeded only by those of *carbon, *hydrogen, and *oxygen. One consequence of such short residence times is that the concentrations of many species are highly variable, and concentrated near their sources. The main exception to this general rule is for carbonyl sulfide, COS, which reacts only slowly. It is produced in very small quantities, in biological decay and burning, but because of its much longer residence time, it has the highest average atmospheric concentration of any sulfur compound.

Figure II.41 shows the main routes by which sulfur compounds are oxidised in the atmosphere.[83] The reactive species chiefly involved are hydroxyl radicals (OH), and oxides of nitrogen (NO_2). The stronger bonds in COS are less susceptible to attack, and some of this compound enters the stratosphere where it is rapidly removed by photochemical reactions.

Oxidation of DMS is especially interesting, as it forms the major biogenic sulfur emission. NO_2 is involved in the major route when NO_x species are present in

Table II.52 Atmospheric sulfur compounds. Most figures, especially for annual inputs and residence times, are very approximate. (Based on data in Schlesinger, 1991)

Compound	Annual input 10^9 kg S $year^{-1}$	Average concentration (ppb)	Atmospheric residence time (days)	Main sources
SO_4^{2-}	165	–	1–10 ?	sea salt, desert sand
SO_2	110	0.2	14	industrial, volcanoes
H_2S	20 ?	0.01	4	swamps, soils
$(CH_3)_2S$ (DMS)	40	0.01	1	ocean biota
CS_2	4	0.01	10	swamps
COS	3	0.5	>300	swamps

83 See Wayne (1991) for more details.

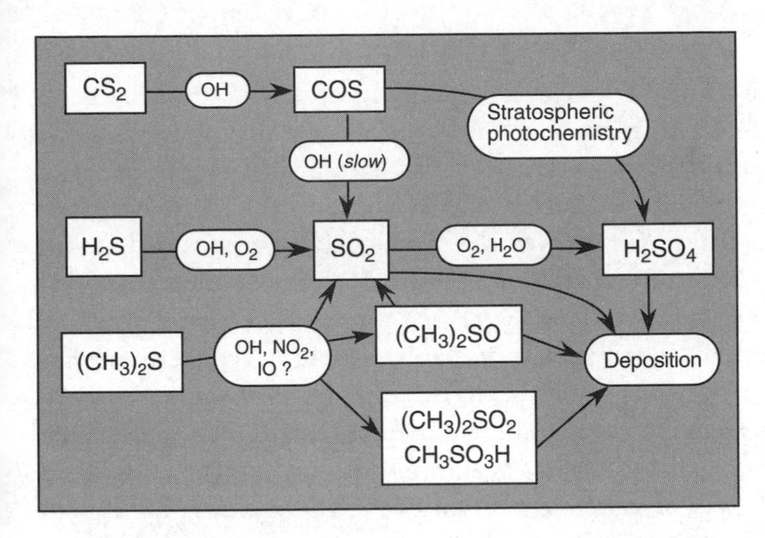

Fig. II.41 Routes for atmospheric oxidation of sulfur compounds, showing the main reactive species (round-cornered boxes) responsible for different conversions.

sufficient concentration. Some oxidation of DMS may proceed to SO_2 and thence to sulfate, but the incomplete oxidation products $(CH_3)_2SO_2$ and CH_3SO_3H (methane sulfonic acid) may be deposited directly. The relative proportions of these different products are not well known; this is one of the difficulties which make it hard to assess the possible role of DMS in climate control, discussed below.

Atmospheric SO_2 comes partly from oxidation of other compounds, and partly as the normal product of burning sulfur and organic sulfur compounds in air. There is some natural input from volcanoes, but it is very variable and tends to be concentrated during major eruptions. Such natural sources are dwarfed by the anthropogenic inputs of SO_2, chiefly from burning fossil fuels. These fuels contain sulfur derived directly from organic compounds; in the case of coal, there is another contribution from sulfide minerals such as iron pyrites, deposited along with the organic matter which forms the coal.

SO_2 is one of the most serious atmospheric pollutants. In concentrations above 1 ppm it has an irritant effect on the membranes of the nose and lungs, which seems to be greatly exacerbated by the present of smoke particles. Many people died from this combination in the notorious London smogs of the past, and following 4000 deaths in one episode in 1952 the Clean Air Act was introduced which forbade the burning of coal in certain areas.

Oxidation of SO_2 proceeds relatively slowly, and normally involves the intervention of water, to produce sulfuric acid, as H^+ and SO_4^{2-}. When cloud drops are not already present, sulfuric acid may be important as a nucleating agent, facilitating the condensation of water. Thus it can have two distinct roles in atmospheric processes, as a major component of acid rain (see *hydrogen), and as an influence on the degree of cloud cover. These are discussed below in connection with the sulfur cycle.

The sulfur cycle

The environmental cycling of sulfur is unusually complex. Figure. II.42 shows the principal types of chemical reaction involved, most of which have been discussed in the preceding sections. It is clear that life plays a dominant role in many of these reactions: direct effects include oxidation and reduction cycles involving the uptake and excretion of biological sulfur, as well as energy-yielding reactions without any assimilation of sulfur; indirect effects of life include the atmospheric reactions resulting from the biogenic gas O_2.

In the early environment, nearly all sulfur was probably present in reduced form, as sulfide minerals. Photosynthetic reactions producing sulfur may have started early in the history of life; it has been suggested that sulfide minerals such as FeS might even have been involved in the origin of life itself. Small amounts of *oxygen, produced abiotically through the photolysis of water vapour in the atmosphere, reacted with surface rocks to produce sulfates. The SO_2 and sulfuric acid present in the atmosphere of Venus must come from this type of reaction—the presence of liquid water on

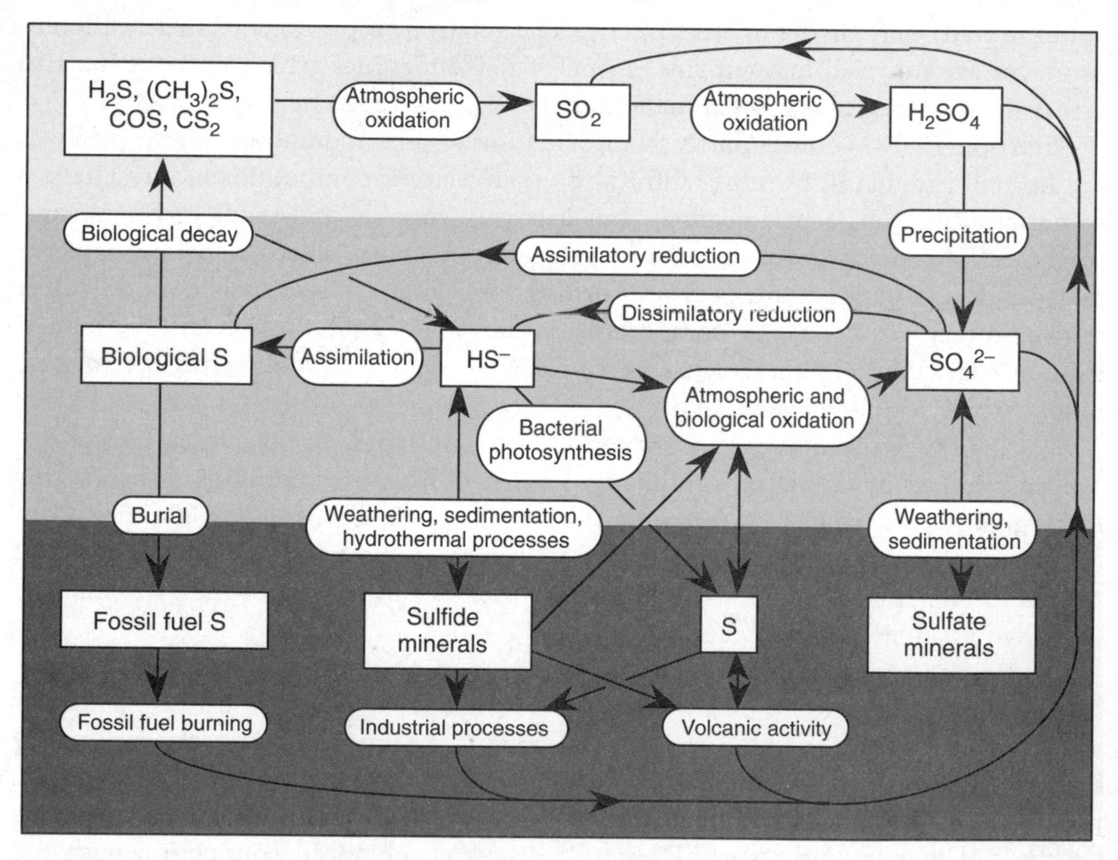

Fig. II.42 Chemical transformations in the sulfur cycle.

Earth has prevented a similar accumulation of sulfur compounds in our own atmosphere. There is evidence from the isotopic composition of sulfur-containing minerals that the dissimilatory reduction of sulfate also started early in the history of life.

With the advent of larger quantities of oxygen from water-based photosynthesis, oxidation of sulfides became more rapid. This reaction, together with the oxidation of *iron, has taken up nearly all the net oxygen ever produced by life: only about 4 per cent remains in the atmosphere. Thus sulfur has been of crucial importance in buffering the redox state of the Earth's surface, and remains so today, especially through its use in life.

Figure II.43 shows an estimate of the fluxes in the present-day sulfur cycle. Some of these numbers are very uncertain, and the very rapid cycling of sulfur through the

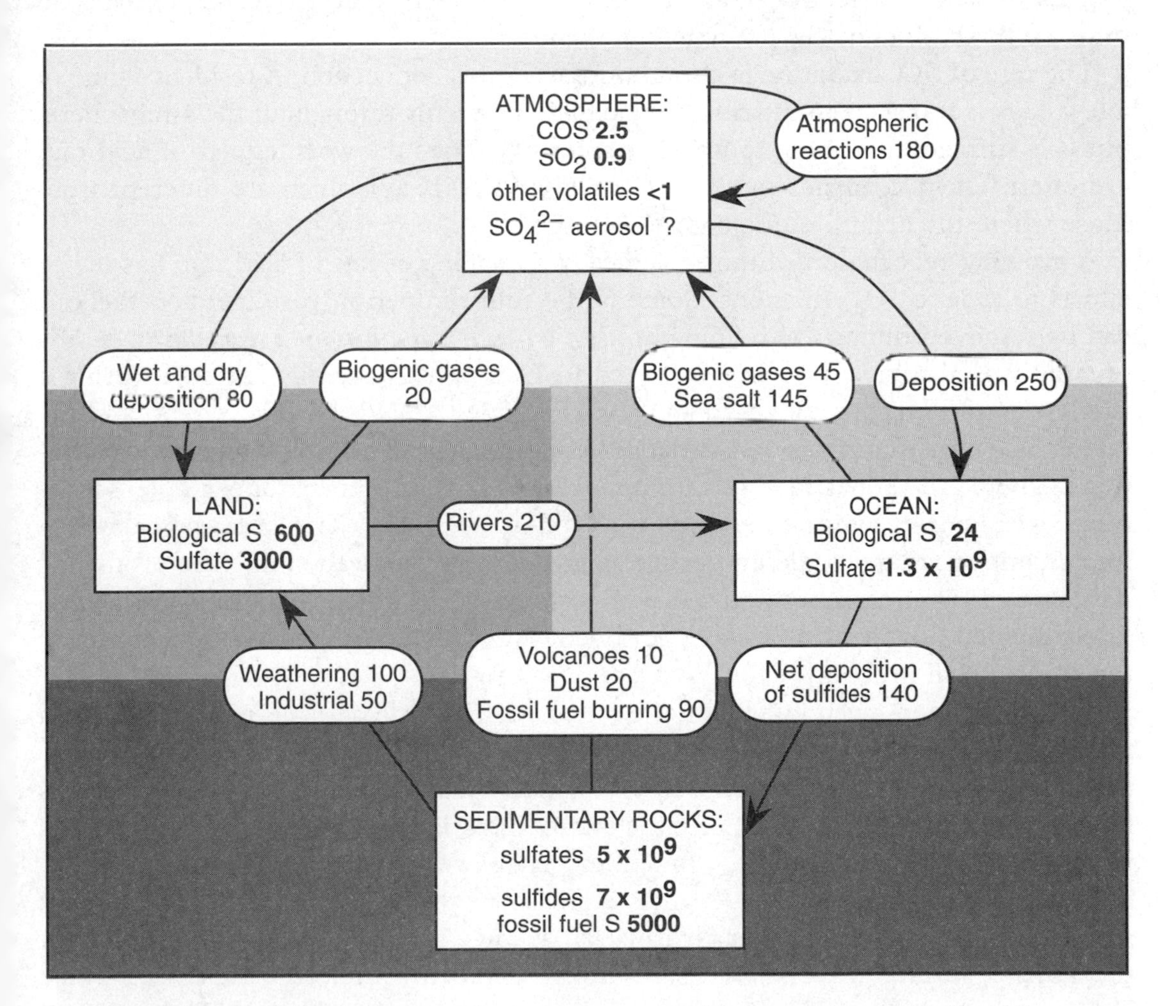

Fig. II.43 The sulfur cycle. Reservoirs (square-cornered boxes) and annual fluxes (round-cornered boxes) are given in units of 10^9 kg S. (Based on data in Schlesinger, 1991.)

atmosphere makes it particularly hard to estimate the magnitudes involved. It can be seen however that human activities, especially fossil fuel burning, contribute an input to the atmosphere which is comparable to the natural sources. It is particularly significant in the formation of acid rain (see under *hydrogen, especially Table II.23). Such acidity arises when SO_2 is oxidised, the net reaction being:

$$2SO_2 + O_2 + 2H_2O \rightarrow 4H^+ + 2SO_4^{2-}.$$

Sulfate entering the atmosphere directly by the evaporation of sea salt does not contribute to acidity in this way, as it is derived from a solution at nearly neutral pH, where each SO_4^{2-} is accompanied by non-acid cations such as Mg^{2+} to balance the charge. When this fact is taken into account, it appears from Fig. II.43 that anthropogenic SO_2 emissions may make a contribution to global acidity which exceeds all other sulfur-based sources put together.

The rate of SO_2 oxidation is also important in this connection. A residence time of a few days is too short to distribute the emissions evenly throughout the atmosphere, but it is sufficient for them to travel appreciably. Thus the worst effects of acid rain are often felt in countries (such as Canada and Norway) which are different from those where the pollution originates (USA, UK, etc.).

A growing recognition of the ecologically damaging effects of acid rain has led to efforts to reduce SO_2 emissions. Some of the sulfur in petrol (gasoline) and fuel oils can be removed during the refining process by *hydrodesulfurisation reactions*, whereby organic sulfur compounds are converted to H_2S (which is easily removed) using a catalyst containing molybdenum and cobalt sulfides. Coal is harder to treat, and the major SO_2 emissions from industrialised countries come from coal-fired power stations. There are various methods for removing SO2 from smoke, for example 'scrubbing' with basic compounds such as $CaCO_3$ or MgO. They are expensive to implement on a large scale and progress in this area has been slow. One promising development is the use of 'fluidised-bed' technology, in which coal can be burnt as a finely divided powder. $CaCO_3$ can be added, to react with SO_2 and remove it during the combustion process.

There has been much interest in the role of sulfuric acid aerosols in acting as *cloud condensation nuclei*, which promote the growth of water droplets, and thus increase the cloud cover (see *hydrogen). It is very likely that atmospheric sulfur emissions have some influence on the Earth's climate. This is particularly true for aerosols which form in the stratosphere, where they may be very long lived because of the dry and stable conditions there. A small increase in the scattering and reflection of incoming solar radiation can easily cause an average cooling of one or two degrees at the Earth's surface. Except for the relatively unreactive molecule COS (see Table II.52), sulfur compounds do not normally reach such altitudes. Violent volcanic eruptions may however eject gases including SO_2 into the upper atmosphere, and so lead to an increase in the stratospheric aerosol content. The cooling effect of some eruptions has been well established, and may last for several years. It has also been suggested that anthropogenic sulfur emissions may have a similar influence, and so could in some ways be mitigating the global warming produced by *carbon compounds.

Proponents of the Gaia hypothesis have suggested that the natural cycling of sulfur, through DMS emissions from the ocean, might provide a mechanism for regulating the global climate.[84] Figure II.44 shows a simplified version of what has been called the CLAW cycle after its originators.[85] The idea is that DMS produced by marine plankton enters the atmosphere and is oxidised to sulfate aerosol. The resulting cloud condensation nuclei increase the cloud cover, which reflects more sunlight and so cools the Earth's surface. If a general rise in temperature stimulated more

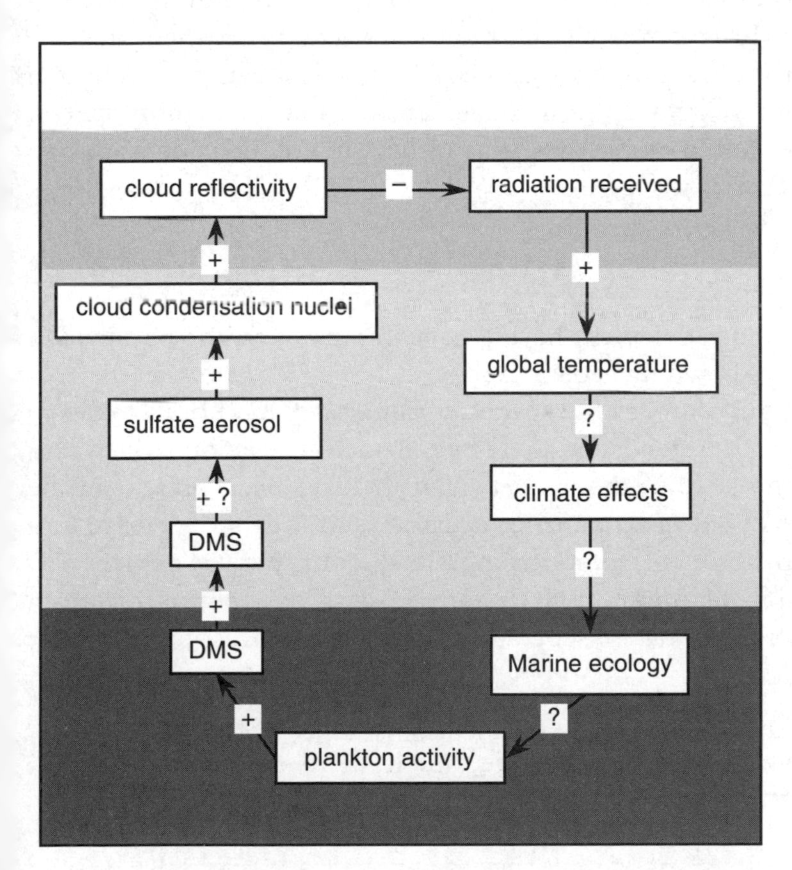

Fig. II.44 Proposed feedback cycle involving the production of dimethyl sulfide (DMS) by marine plankton, and its influence on climate. The + and – signs indicate the sign of the effect that each component of the cycle is expected to have on the next one. An overall negative feedback would require there to be an odd number of – influences; some of the effects, however, are uncertain, and marked ?.

84 See Lovelock (1988) for a readable exposition of the Gaia hypothesis in general, and of the proposal concerning DMS.

85 Charlson, Lovelock, Andreae, and Warren (1987); see articles in Schneider and Boston (1991) for further discussion, some of it critical.

DMS production, the cycle could provide a negative feedback loop overall, serving to limit temperature changes caused by other factors. In order for this to work, the cycle drawn in Fig. II.44 must contain an odd number of negative (–) influences, the remainder being positive (+). As indicated, however, there are several uncertainties (marked ?). Sulfate is not the only possible oxidation product of DMS (see Fig. II.41), and the amount produced is not known. There is also no evidence that DMS production is actually correlated with ocean temperatures. The CLAW hypothesis must therefore be regarded as unproven. Other possible Gaia regulation cycles are discussed in connection with the *carbon cycle.

Tantalum ($_{73}$Ta)

Tantalum is a fairly rare element, chemically similar to *niobium. It occurs in the mineral *tantalite*, (Fe,Mn)(Nb,Ta)$_2$O$_6$ (also known as *columbite* when niobium predominates). Although the metal is very electropositive, it forms an oxide film which is extremely resistant to any chemical attack. It is used in the manufacture of chemical plant, in surgery, and in electronics. The insolubility of its oxide leads to a very low abundance in natural waters.

Technetium ($_{43}$Tc)

Technetium is the lightest element having no stable isotopes, the half-lives of the longest-lived ones being shown in Table II.53.

Technetium does occur in nature, as the spectra of some stars have been found to include lines from Tc atoms. It is generated there by the nuclear reactions which synthesise other heavy elements. Any technetium that might have been present in the formation of the Earth, however, has long since decayed, and it is not regarded as a 'natural' element in the environment, although minute amounts may be present as a result of the natural fission of uranium. It is formed as a fission product from *uranium in nuclear reactors, and can be separated during the reprocessing of nuclear

Table II.53 Half-lives of some technetium isotopes. ^{99m}Tc is a 'metastable' excited nuclear state, which decays to ^{99}Tc

Isotope	Half-life
^{97}Tc	2.6×10^6 years
^{98}Tc	4.2×10^6 years
^{99}Tc	2.1×10^5 years
^{99m}Tc	6.01 h

fuel. It has come to be used as an important radioactive tracer, especially for medical diagnosis. Particularly useful is the short-lived ^{99m}Tc, the notation standing for a 'metastable' state of the nucleus that decays to the normal ^{99}Tc by emitting γ radiation. This decay is so rapid that ^{99m}Tc must be made by nuclear bombardment in a particle accelerator. It is rapidly separated, converted into a chemically suitable form, and injected into the body. Different chemical compounds have been found to 'target' particular organs such as the heart or the brain. A γ-ray detector can then be used to scan the relevant area, and abnormal conditions such as a failure of blood flow to part of the brain show up in the distribution of radioactivity. The long-lived ^{99}Tc is excreted, and although the body is exposed to some radiation as a result of the procedure, this compares favourably with the exposure resulting from medical X-rays.

Some technetium is released into the environment, although it is not normally regarded as a significant contaminant compared with other radioactive elements. The major problem with technetium is that it makes a contribution to the long-term radioactivity of spent nuclear fuel, the disposal of which is a major environmental issue.

Tellurium ($_{52}Te$)

A very rare element in the Earth's crust, tellurium is strongly chalcophilic in its geochemical behaviour, and is found in the sulfide ores of elements such as copper. It is used as a constituent of some steels and other alloys, and in small amounts for tinting glass and in the chemical industry. The dissolved concentration of tellurium in natural waters is extremely low. It is not essential for life, and is fairly toxic, although less so than some other heavy elements because of the low solubility of compounds such as TeO_2. Industrial exposure has caused symptoms similar to those of *selenium poisoning, including a 'garlic odour' in the breath, probably from the formation of the dimethyl compound, $(CH_3)_2Te$.

Terbium ($_{65}Tb$)

A lanthanide (rare earth) element, discussed under *lanthanum.

Thallium ($_{81}Tl$)

Thallium is a rare element. Like its neighbours in the periodic table, mercury and lead, it is a chalcophile, and it is obtained as a by-product of the production of zinc and lead. However, the Tl^+ ion also has chemical similarities with potassium K^+, so that thallium occurs in low concentrations in potassium-containing minerals such as feldspars. Some typical environmental concentrations are shown in Table II.54.

Table II.54 Thallium concentrations in the environment

Location	Concentration
Crust	0.6 ppm
Sea water	0.001 ppb
Fresh waters	0.004 ppb
Soils	0.1–1 ppm
Human body	0.1 ppm

Thallium compounds were previously used as a poison for rats and ants, and for the removal of hair, but these uses have been abandoned.

The toxicity of thallium is well known, and it is one of the few elements apart from *arsenic to have appeared as a poison in popular detective fiction.[86] It is especially dangerous as its compounds are soluble in water, and tasteless. It may interfere with the action of *potassium in nerve cells, and indeed increased potassium intake does to some extent counter the effects of thallium poisoning. But the chalcophilic nature of thallium, combining like other heavy elements with sulfur, is probably also important. Thallium is retained in the body, and toxic symptoms appear slowly. These include abdominal pains and vomiting, nervous problems such as tingling pain in the extremities, and loss of hair. Exposure during pregnancy is also thought to cause birth abnormalities. Like *mercury and *lead, thallium is a very dangerous element; it is much less important as an environmental pollutant, however, because it is little used in industry.

Thorium ($_{90}Th$)

Thorium is a long-lived radioactive element, the major naturally occurring isotope ^{232}Th having a half-life of 1.4×10^{10} years. Like uranium, it makes a contribution to the natural radioactivity in the crust, and to the heat supply which fuels the tectonic processes within the Earth (see §1.3). Its decay leads in turn to the formation of a series of short-lived radioactive elements, and ultimately to the stable *lead isotope ^{208}Pb, as shown in Table II.55. The half-lives of the isotopes in the thorium decay series are generally shorter than those from ^{238}U however, and so the concentrations of these elements in thorium minerals are much less than is found with *uranium. Thorium is found in association with the lanthanide (rare earth) elements in the mineral *monazite*, $(La,Ce,Th)PO_4$, and with uranium in *uranothorite*, a mixed silicate. It is extracted in small amounts, the major use being as the highly refractory oxide ThO_2 in furnaces and in gas mantles. Thorium is not a very rare element in the

86 Christie (1961).

Table II.55 Isotopes of the thorium decay series

Isotope	Half-life	Decay mode
^{232}Th	1.41×10^{10} years	a
^{228}Ra	5.76 years	β
^{228}Ac	6.13 h	β
^{228}Th	1.91 years	α
^{224}Ra	3.66 d	α
^{220}Rn	55.6 s	α
^{216}Po	0.15 s	α, β (0.014%)
^{216}At	0.30 ms	α
^{212}Pb	10.6 h	β
^{212}Bi	60.6 min	β (66%), α (34%)
^{212}Po	0.30 μs	a
^{208}Tl	3.05 min	β
^{208}Pb	stable	–

crust (having an abundance comparable to that of tin and bromine), but its compounds are very insoluble in water, and their dissolved concentrations are extremely low. It is not essential for life, and apart from the danger from its radioactivity, it is non-toxic.

Although thorium does not undergo fission like ^{235}U and ^{239}Pu (see *uranium), there is a possibility that it could be used as a nuclear fuel. Under neutron irradiation, the naturally occurring isotope ^{232}Th is converted to ^{233}Th, which rapidly decays to the fissile isotope ^{233}U:

$$^{232}\mathrm{Th} + \mathrm{n} \rightarrow {}^{233}\mathrm{Th} \rightarrow {}^{233}\mathrm{Pa} \rightarrow {}^{233}\mathrm{U}.$$

^{233}U could be used in nuclear reactors, which might in turn 'breed' more of this isotope from ^{232}Th. The ready availability of thorium, compared with that of the rare isotope ^{235}U, thus gives the possibility of extending the lifetime of nuclear fuels. This has been little explored so far, compared with the alternative 'breeding' scheme which produces ^{239}Pu from ^{238}U (see also *plutonium).

Thulium ($_{69}$Tm)

An element of the lanthanide series, discussed under *lanthanum.

Tin ($_{50}$Sn)

An element of rather low abundance in the crust, tin differs from its neighbouring elements in the periodic table in being more lithophilic. The main mineral is the oxide *cassiterite*, SnO_2; the fact that this can be reduced to metallic tin by glowing coals accounts for its familiarity in the ancient world. Prior to the twentieth century an important source was Cornwall in England, but the Cornish tin mines became increasingly uneconomic, and none remain in operation. Metallic tin is used as a corrosion-resistant and non-toxic coating for other metals, especially in 'tins' used for canning food and soft drinks, and in numerous alloys, including bronze (with copper), solder, and pewter. A more recent application of tin is in organometallic compounds. They are used as stabilisers for plastics, and as fungicides for food crops such as potatoes, and in marine 'antifouling' paints which resist attack by barnacles and other creatures.

Tin oxide is very insoluble, and the dissolved concentration of the element in natural waters is very low. Because of its insolubility, and also its lower affinity for sulfur, it differs from many 'heavy metals' such as *cadmium and *lead in being generally non-toxic. It is thought that it may be essential for life in some species, including possibly humans, although its role is obscure. Some organotin compounds are fairly toxic however, and the use of tributyl tin (TBT) compounds such as $(C_4H_9)_3SnOCO.CH_3$ in marine paints has attracted criticism.[87] They are very harmful to plankton and mollusc larvae. Other effects include a thickening of the shells of oysters, and a bizarre condition known as 'imposex' in some marine snails: this involves the development of a penis and a sperm duct in females, which blocks the oviduct and prevents eggs being laid. Because of these problems the use of TBT in coastal waters and on small boats has been banned in many countries, although its use on large ocean-going vessels continues. Organotin compounds are thought to decompose fairly quickly in the ocean to harmless inorganic forms.

Titanium ($_{22}Ti$)

Titanium is a very common element, being widely distributed in silicates, and occurring in the important oxide minerals *ilmenite* ($FeTiO_3$) and *rutile* (TiO_2). The extraction of the metal is expensive, as the element is extremely electropositive and its oxide hard to reduce. It is normally manufactured by the reduction of $TiCl_4$ with metallic magnesium or sodium, and is widely used as a light and strong metal in aircraft and other industries. The oxide TiO_2 is also used in paints and paper manufacture as a white pigment, having largely replaced other more toxic compounds (such as those of lead) for this purpose.

The oxide minerals of titanium are very insoluble, and its concentration in natural waters is very low. It is not essential for life and is non-toxic, although some carcinogenic properties have been suspected. The main environmental problem is connected

87 Clark (1989).

with the iron oxides that must be separated during the purification of titanium oxide. Large quantities of so-called 'acid iron wastes' have been dumped in coastal waters, for example in the North Sea, the Baltic Sea, and the New York Bight.[88] The effect of this dumping has been controversial, some claims being made that no long-term harmful effects occur. However, in the USA and Europe coastal dumping of iron wastes has been curtailed, or replaced by dumping in deeper waters.

Tungsten ($_{74}W$)

Tungsten is a rather rare element, occurring principally in the oxide minerals *scheelite*, $CaWO_4$, and *wolframite*, $(Fe,Mn)WO_4$. The metallic element has an exceptionally high melting point, and is used as a filament in electric light bulbs. The other main use of tungsten is in the form of the carbide, WC, in cutting tools. The dissolved concentration in natural waters is very low. Tungsten is non-toxic, and was until recently not thought to have any biological role. However, a few enzymes containing tungsten have been found in some anaerobic bacteria such as *Clostridium*. The role of the tungsten seems to be to assist in oxidation/reduction reactions, probably involving two-electron steps; but why tungsten should be used in preference to the chemically similar and more abundant element *molybdenum is not known.[89]

Uranium ($_{92}U$)

Uranium is the most important long-lived radioactive element. The naturally occurring isotopes are shown in Table II.56. ^{235}U and ^{238}U have half-lives comparable to the age of the Earth (4.6×10^9 years), and their abundance, especially that of the shorter-lived ^{235}U, was originally larger than at present. ^{234}U is too short-lived to have survived this time, and is present only because it is formed in the decay series discussed below.

Table II.56 Naturally occurring uranium isotopes

Isotope	Half-life	Natural abundance
^{234}U	2.45×10^5 years	0.005 %
^{235}U	7.04×10^8 years	0.720 %
^{238}U	4.46×10^9 years	99.275 %

88 See Clark (1989).

89 Da Silva and Williams (1991).

Uranium is a strongly lithophilic element, widespread in low concentrations in oxide minerals, especially in igneous rocks such as granite (Table II.57). The main ores from which the element is obtained are *uraninite* (also known as *pitchblende*), U_3O_8, and *carnotite*, $K(UO_2)(VO_4)H_2O$. Uranium has some minor applications in glazes for pottery and as a very hard dense metal in armour-piercing weapons, but its major use by far is in the manufacture of nuclear weapons and as a fuel for nuclear power stations. These applications, and some background to the environmental disputes arising from them, are discussed below.

The very insoluble U_3O_8 is oxidised in air to the soluble *uranyl* ion, UO_2^{2+}, which occurs in sea water. Uranium is not essential for life, and is not chemically toxic, although it is harmful owing to its radioactivity. Apart from the problems related to its use in nuclear power production and weapons, the main environmental problem associated with uranium is the presence of the radioactive gas ⋆radon in the neighbourhood of uranium-containing rocks such as granite, and especially in mines where uranium ores are obtained.

The uranium decay series

Both ^{235}U and ^{238}U decay into isotopes that are themselves radioactive, and so form the first members of *decay series* of short-lived radioactive elements, eventually ending up as stable isotopes of lead. The two series are shown in Tables II.58 and II.59.

Two types of radioactive decay mode can occur, as shown in the tables. In α decay an unstable nucleus emits a ^4He nucleus, known as an α particle. In consequence, the atomic number is reduced by two units, and the mass number by four, corresponding to the charge and mass of ^4He. For example, ^{238}U decays to ^{234}Th in the first step of the ^{238}U series. The other type of decay is the emission of an electron, a 'β particle'. The light electron does not change the mass number, but its negative charge leads to an increase by one unit in the atomic number, for example ^{234}Th becomes ^{234}Pa. The reason why β decay occurs in the series is that the relative

Table II.57 Uranium concentrations in the environment

Location	Concentration
Crust:	
average	2.4 ppm
granite	3.4 ppm
Sea water	3 ppb
Fresh waters	1 ppb
Human body	1 ppb

Table II.58 Isotopes of the ^{235}U decay series. Some isotopes can decay by more than one route; the numbers in parentheses show the relative probabilities

Isotope	Half-life	Decay mode
^{235}U	7.04×10^8 years	α
^{231}Th	25.5 h	β
^{231}Pa	3.28×10^4 years	α
^{227}Ac	21.8 years	β (98.8 %), α (1.2 %)
^{227}Th	18.7 d	α
^{223}Fr	21.8 min	β
^{223}Ra	11.4 d	α
^{219}Rn	3.96 s	α
^{215}Po	1.78 ms	α, β (0.0005 %)
^{215}At	0.10 ms	α
^{211}Pb	36.1 min	β
^{211}Bi	2.15 min	β (99.7 %), α (0.3 %)
^{211}Po	0.52 s	α
^{207}Tl	4.77 min	β
^{207}Pb	stable	–

Table II.59 Isotopes of the ^{238}U decay series. Some isotopes can decay by more than one route; the numbers in parentheses show the relative probabilities

Isotope	Half-life	Decay mode
^{238}U	4.47×10^9 years	α
^{234}Th	24.1 d	β
^{234}Pa	6.75 h	β
^{234}U	2.45×10^5 years	α
^{230}Th	8.0×10^4 years	α
^{226}Ra	1600 years	α
^{222}Rn	3.82 d	α
^{218}Po	3.05 min	α (99.96 %), β (0.04 %)
^{218}At	2 s	α
^{214}Pb	26.8 min	β
^{214}Bi	19.7 min	β (99.96 %), α (0.04 %)
^{214}Po	0.164 ms	α
^{210}Tl	1.30 min	β
^{210}Pb	22.3 years	β
^{210}Bi	5.01 d	α
^{206}Pb	stable	–

numbers of protons and neutrons in stable nuclei depend on the mass. In order to end up with a stable nucleus from ^{235}U or ^{238}U, it is necessary not only to lose mass (by α emission), but also to change some of the remaining neutrons into protons, and this is effectively what β decay does.

As shown in the tables, some nuclei have alternative decay modes; for example ^{218}Po in the ^{238}U series undergoes α decay to ^{214}Pb usually, but occasionally (in 0.04 per cent of cases) gives ^{218}At by β decay.

Most radioactive decays, by either α or β routes, are also accompanied by γ rays. These consist of high-energy electromagnetic radiation similar to X-rays, and are important because this radiation is much more penetrating than α or β rays.

One consequence of the decay series is that natural uranium minerals contain small amounts of the other elements involved in the series. This is the main reason why elements such as *radium and *radon exist on Earth, as their half-lives are very much shorter than the age of the Earth, so that any which might have been present from the origin of the elements in stars must have long ago decayed. The amount of each radioactive isotope present in steady state should be proportional to its half-life. Thus one can estimate that uranium ore should contain 0.3 mg of ^{226}Ra and 2 ng of ^{222}Rn for each kilogram of ^{238}U present. But there are factors which complicate this calculation. ^{222}Rn is an isotope of the noble gas *radon, which is chemically inert and so does not remain combined in the oxide uranium minerals. Small amounts may diffuse out and be lost from the rocks. For this reason radon, in spite of its short half-life and the very small amount present, is now thought to constitute the major radioactive hazard arising from natural uranium-containing minerals; these are present for example in granite. A second complication in the decay series comes from the alternative routes of decay. Isotopes of *francium (Fr) and *astatine (At) are only produced by minor, uncommon, 'branches' in the decay series. Given their very short half-lives as well, this means that their abundance is exceptionally low, and they are only known from artificial routes.

The ultimate products of decay of uranium are the stable isotopes ^{206}Pb and ^{207}Pb of *lead. The amounts of these present in uranium minerals depends on the age since they solidified, and can be calculated on the assumption that no lead or intermediate elements such as radon have been lost. In practice these assumptions are rarely true, but in spite of this difficulty the proportion of different lead isotopes present in minerals can be used for geological dating. It is on this basis, for example, that the age of the Earth is known fairly reliably as 4.6 billion years.

Uranium fission: nuclear power and weapons

In 1939 it was discovered that lighter elements such as barium are produced when uranium is bombarded with neutrons. The explanation is that some heavy nuclei become so unstable when they capture a neutron that they undergo *fission* into lighter fragments.[90] In this process, a very large amount of energy is liberated, equivalent to 2×10^{13} joules per mole of uranium consumed, and vastly greater than can be obtained from chemical reactions. Another feature of the fission event is that more neutrons are liberated, either simultaneously or after a short delay. Thus one can get a *chain reaction*, with neutrons multiplying through successive fission events, and an enormous release of energy. There are several reasons why this does not happen easily with natural uranium: only the less common isotope ^{235}U normally undergoes fission with slow neutrons; many nuclei which might also be present absorb neutrons, and so reduce the number available to cause further fission; and neutrons may also escape from the surface of the sample. A self-sustaining chain reaction requires a critical mass of material, the value of which depends on various factors. For nearly pure ^{235}U it is only a few kilograms. With natural uranium containing only 0.72 per cent ^{235}U a chain reaction can only be produced with a very large mass, and even then only if a *moderator* is present. This is a material which slows down the neutrons coming from fission, and thereby increases the likelihood that they will be captured by other ^{235}U nuclei. Water is a good moderator, and in a concentrated solution of pure ^{235}U, only 0.8 kg is required for a critical mass.

A nuclear weapon works on the principle of assembling a critical mass of nearly pure ^{235}U very quickly. In the simplest design, used in the bomb dropped on Hiroshima in 1945, two pieces of uranium, each of more than half the critical mass, are shot towards each other by explosive charges. A more sophisticated method uses a hollow sphere of material, which is 'imploded' into a critical concentration by several explosive charges detonated simultaneously.

The development of nuclear weapons based on ^{235}U depends on methods for separating this isotope from the commoner ^{238}U. The first method employed was electromagnetic, essentially consisting of a large-scale mass spectrometer, where ions travelling in a vacuum chamber are separated according to their mass. This was succeeded by techniques which use the volatile compound UF_6. Gas diffusion methods depend on the slightly faster diffusion of the lighter isotope; more recent developments make use of high-speed gas centrifuges. Many stages of separation are needed in these methods, as each one only produces a small degree of fractionation. The plants are very large and use substantial amounts of electric power. Laser spectroscopy methods, utilising small differences in the spectra of the isotopes, have also been investigated.

Nuclear reactors were first built for research purposes (and are still used as a source of neutrons), and for the production of *plutonium for nuclear weapns. Most

---

90 See Choppin and Rydberg (1980) and Knief (1992) for accounts of nuclear fission and its applications.

reactors in operation in the world today are intended for electric power generation. The fuel of uranium metal, or more commonly the oxide UO_2, is enclosed in stainless steel or zirconium alloy cladding. Slightly enriched uranium, containing 2–3 per cent ^{235}U, improves the efficiency of operation. Graphite or water is commonly used as a moderator, with cooling and heat transfer to the steam turbines provided by water, or in some designs by gases such as CO_2. The fission reaction is controlled by rods of some neutron-absorbing material that can be lowered into the reactor; boron, indium, and cadmium are particularly effective elements for this. Such control is made much easier by the fact that some neutrons from the fission of ^{235}U are emitted with a delay of up to several minutes. This means that the natural response time of the chain reaction under operating conditions is rather slow.

The uranium fuel rods must be replaced regularly, as the efficiency of fission declines with use. The ^{235}U is slowly consumed, but only a fraction can be used in practice because of the accumulation of fission products, many of which absorb neutrons and so 'poison' the reactor. In a typical operation, a fifth of the fuel rods are replaced every year. The lifetime of uranium reserves for power generation can in principle be extended by reprocessing the fuel (see below), or by 'breeding' new fissile nuclei, through the neutron bombardment of natural material in a reactor. The two most promising candidates for breeding are ^{239}Pu, formed from ^{238}U (see *plutonium), and the fissile isotope ^{233}U (which has a half-life of 162 000 years), which can be bred from natural thorium. The reactions are

$$^{232}Th + n \rightarrow {}^{233}Th \rightarrow {}^{233}Pa \rightarrow {}^{233}U,$$

where ^{233}Th and ^{233}Pa decay fairly rapidly by β decay.

Ionising radiation resulting from nuclear fission comes in two ways. Firstly, the fission event itself produces penetrating γ radiation, which accounts for part of the damaging effect of nuclear weapons. Secondly, many of the products of fission are themselves radioactive, some persisting for long after the fission has ceased. Figure II.45 shows some of the most important radioactive elements formed, classified according to their origin.

Fission products are the elements formed when the ^{235}U nucleus splits in two. Rather than two equal fragments with mass around 116, the commonest products are one lighter nucleus (mass range 90–105, corresponding to elements between krypton and rhodium), and a heavier one (mass range 130–145, giving elements between tellurium and europium). Many of the immediate products of fission still have an excess of neutrons, and so undergo β decay (see §2.4) with a wide range of half-lives.

Other radioactive elements come from the neutron bombardment of materials present in the reactor core. They include actinide elements such as plutonium, americium, and curium, formed by successive neutron capture processes starting

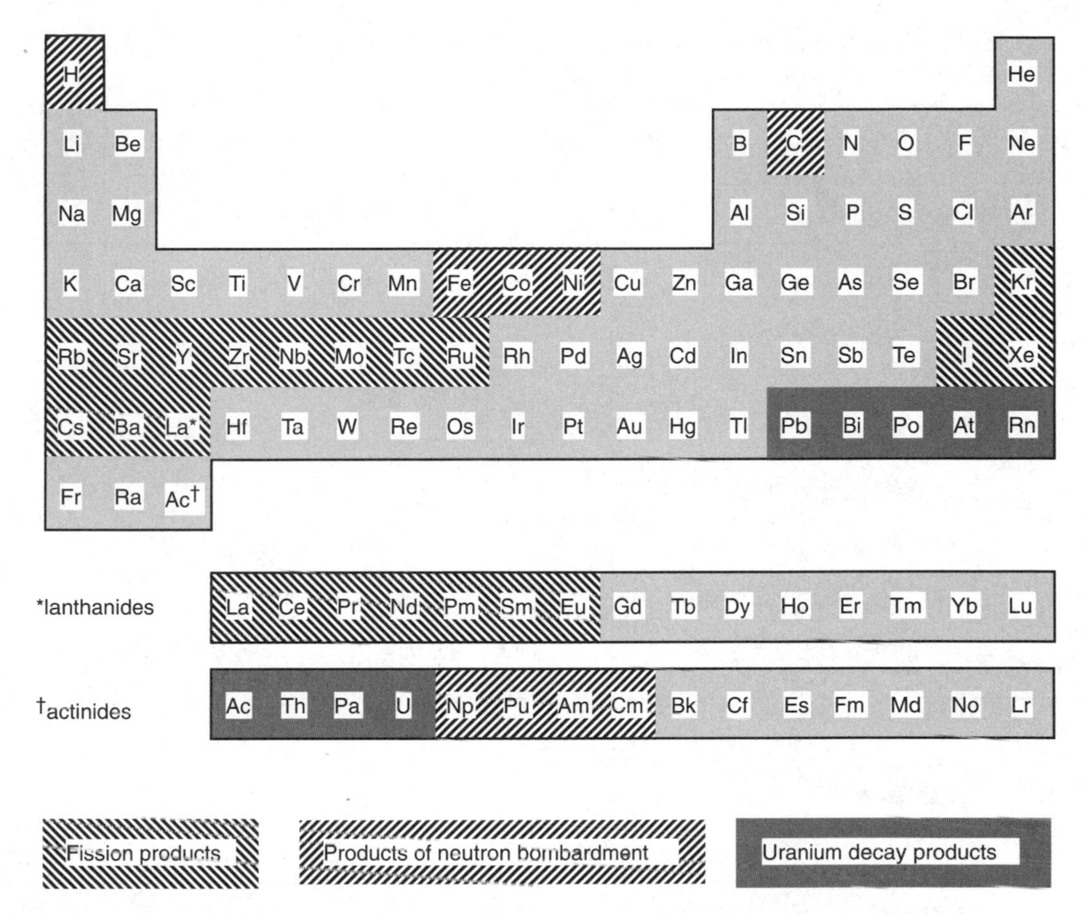

Fig. II.45 Radioactive elements resulting from nuclear power generation, classified according to their origin.

with uranium. Other such products include tritium (^{3}H, by neutron capture from deuterium), and ^{60}Co from irradiation of fuel cladding. A final class shown in Fig. II.45 comprises products formed in the normal radioactive decay of uranium and of heavier actinides.

Table II.60 lists a selection of the radioactive isotopes found in spent nuclear fuel, with an indication of how much radioactivity each one contributes. These values vary significantly according to the reactor type, and the extent to which the fuel has been used (known technically as the 'burn-up'). Nevertheless they give some guide as to what can be expected. The immediate activity, in the core and just after removal of a fuel rod, comes mostly from short-lived fission products such as ^{89}Sr, ^{91}Y, ^{95}Nb, and ^{131}I. After a few years' storage, these have decayed, and the longer-lived isotopes ^{90}Sr, ^{137}Cs, and ^{241}Pu now make the largest contribution. ^{241}Pu decays to ^{241}Am, which in the period 100–1000 years after use is predicted to be the most

Table II.60 Selected radioactive isotopes found in spent nuclear fuel. The activities vary considerably according to the operating conditions

Isotope	Decay mode	Half-life (years)	Activity (Ci kg^{-1} U)[a]	
			immediately	after 10 years
^{3}H	β	12.3	0.2	0.1
^{85}Kr	β, γ	10.7	2.2	1.2
^{89}Sr	β, γ	0.14	400	–
^{90}Sr	β	29	18	13
^{91}Y	β, γ	0.16	600	–
^{95}Zr	β, γ	0.18	800	–
^{95}Nb	β, γ	0.10	800	–
^{99}Tc	β, γ	210 000	3×10^{-3}	3×10^{-3}
^{106}Ru	β	1.0	200	0.2
^{131}I	β, γ	0.02	500	–
^{134}Cs	β, γ	2.2	17	0.5
^{137}Cs	β, γ	30	25	20
^{144}Ce	β, γ	0.8	400	0.1
^{147}Pm	β, γ	2.6	60	5
^{239}Pu	α, γ	24 400	0.15	0.15
^{240}Pu	α, γ	6600	0.24	0.24
^{241}Pu	β, γ	15	23	14
^{241}Am	α, γ	432	0.01	0.3
^{242}Cm	α, γ	0.45	3	–

[a]One curie (Ci) represents 3.7×10^{10} radioactive disintegrations per second (Bq), and is often used as a more practical unit for large amounts of radioactive material.

significant isotope; when it in turn has decayed, the remaining activity will come mostly from ^{239}Pu.

A similar range of radioactive elements is generated during a nuclear explosion, although their relative amounts are somewhat different. These products are entirely vaporised, and when a bomb is exploded above ground, they enter the atmosphere, and thence are slowly precipitated as 'fallout'. The most significant isotopes in fallout from the nuclear tests of the 1950s were ^{90}Sr and ^{137}Cs; the former of these is especially dangerous as it is strongly retained by the body (see *strontium). Most atmospheric testing ceased following the Partial Test-Ban Treaty of 1963, and was replaced by underground testing at suitably remote sites, where the radioactive products can be contained.

The design of nuclear reactors is intended to ensure that radioactive elements do not escape into the environment, but a number of accidents have nevertheless led to some release. Partial failure of the cooling and/or control systems can result in an uncontrolled temperature rise, as happened at Three Mile Island in Pennsylvania in 1979, or even a fire, as at Windscale (now Sellafield) in England in 1957, and at Chernobyl in Ukraine in 1986. The release of elements in such circumstances depends on their volatility. Tritium and the noble gases, krypton and xenon, are the hardest to contain, and are most likely to be released on overheating, as at Three Mile Island. The fire at Windscale also liberated a significant amount of iodine, and caesium was also released in the much more serious accident at Chernobyl (see Table II.61). Some workers at the site were exposed to lethal doses of radiation directly from the core. Apart from this, the most serious source of radiation to local populations near Chernobyl came from ^{131}I, which has a half-life of 8 days and can cause thyroid cancer (see *iodine); over a longer time-scale, and longer distances, ^{137}Cs has been most important.

Apart from the possibility of accidents, the major worries with nuclear power are concerned with the ultimate disposal of the radioactive material. Immediately after removal from the reactor, the fuel is so radioactive that it must be stored for at least a year. The radioactive decay also generates a large amount of heat, and so fuel elements are normally stored in water. In some countries, it is intended to leave spent fuel elements in this condition (although they must be re-clad), but others have built plants for reprocessing the fuel. The fuel rods and their cladding are dissolved in nitric acid, and a sequence of chemical processes is used to separate unused uranium, plutonium, and some selected fission products. Reprocessing was first motivated by military concerns, to extract plutonium for atomic weapons. The 'peaceful' arguments for it are contentious. It allows uranium reserves to be extended, by recycling unused uranium, and by providing plutonium which can in principle be used as an alternative fuel. Some of the fission products are useful, for example ^{137}Cs for the irradiation of food. Separation of the very long-lived ^{239}Pu will also reduce the activity of the remaining waste in the long term. Against these arguments in favour of reprocessing, however, there are others. Plants to treat highly radioactive materials are extremely expensive, especially if adequate precautions are taken against leakage. In spite of all the safety measures, there is always the possibility of accidental release. It may also be undesirable to separate plutonium, which

Table II.61 Release of elements from reactor core at Chernobyl, April–May 1986

Fraction released (%)	Elements
100	Kr, Xe
10–20	Te, I, Cs
3–4	most others

could possibly be seized by terrorists to make nuclear bombs. Finally, although the activity of the waste may ultimately be decreased by removal of ^{239}Pu, this is only really significant after hundreds of years, and meanwhile the volume of waste has become much larger: the fission products end up dissolved in acid solution, and large quantities of other reagents used in the process have become contaminated. For the moment, these waste solutions are stored in tanks, awaiting a decision on what to do with them. At Hanford in the USA, hundreds of millions of gallons have accumulated since the 1940s, when reprocessing started. Some of the tanks are leaking, and there is a serious prospect of their contents reaching the Columbia River, a major water source for the states of Washington and Oregon. British and French reprocessing plants are on the coast (Sellafield and Cap la Hague, respectively) and accidental leakages are quickly diluted. They do, however, contribute to the level of radioactivity in the immediate environment.

Most proposals for the ultimate disposal of high-level nuclear waste envisage burying it underground, either in existing mines, or ones specially made.[91] Such sites should be geologically stable, with little prospect of earthquakes or erosion over periods of up to a million years. If the waste can be incorporated into a highly insoluble glass, from which the radioactive elements could not be leached, then the sea bed would also be a suitable site.

Natural uranium reactors

Given the technical difficulties involved in making a nuclear reactor, it might seem surprising that a chain reaction can occur under natural conditions in a geological deposit. Nevertheless, this seems to have happened at least once in the history of the Earth. Uranium ore mined at Oklo in Gabon has been found to have an anomalous isotopic composition, with considerably less than the normal 0.72 per cent ^{235}U. Evidence that some of this isotope was consumed in a natural fission reactor comes from the presence of fission products in the deposit.[92] For example, samples of the lanthanide element neodymium have an isotopic composition very different from that normally found, but much closer to that expected for material derived from ^{235}U fission.

The Oklo deposit was formed about 2 billion years ago as a result of changing chemical conditions on the Earth. With the increase of atmospheric O_2 content from photosynthesis (see ⋆oxygen), UO_2 was oxidised to the soluble uranyl form, UO_2^{2+}. A concentrated deposit formed where this was reduced again under anaerobic conditions. Water acted as a moderator, and it is thought that the chain reaction started and stopped several times over a period of a few thousand years.

---

91 See Roxburgh (1987) for a discussion.

92 Cowan (1976).

Although it is unlikely that conditions at Oklo were unique, there is no evidence that any other uranium deposits have undergone a chain reaction. The same thing could not happen again today, because the natural ^{235}U content is now too low to support a chain reaction under natural conditions. At the time of the Oklo reactor, the abundance of this shorter-lived isotope was higher, around 3 per cent and so similar to the composition of the enriched fuel used in many types of present-day reactor.

Vanadium ($_{25}V$)

A transition element of moderate abundance in the crust, vanadium has a versatile chemistry. It is widely distributed in oxide minerals, where it can have the oxidation state +3, +4, or +5, and also occurs in sulfides. Important minerals are *vanadinite*, $Pb_5(VO_4)_3Cl$, *carnotite*, $K_2(UO_2)_2(VO_4)_2.3H_2O$, and *patronite*, VS_4. Vanadium is also occurs in some crude oils, especially those from Venezuela, as porphyrin complexes of the vanadyl ion, $[VO]^{2+}$. Its chief use is as a component of steels.

The concentration of vanadium in natural waters is low (see Table II.62), but it is common in soils, where it is derived from the weathering of rocks. It is an essential element for many species, including humans, but its biological functions are not well understood. In some bacteria, it may perform a nitrogen-fixation role similar to that of *molybdenum; in other cases, vanadium may be a component of enzymes which perform redox reactions involving hydrogen peroxide. High concentrations of vanadium are found in some toadstools and marine worms, most remarkably in *Ascidia nigra*, where the blood cells contain up to 1.5 per cent of the element. The reasons for this are unknown.

Xenon ($_{54}Xe$)

One of the chemically inert noble gas elements, xenon occurs at a concentration of 0.09 ppm by volume in the atmosphere. It is extracted in small quantities along with other noble gases, but has very little application outside scientific

Table II.62 Vanadium concentrations in the environment

Location	Concentration
Crust	160 ppm
Sea water	1 ppb
Fresh waters	< 1 ppb
Soils	100–1000 ppm
Human body	1.5 ppb

research. The radioactive isotope ^{133}Xe has a half-life of 5.2 days, and is formed as a fission product from *uranium in nuclear reactors. Because of its chemical inertness, xenon is difficult to contain, and small amounts are liberated from nuclear power stations and during reprocessing of spent fuel elements. ^{133}Xe has contributed a significant amount of the radioactivity released during nuclear accidents, such as the fire in a nuclear reactor at Chernobyl in Ukraine in 1986. Most of the element dispersed rapidly in the atmosphere, however, so that it probably constituted a less important hazard than some other isotopes such as ^{131}I and ^{137}Cs.

Ytterbium ($_{70}Yb$)

A lanthanide element, discussed under *lanthanum.

Yttrium ($_{39}Y$)

Although not one of the lanthanide elements (see *lanthanum), yttrium is chemically very similar to them, and often classified as a 'rare earth element'. It occurs along with the lanthanides, in minerals such as *xenotime*, $(Y,Ln)PO_4$. Yttrium has a number of specialised small-scale uses, especially in the form of yttrium aluminium garnet (YAG; $Y_3Al_5O_{12}$) and yttrium iron garnet (YIG; $Y_3Fe_5O_{12}$) in TV screens, as an artificial gemstone, and for electronics applications. The dissolved concentrations of yttrium compounds in natural waters are very low, and it has no known biological role.

Zinc ($_{30}Zn$)

Zinc is an element of moderate abundance in the crust. It is chalcophilic in nature, and is found in sulfide minerals, especially *zinc blende*, ZnS (known in the USA as *sphalerite*). Surface oxidation liberates the soluble Zn^{2+} ion, which forms occasional carbonate and silicate minerals. ZnS also forms a source of some other chalcophilic elements, especially *cadmium, and the extraction of metallic zinc is sometimes responsible for environmental contamination by this. The metal is used for anticorrosion coatings ('galvanised iron'), roof-cladding, batteries, and some specialised alloys.

As shown in Table II.63, the dissolved concentration of Zn^{2+} in natural waters is low, although it is not rare in soils. Zinc is essential for all life, being one of the most important trace elements. An adult of average weight has about 2.5 g of zinc, mostly forming part of specialised enzymes and other proteins; a selection of these is shown in Table II.64.

Table II.63 Zinc concentrations in the environment

Location	Concentration
Crust	75 ppm
Sea water:	
surface	0.05 ppb
deep	0.5 ppb
Fresh waters	10 ppb
Soils	30–1000 ppm
Human body:	
average	30 ppm
blood	7 ppm
muscle	240 ppm

Table II.64 Some zinc-containing enzymes

Name	Function
Carbonic anhydrase	$HCO^{3-} \rightarrow CO_2$
Alcohol dehydrogenase	$R_2CHOH \rightarrow R_2CO$ e.g. $C_2H_5OH \rightarrow CH_3CHO$
RNA polymerase, DNA polymerase, Reverse transcriptase	nucleic acid synthesis
Carboxypeptidase, Aminopeptidase, Collagenase	digestion
Superoxide dismutase (with Cu)	$O_2^- \rightarrow O_2 + H_2O_2$

Zinc has no redox activity, occurring in biology, as in the inorganic environment, exclusively as Zn^{2+}. It is frequently bound to nitrogen- and sulfur-containing groups in proteins, and has a coordinating ability that facilitates acid–base reactions of diverse kinds. One example is in *carbonic anhydrase*, a zinc-containing enzyme that catalyses the reaction

$$HCO_3^- \rightarrow CO_2 + OH^-,$$

which is otherwise rather slow. Other examples of dehydration and hydration reactions catalysed by zinc-containing enzymes are the condensation polymerisation of DNA and RNA, and the hydrolysis of organic polymers during the digestion of food. Zinc also occurs in many non-enzymic proteins, some of which have important control functions. *Zinc finger* proteins, so called after a characteristic loop or 'finger' of protein chain associated with a zinc binding site, are involved in the expression of

DNA, the sequence of reactions whereby proteins are synthesised according to the instructions of the genetic code.

An adequate dietary supply of zinc is essential for normal growth and maturation. Although nearly all foods contain sufficient, it appears that the zinc in some cereals is in an insoluble form which is not absorbed by the body. Deficiency in zinc absorption has been linked to the incidence of dwarfism in some countries.[93] The use of dietary supplements containing zinc sulfate has been advocated, although a normal Western diet should contain an adequate amount. An excess can be toxic, although much less so than for many other elements.

Zirconium ($_{40}Zr$)

Zirconium is a moderately common element, its abundance in the crust being comparable to those of barium and chlorine. It is strongly lithophilic and is widespread in low concentrations in silicates of other elements, as well as forming the minerals *zircon*, $ZrSiO_4$, and *baddeleyite*, ZrO_2. The former has been prized as a gemstone, and synthetic ZrO_2 is used as a cheaper substitute for diamonds in jewellery. Because of its highly electropositive character, the element itself is expensive to extract. It is used in some high-performance alloys, and as a cladding material for uranium oxide fuel elements in nuclear reactors. Zirconium always occurs in association with *hafnium, which is very similar chemically. For many purposes it is unnecessary to separate these elements, but this must be done for nuclear power applications: in spite of their chemical similarity, zirconium and hafnium have quite different nuclear properties, zirconium being a very weak absorber of neutrons (which is why it is used), but hafnium a very strong one.

Zirconium compounds are very insoluble in water at neutral pH, and the concentrations in rivers and sea water are extremely low. It has no known role in life.

93 See Lenihan (1988).

References

As a guide to readers, the references have been classified as follows:

(1) no mark: general or 'popular' accounts assuming little scientific background;

(2) single asterisk (*): technical books or reviews assuming a level of understanding similar to that needed for this book;

(3) double asterisk (**): more technical papers, reviews or books, assuming specialised knowledge.

Alvarez, L. W. (1987), Mass extinctions caused by large bolide impacts, *Physics Today*, **40** (7), 24.

Andouze, J. and Israel, G. (1985), *The Cambridge atlas of astronomy*, Cambridge University Press.

*__Ashby, J. R. and Craig, P. J.__ (1990), *Organoelement compounds in the environment*, in R.M. Harrison (ed.), *Pollution: causes, effects, and control*, 2nd edition, Royal Society of Chemistry, Cambridge, pp. 309–42.

Atkins, P. W. (1984), *The Second Law*, Scientific American Books, New York.

*__Atkins, P. W.__ (1990), *Physical chemistry*, 4th edition, Oxford University Press.

*__Barrow, J. D. and Tipler, F. J.__ (1986), *The anthropic cosmological principle*, Oxford University Press, Oxford.

**__Berry, F. J. and Vaughan, D. J.__ (1985), *Chemical bonding and spectroscopy in mineral chemistry*, Chapman and Hall, London.

*__Bridgman, H.__ (1990), *Global air pollution: problems for the 1990s*, Belhaven Press, London.

Brimhall, G. (1991), The genesis of ores, *Scientific American*, **264** (5), 48.

Brock, W. H. (1992), *The Fontana history of chemistry*, Fontana Press, London.

*__Brown, G. C. and Musset, A. E.__ (1981), *The inaccessible Earth*, George Allen and Unwin, London.

*__Cairns-Smith, A. G.__ (1982), *Genetic takeover and the mineral origins of life*, Cambridge University Press, Cambridge.

*__Charlson, R. J., Lovelock, J. E., Andreae, M. O. and Warren, S. G.__ (1987), Oceanic phytoplankton, atmospheric sulfur, cloud albedo, and climate, *Nature*, **326**, 655.

*__Choppin, G. R. and Rydberg, J.__ (1980), *Nuclear chemistry: theory and applications*, Pergamon Press, Oxford.

Christie, A. (1961), *The pale horse*, Collins, London.

Clark, R. B. (1989), *Marine pollution*, 2nd edition, Oxford University Press.

Close, F. (1990), *Too hot to handle: the race for cold fusion*, W.H. Allen, London.

Cowan, G. A. (1976), A natural fission reactor, *Scientific American*, **235**, 36–40.

*__Cox, P. A.__ (1989), *The elements: their origin, abundance and distribution*, Oxford University Press.

Crowson, P. (1992) *Minerals handbook 1992–3*, Macmillan, London.

*__D'Abro, A.__ (1951), *The rise of the new physics: its mathematical and physical theories*, Vol. 2, Dover Publications, New York.

*__Da Silva, J. J. R. F. and Williams, R. J. P.__ (1991), *The biological chemistry of the elements*, Clarendon Press, Oxford.

*__Deer, W. A., Howie, R. A. and Zussman, J.__ (1966), *An introduction to the rock-forming minerals*, Longman, London.

*__Dorman, J. R. and Woolfson, M. M.__ (1989), *The origin of the Solar System: the capture theory*, Ellis Horwood, Chichester.

*__Duley, W. W. and Williams, D. A.__ (1984), *Interstellar chemistry*, Academic Press, London.

*__Elsom, D.__ (1987), *Atmospheric pollution*, Blackwell, Oxford.

*__Emsley, J.__ (1991), *The elements*, 2nd edition, Clarendon Press, Oxford.

*__Evans, A. M.__ (1987), *An introduction to ore geology*, Blackwell, Oxford.

*__Faure, G.__ (1986), *Principles of isotope geology*, 2nd edition, Wiley, New York.

*__Fergusson, J. E.__ (1990), *The heavy elements: chemistry, environmental impact and health effects*, Pergamon Press, Oxford.

*__Freedman, B.__ (1989), *Environmental ecology*, Academic Press, San Diego.

*__Greenwood, N. N. and Earnshaw, A.__ (1984), *Chemistry of the elements*, Pergamon Press, Oxford.

*__Harrison, R. M.__ (ed.) (1990), *Pollution: causes, effects, and control*, Royal Society of Chemistry, Cambridge.

*__Harrison, R. M.__ (ed.) (1992), *Understanding our environment: an introduction to environmental chemistry and pollution*, Royal Society of Chemistry, Cambridge.

*__Henderson, P.__ (1982), *Inorganic geochemistry*, Pergamon Press, Oxford.

*__Henderson-Sellers, A. and Robinson, P. J.__ (1986), *Contemporary climatology*, Longman, Harlow.

Houghton, J. T., Jenkins, G. J. and Ephraums, J. J. (ed.) (1990), *Climate change: the IPCC scientific assessment*, Cambridge University Press.

Hoyle, F. and Wickramasinghe, N.C. (1978), *Life cloud*, J.M. Dent, London.

***Hutchison, C. S.** (1983), *Economic deposits and their tectonic setting*, Macmillan, Basingstoke.

Ihde, A. J. (1964), *The development of modern chemistry*, Harper and Row, New York.

****IUPAC** (1983), Atomic weights of the elements, 1981, *Pure and Applied Chemistry*, **55**, 1101.

***IUPAC** (1989), *Nomenclature of inorganic chemistry*, Butterworths, London.

***Johnson, D. A.** (1982) *Some thermodynamic aspects of inorganic chemistry*, 2nd edition, Cambridge University Press.

***Knief, R. A.** (1992), *Nuclear Engineering*, Hemisphere Publishing Corporation, Washington.

Lenihan, J. (1988), *The crumbs of creation: trace elements in history, medicine, industry, crime, and folklore*, Adam Hilger, Bristol.

***Levett, P. N.** (1990), *Anaerobic bacteria*, Open University Press, Milton Keynes.

***Lindsay, W. L.** (1979), *Chemical equilibria in soils*, Wiley, New York.

Lovelock, J. (1982), *Gaia: a new look at life on Earth*, Oxford University Press.

Lovelock, J. (1988), *The ages of Gaia*, Oxford University Press.

***Manahan, S. E.** (1991), *Environmental chemistry*, 5th edition, Lewis Publishers, Chelsea, Michigan.

Martin, A. and Harbison, S. (1986), *An introduction to radiation protection*, Chapman and Hall, London.

***Mason, B. and Moore, C. B.** (1982), *Principles of geochemistry*, John Wiley, New York.

***Mason, S. F.** (1991), *Chemical evolution*, Clarendon Press, Oxford.

Meyer, C. (1985), Ore metals through geologic history, *Science*, **227**, 1421.

***Mitchell, J. F. B.** (1989), The 'greenhouse' effect and climate change, *Reviews of Geophysics*, **27**, 115–39.

NRPB (1986), *Living with radiation*, HMSO, London.

NRPB (1988), *Radiation exposure of the UK population: 1988 review*, HMSO, London.

Prigogine, I. and Stengers, I. (1984), *Order out of chaos*, Heinemann, London.

****Ramanathan, V., Callis, L., Cess, R., Hansen, J., Isaksen, I., Kuhn, W.,** *et al.* (1987), Climate–chemical interactions and effects of changing atmospheric trace gases, *Reviews of Geophysics*, **25**, 1441–82.

***Roth, E. and Poty, B.** (1989), *Nuclear methods of dating*, Kluwer Academic Publishers, Dordrecht.

⋆**Roxburgh, I. S.** (1987), *Geology of high-level nuclear waste disposal: an introduction*, Chapman and Hall, London.

⋆**Schlesinger, W. H.** (1991), *Biogeochemistry: an analysis of global change*, Academic Press, San Diego.

⋆**Schneider, S. H. and Boston, P. J.** (ed.) (1991), *Scientists on Gaia*, MIT Press, Cambridge, MA.

⋆⋆**Shannon, R. D. and Prewitt, C. T.** (1969), Revised values of effective ionic radii, *Acta Crystallographica B*, **26**, 1046.

⋆⋆**Shannon, R. D. and Prewitt, C. T.** (1970), Effective ionic radii in oxides and fluorides, *Acta Crystallographica B*, **25**, 925.

Sherwood Taylor, F. (1957), *A history of industrial chemistry*, Heinemann, London.

⋆**Shriver, D. F., Atkins, P. W. and Langford, C. H.** (1990), *Inorganic chemistry*, Oxford University Press.

⋆**Smith, G. R.** (ed.) (1981), *The Cambridge encyclopedia of earth sciences*, Cambridge University Press.

⋆**Stryer, L.** (1988), *Biochemistry*, 3rd edition, W.H. Freeman, New York.

Tylecote, R. F. (1992), *A history of metallurgy*, 2nd edition, Institute of Materials, London.

⋆**Wayne, R. P.** (1991), *Chemistry of atmospheres*, 2nd edition, Clarendon Press, Oxford.

Weeks, M. E. and Leicester, H. J. (1968), *Discovery of the elements*, 7th edition, Journal of Chemical Education.

⋆**Williams, R. J. P. and Da Silva, J. R. R. F.** (ed.) (1978), *New trends in bioinorganic chemistry*, Academic Press, London.

White, I. D., Mottershead, D. N. and Harrison, S. J. (1984), *Environmental systems*, Allen and Unwin, London.

Young, S. (1993), The body's vital poison, *New Scientist*, **137** (1864), 36.

Index

For individual chemical compounds, see the entry in Part II for the corresponding element. (For example carbon dioxide is listed under carbon.)